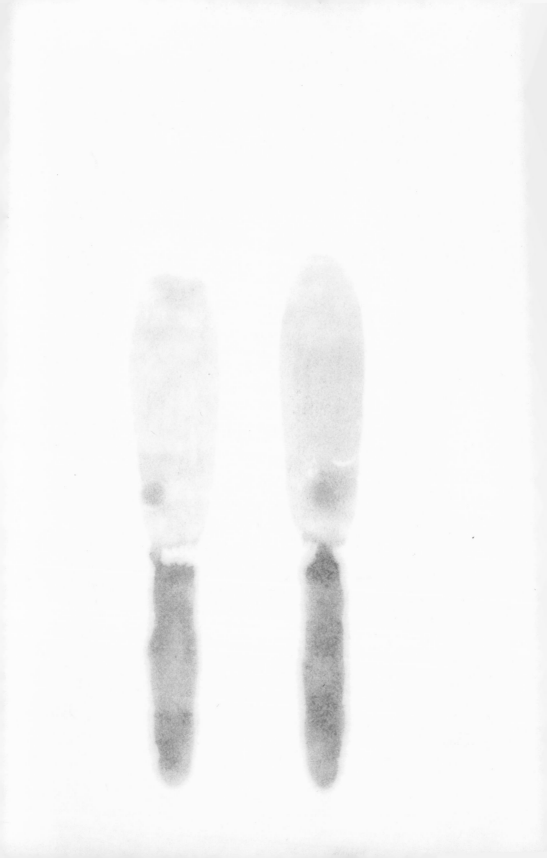

AGRICULTURAL AND FOOD CHEMISTRY: PAST, PRESENT, FUTURE

AGRICULTURAL AND FOOD CHEMISTRY: PAST, PRESENT, FUTURE

Edited by
Roy Teranishi, Ph.D.
U.S. Department of Agriculture

AVI PUBLISHING COMPANY, INC.
Westport, Connecticut

Library of Congress Cataloging in Publication Data
Main entry under title:

Agricultural and food chemistry.

Includes index.
1. Agricultural chemistry—Congresses. 2. Food—Composition—Congresses. 3. Food—Analysis—Congresses. 4. Food adulteration and inspection—Congresses. 5. Nutrition—Congresses. I. Teranishi, Roy, 1922-   II. American Chemical Society. Division of Agricultural and Food Chemistry.
S583.2.A37   630'.24   77-10327
ISBN 0-87055-231-7

Printed in the United States of America

Contributors

DR. JAMES J. ALBRECHT, Ph.D., Vice President, Research, Nestlé Enterprises, White Plains, N.Y.

DR. B. BORENSTEIN, Ph.D., Corporate Director. Consumer R&D, CPC International, Inc., Englewood Cliffs, N.J.

DR. T. C. BYERLY, Ph.D., USDA, ARS, Washington, D.C.

DR. A. S. CLAUSI, Ph.D., Vice President, Director of Technical Research, General Foods Corporation, White Plains, N.Y.

DR. SAMUEL B. DETWILER, JR., Ph.D., USDA, ARS, Washington, D.C.

DR. F. J. FRANCIS, Ph.D., Professor and Chairman, Department of Food Science and Nutrition, University of Massachusetts, Amherst, Massachusetts.

DR. R. A. FREDERIKSEN, Ph.D., Professor, Department of Plant Sciences, Texas A&M University, College Station, Texas.

DR. WARREN H. GABELMAN, Ph.D., Professor, Department of Horticulture, University of Wisconsin, Madison, Wisconsin.

DR. SAMUEL A. GOLDBLITH, Ph.D., Underwood-Prescott Professor of Food Science, Department of Nutrition and Food Science, and Director of Industrial Liaison, Massachusetts Institute of Technology, Cambridge, Massachusetts.

DR. WILLIS A. GORTNER, Ph.D., National Program Staff, USDA, ARS, Beltsville, Maryland.

DR. RICHARD L. HALL, Ph.D., Vice President, Science and Technology, McCormick and Company, Ltd., Hunt Valley, Maryland.

MR. WARREN E. HARTMAN, Vice President, Scientific Affairs, Worthington Foods, Worthington, Ohio.

DR. VIRGIL W. HAYS, Ph.D., Chairman, Department of Animal Science, University of Kentucky, Lexington, Kentucky.

DR. SAM R. HOOVER, Ph.D., USDA, ARS, Washington, D.C.

DR. GEORGE E. INGLETT, Ph.D., Chief, Cereal Science and Foods, NRRC, USDA, ARS, Peoria, Illinois.

DR. GEORGE W. IRVING, Ph.D., Research Associate, Life Sciences Research Office, Federation of American Societies for Experimental Biology, Bethesda, Maryland.

DR. PETER R. JENNINGS, Ph.D., Agriculturalist, The Rockefeller Foundation, New York, N.Y.

DR. J. W. JOHNSON, Ph.D., Associate Professor, Texas Agricultural Experiment Station, Lubbock, Texas.

DR. STANLEY J. KAZENIAC, Ph.D., Campbell Institute for Food Research, Camden, N.J.

DR. RICHARD J. MAGEE, Ph.D., Agricultural Center, American Cyanamid Company, Princeton, N.J.

DR. CARL S. MENZIES, Ph.D., Professor, Texas A&M University, San Angelo, Texas.

DR. EMIL M. MRAK, Ph.D., Chancellor Emeritus, University of California, Davis, California.

DR. H. E. NURSTEN, Ph.D., D.Sc., F.R.I.C., F.I.F.S.T., Reader, Procter Department of Food and Leather Science, the University of Leeds, England, now Professor and Head of Department of Food Science, University of Reading, England.

DR. J. E. OLDFIELD, Ph.D., Professor and Chairman, Department of Animal Science, Oregon State University, Corvallis, Oregon.

DR. HERMAN OLSMAN, Ph.D., Unilever Research Duiven, The Netherlands.

MR. ROBERT L. OPILA, Vice President and General Manager, Globe Engineering Company, Chicago, Illinois.

DR. BERNARD L. OSER, Ph.D., Bernard L. Oser Associates, Inc., Forest Hills, N.Y.

DR. GODEFRIDUS A. M. VAN DEN OUWELAND, Ph.D., Unilever Research Duiven, The Netherlands.

DR. HEIN G. PEER, Ph.D., Unilever Research Duiven, The Netherlands.

DR. HARRY J. PREBLUDA, Ph.D., Consultant, Trenton, N.J.

DR. J. T. REID, Ph.D., Professor and Chairman, Department of Animal Science, Cornell University, Ithaca, N.Y.

DR. D. REYMOND, Ph.D., Manager, Fundamental Research, Nestlé Technical Assistance Co., Ltd., Lausanne, Switzerland and Professor of Industrial Biochemistry, University of Geneva, Switzerland.

DR. HOWARD R. ROBERTS, Ph.D., Acting Director, Bureau of Foods, Food and Drug Administration, Washington, D.C.

DR. D. T. ROSENOW, Ph.D., Associate Professor, Texas Agricultural Experiment Station, Lubbock, Texas.

DR. LAWRENCE ROSNER, Ph.D., President, Rosner-Hixson Laboratories, Chicago, Illinois.

DR. JAMES SCALA, Ph.D., Director of Nutrition and Health Sciences, General Foods Corporation, White Plains, N.Y.

DR. MAURICE E. SHILS, M.D., Sc.D., Director of Nutrition and Attending Physician, Memorial Hospital, N.Y.

DR. G. F. SPRAGUE, Ph.D., Professor, Department of Agronomy, University of Illinois, Urbana, Illinois.

DR. MILTON L. SUNDE, Ph.D., Professor, Department of Poultry Science, University of Wisconsin, Madison, Wisconsin.

DR. G. L. TEETES, Ph.D., Associate Professor, Department of Entomology, Texas A&M University, College Station, Texas.

DR. ROY TERANISHI, Ph.D., Research Leader, Food Quality, USDA, ARS, WRRC, Albany, California.

DR. CLAIR E. TERRILL, Ph.D., National Program Staff, USDA, ARS, Beltsville, Maryland.

DR. OTTILIE D. WHITE, Ph.D., Professor, Department of Animal Science, Cornell University, Ithaca, N.Y.

DR. HAROLD L. WILCKE, Ph.D., Ralston Purina Company, St. Louis, Missouri.

Dedicated to
Peace

Preface

The food we eat is necessary for our health and is as important to our well-being as the genetic entity we have inherited and as the exercise we engage in to keep the body in good physical condition. The people in the United States enjoy plentiful, varied and high quality food. Not only are we blessed with sufficient fertile soil and favorable climate, but also we have energetic and educated farmers, who comprise less than 5% of our total population but produce more abundantly than farmers in any other country. The advanced technology in production, transportation, storage, processing, and utilization of various farm products has come about because of the forefathers who had enough vision to initiate research and to disseminate knowledge to farmers by establishing Land Grant Colleges and the U.S. Department of Agriculture.

These are the AGFD papers presented at the Agricultural and Food Chemistry sessions of the American Chemical Society's Centennial Meeting, New York City, April 4–9, 1976. We gathered experts to give their impressions as to the great developments in production, processing, and utilization (which includes nutrition and safety) in the last 100 years. Because each topic is so expansive, each contributor could fill volumes; it therefore is obvious that only highlights could be covered.

The AGFD Division is very appreciative that these contributors have addressed themselves to this enormous task and have given their time and talent to make this AGFD Centennial Meeting such a success.

In view of the looming crisis of feeding the increasing world population, we need to build on every aspect discussed at this historic meeting. Contributions from those engaged in agricultural and food chemistry will be needed even more in the years to come.

ROY TERANISHI
Chairman, AGFD
Western Regional Research Center
Agricultural Research Service
U.S. Department of Agriculture
May 1, 1977   *Albany, California*

Contents

Section 1

Highlights in U.S. Agriculture and Food Chemistry

Introduction

Harry J. Prebluda

Thomas Jefferson once said, "A morsel of genuine history is a thing so rare as to always be valuable." It is appropriate, then, that our Agricultural and Food Division pay tribute to the U.S. Department of Agriculture during this American Chemical Society Centennial celebration. A hundred years ago agriculture was more of an art than a science. Common sense and tradition were woven into the "trial and error" research of that time. In the last century more agricultural research has been carried out under the United States Government and industrial sponsorship than through any other nation in the world. The idea for a Department of Agriculture began with George Washington and Thomas Jefferson, who wanted Congress to create a National Board of Agriculture; this never came to a vote. The U.S. Department of Agriculture had its strange beginnings when $1000 was appropriated by Congress in 1839, at the request of President Van Buren, for collection by the Patent Office of agricultural statistics, "and other purposes." Up to that time Patent Commissioner Henry L. Ellsworth had distributed seeds and plants to farmers at his own expense and without specific authority.

In 1862, a year after Abraham Lincoln became President, four Acts were signed into law to encourage agricultural research and experimentation. The first of these established a U.S. Department of Agriculture. The second, called the Homestead Act, provided 160 acres of land for heads of families who would improve it and live on it for 5 years. The third Act, called the Transcontinental Railroad Act, provided financing to the railroads and helped take farm-grown crops to market. The fourth of the Agricultural Reform Acts was called the Morrill Land Grant College Act. It granted land to each state for colleges of agriculture and the mechanical arts.

The chapter following this one is by our former Division Chairman, Dr. Richard J. Magee. He is a well-known scientist with the American Cyanamid Company and describes the role of chemists in agricultural

and food research progress. He also traces the important developments of our Division of Agricultural and Food Chemistry within the
American Chemical Society.

Dr. Ted C. Byerly later salutes the many contributions of the U.S.
Department of Agriculture in production research. These accomplishments have catalyzed productivity and efficiency of farming in
the United States.

The recent emphasis on agricultural utilization research is covered
by Drs. Sam R. Hoover and Samuel B. Detwiler. They point out a
great many break-through contributions from USDA's Regional
Research Laboratories.

The important achievements of the U.S. Department of Agriculture in Nutrition and Food Research are covered by Dr. Willis A.
Gortner of the National Program Staff on the Agricultural Research
Service. Dr. Gortner's presentation provides a proper prologue to
the great highlight of the first section, the Eighth Annual W. O.
Atwater Memorial Lecture of Dr. Emil Mrak, Chancellor Emeritus
of the University of California at Davis. The lectureship provides
special recognition for individuals who have made unusual contributions to the many fields of science broadly related to human
nutrition and world food needs. The title is "Food Science and
Technology: Past, Present and Future." This chapter will help all
of us to understand the role of chemistry in meeting our world
food needs.

Appraisal for the 1976 American Chemical Society Centennial

Richard J. Magee

The emblem of the American Chemical Society displays a phoenix being transformed by fire. Less conspicuously, there appears an unfamiliar triangular, bulbous tube. This piece of obsolete laboratory equipment so prominently displayed is a Liebig tube. It commemorates the name of one of the greatest chemists of all time, the man who is, and who should be more widely known as, the father of agricultural chemistry—Justus von Liebig.

To Liebig's laboratory at the University of Giessen came science students from the United States and elsewhere to be trained by the man who in 1842, at the age of 37, had published a pioneering book entitled *Organic Chemistry in its Application to Agriculture and Physiology*. One of these students, the chemist C. M. Wetherwill, was to become the first scientist in the United States Department of Agriculture, a distinction received from President Abraham Lincoln in 1862. In the annals of agricultural chemistry for the Civil War period, Wetherwill shares the spotlight with another chemist, Edward Ruffin of South Carolina, a prominent publisher of scientific agriculture journals who is at least as well known for his claim that he fired the first shot on Fort Sumter in 1861.

The American Chemical Society was not established until 1876. But the period before and surrounding its founding reflected the fact that the United States was on the threshold of the era of science. The Morrill Act of 1862 encouraged the creation of colleges for teaching of agricultural and mechanical arts, and the Hatch Act of 1887 provided funds for the establishment of state experiment stations as model farms. The latter bill was probably based on the success of the Connecticut Agricultural Experiment Station, the first permanently established experiment station in the United States, founded in 1875. S. W. Johnson of Yale, another of Liebig's students and the most influential agricultural scientist of his day, first convinced the Connecticut legislature and then led the movement of experiment stations in every state. W. O. Atwater, a former pupil of Johnson's, later known for his outstanding work in respiratory calorimetry, was the first director of the Connecticut station originally located at Wesleyan University. In 1877, Johnson himself became director of a new and larger Connecticut station in New Haven, and T. B. Osborne, his son-in-law and later a giant in the field of plant protein chemistry, was his assistant. In 1878, Johnson also served as president of the American Chemical Society.

Early agricultural science concerned itself largely with soil chemistry and the benefits of fertilizer, and chemists dominated the field. State chemists were first appointed to provide controls on fertilizer quality to prevent fraud, but shortly the methods of analytical chemistry were being extended to the quality of produce as well. The Kjeldahl method of nitrogen analysis became available in 1883. The Babcock butterfat test was developed in 1890. A law regulating the quality of milk appeared in Connecticut in 1882, and largely through the persistent efforts of Dr. Harvey Wiley, the famed USDA chemist who was president of ACS in 1893, the Federal Pure Food and Drug Law was passed in 1906. The AOAC, the Association of Official Agricultural Chemists, was founded in 1884—with the energetic, ubiquitous S. W. Johnson as its first president.

Liebig had classified food constituents as proteins, carbohydrates, fats, and minerals, but not until the turn of the century was much study aimed at understanding the specific chemical requirements of nutrition. Eijkman's observation on polished rice was made in 1897. The use of laboratory animals in nutrition studies was an innovation which quickly led to the realization that in food there were many unidentified chemical substances essential to growth.

The ACS Division of Agricultural and Food Chemistry was founded in 1908. The year 1908 was also the year of F. G. Hopkins' classic paper proposing the existence of accessory food factors, substances which were named "vitamines" by Casimer Funk in 1911. The three decades that followed saw the unfolding of the chemical structures of vitamins and the chemical reactions underlying sound nutrition. Many chemists working in agricultural laboratories contributed to this progress; we need only think of the major contributions from the Department of Agricultural Chemistry at the University of Wisconsin, and from the College of Agriculture of the University of California at Berkeley and Davis. In the race to find and characterize the vitamins, absolute priorities of discovery are sometimes hard to discern. Among the most prominent names are E. V. McCollum, fat-soluble Vitamin A (1912) and water-soluble vitamins (1915); E. V. McCollum (1922) and H. Steenbock (1925), Vitamin D; C. G. King and A. Szent-Györgyi, identification of Vitamin C as ascorbic acid (1931–1932); R. R. Williams, Vitamin B-1 (1936); C. Elvehjem, identification of the pellagra-preventive factor as nicotinic acid (1937); Lepkovsky (1938) and Kuhn (1939), Vitamin B-6 as pyridoxine; Almquist, synthesis of Vitamin K (1939); T. H. Jukes (1939) and K. Folkers (1940), recognition and synthesis of pantothenic acid.

Vitamin research was not the only area of progress during this period. The nutritional essentiality of trace elements was recognized about 1928, and that of essential fatty acids in 1932. While Osborne and Mendel had reported "dietary indispensible amino acids" in 1913, it was W. C. Rose and his colleagues at Illinois who discovered threonine (1932–1934). Within a decade the approximate requirements of man for the essential amino acids had been determined. The nutritional knowledge gained in this period was not only of direct value to human health but also found important application in the production of meat animals.

Fermentation processes have had an important role in the production of foods and beverages and in the utilization of agricultural crops since ancient times. However, during World War II fermentation technology received real stimulation by the requirement for penicillin production. With the continued discovery and development of antibiotics, considerable progress has been made by microbiologists, biochemists and organic chemists in putting our knowledge and application of fermentation on a scientific basis. Antibiotics have also been proven important in the prevention of disease in meat-producing animals and have become significant as feed additives in animal husbandry.

Other important advances in food technology were made during the 1940's, and chemists and chemical engineers contributed significantly to their success. Advances were made in the preparation of dehydrated foods as the result of wartime demands, and the frozen food industry was founded.

New investigative tools, such as radiotracers, in the hands of biochemists have led to improved understanding of the physiology of plants. The carbon pathways in photosynthesis were detailed by Calvin and co-workers, as were the energy pathways by Arnon and others. Significant advances were made in understanding the chemistry of nitrogen fixation. Applications in agriculture soon followed the discoveries of auxin, the gibberellins, kinetin and ethylene. The identification of *Opaque II*, a type of corn rich in lysine, was the result of work by E. Mertz and co-workers at Purdue University.

Pesticides, soil studies, and fertilizers have played essential roles in fostering agricultural progress. Flavor chemistry, food chemistry and protein chemistry continue to produce new information about the constituents of foods. It would be impossible in a paper of this length to detail every accomplishment. But I would like to note some of the developments within the American Chemical Society that paralleled these advances.

On December 30, 1908, ACS gave approval to the formation of a Division of Agricultural and Food Chemistry. H. J. Wheeler of the University of Rhode Island led the petitioners; W. D. Bigelow was the Division's first chairman. Just to state a fact—with no further implications—the first paper presented on the first program of this new division of 40 members was on the subject of whiskey.

Programming in specialized areas has long played an important part in the Division of Agricultural and Food Chemistry. The Division's first symposium was held in 1922 on the subject of edible oils and fats. The first vitamin symposium was held in 1935; this topic was repeated at least 20 times before 1960, including 5 symposia on Vitamin B-12 between 1949 and 1951.

In addition to symposia, subdivisions have played an important part in the Division's activities. In 1946, a Fermentation Subdivision was formed with C. S. Boruff as chairman. It was extremely active until 1961 at which time it was given independent status as the Division of Microbial Chemistry and Technology. In 1950, the Pesticide Subdivision was formed with J. L. St. John as chairman. It prospered within the Division until 1968, at which time it became the Division of Pesticide Chemistry. Despite the recent spawning of two new Divisions, the membership of the parent Division has remained strong. At the parturition of each of the aforesaid Divisions membership dropped by about 400, but in each case grew steadily upward again; at the present time membership exceeds 1200. Currently, the Division has three subdivisions: the Flavor Subdivision, which was founded in 1965 with I. Hornstein as chairman; the Protein Subdivision, which was founded in 1970 with G. E. Inglett as chairman; and a new subdivision in Nutrition, currently being organized by L. Rosner.

Subjects presented by the Division have always been at the forefront of research, often presaging wider interest by several years. As early as 1952 a paper was presented entitled "Analysis of Human Fat for DDT and Related Substances." A subject discussed in 1958 was

"Control of Physiological Processes in Plants by Chemicals," while in 1960 "Carcinogenic Hazards of Food Additives" and "Production and Use of Enzymes in Agricultural and Food Processing" were symposium topics.

In the United States, research in agricultural and food chemistry has always been characterized by close cooperation among government, university, and industrial scientists. From such interaction have come the major accomplishments of the past, and from such scientists will come the research breakthroughs of tomorrow. In a centennial statement, Charles L. Flint said, ". . . the present is but the dawn of a new era—an era of improvements of which we cannot yet form an adequate conception . . . for the large number of young [people] who will go forth every year. . . , many of them, thoroughly instructed in chemistry and kindred sciences, will give us . . . the conditions for new discoveries which will open the way to higher triumphs, and so lead on to the golden age of American agriculture." Mr. Flint, later president of the Massachusetts Agricultural College, was speaking in 1876 as the Nation was preparing to celebrate its first centennial, when the world population was less than 1.5 billion. His words were prophetic for the hundred years just past. Now that the world population is 4 billion and growing, it is our challenge to see that his prediction of "higher triumphs" remains prophetic for the century to come.

3

Contributions of USDA in Production Research

T. C. Byerly

Efficiency of production of agricultural products has more than doubled during the past hundred years. U.S. Department of Agriculture scientists have made many major research achievements which have contributed to the increase in production and production efficiency. Many equally important achievements have been made by research workers in State Agricultural Experiment Stations (SAES); USDA; and many achievements were accomplished through the joint efforts of scientists in USDA and SAES.

This chapter consists of brief accounts of those achievements in which USDA scientists made important major or sole contributions. The items discussed are a selected sample; there are many other achievements of equal importance. The items reported were selected to illustrate both breadth and depth of USDA achievements in production research. The items include both basic and applied research achievements. USDA scientists have contributed and continue to contribute to science through basic research, from the first discovery of an arthropod-transmitted disease—cattle tick fever—by Smith and Kilbourne (1893) to the development of an effective vaccine from a protein coat moiety of foot-and-mouth virus, free of the virus itself, in 1975 (Bachrach *et al.* 1975).

ANIMAL DISEASE RESEARCH

Cattle Tick Fever

Cattle tick fever was general in Southwestern United States and Latin America prior to 1906 (MacKellar 1942). Trail movement of cattle from the infested areas to northern and western areas was followed by heavy losses of cattle native in those areas. In 1883, USDA conducted a survey to determine the northern boundary of the infested area. This was followed by the first national animal quarantine made by the first Secretary of Agriculture, N. J. Colman, on July 3, 1889.

Research had been initiated in the Bureau of Animal Industry under its first chief, D. E. Salmon. The researchers were Theobald Smith and L. B. Kilbourne. By 1889–1890 they had established the fact that tick infestation is necessary to transmission of the disease; the tick vectors are *Boophilus annulatus* and *B. microplus.* These ticks introduce the causal organism, a parasitic protozoan called *Babesia bigemina.* During the same period, another USDA scientist, Cooper Curtice, established the life history of the tick (Smith and Kilbourne 1893).

The demonstration of arthropod transmission of cattle tick fever by Smith and Kelbourne was the first demonstration that an infectious disease could be transmitted by an arthropod intermediate host or carrier from one animal to another. It was followed by studies in many agencies which demonstrated the similar role of arthropod and other intermediate hosts and carriers in the spread of human, animal and plant diseases; e.g., typhus, bubonic plague, encephalitides, anaplasmosis, and malaria.

A tick eradication program was initiated by USDA in 1906, in cooperation with the infested states and livestock owners. Cattle within the area were dipped at regular intervals. By December of 1940 cattle tick fever had been eradicated, except through the occasional illegal incursions into border counties of tick-infested Mexican cattle. For more than 60 years arsenical dips were used to control ticks on cattle imported legally from infested Mexican areas. These dips were replaced in the 1970's by organic chemicals.

Hog Cholera

In 1878, the Congress appropriated $10,000 to defray the cost of a Commission to Investigate Diseases of Domestic Animals (Cole *et al.* 1962). The Commission gave special attention to diseases of swine, then the most costly of all livestock diseases. One of the members of the Commission, James Law of New York, reported in 1878 that he had inoculated test animals with hog cholera. Another member of the Commission, D. E. Salmon of North Carolina, later became the first Chief of the Bureau of Animal Industry, established in 1884.

Under the participating leadership of Salmon, Smith and Kilbourne research began on hog cholera in 1885. They obtained from Pasteur a vaccine effective against Rouget disease of swine in Europe. It failed to protect swine against hog cholera, thus establishing the fact that hog cholera was not Rouget's disease. Smith and Kilbourne conducted tests with a bacterium now known as *Salmonella cholera suis*, which was then believed to be the pathogen causing hog cholera. Subsequent research by De Schweintz and Dorset established the fact that hog cholera is caused by a virus, not *S. cholera suis* (De Schweinitz and Dorset 1903).

During the next ten years, laboratory research and field trials resulted in the development and application of a system of immunization by simultaneous inoculation of healthy pigs with virulent hog cholera virus and serum from hyper immune pigs. Field tests with about 100,000 pigs were conducted in 1913 and 1914 in counties of 13 States. In the 17 test counties, hog losses in 1912 were at the rate of 178 per 1000; and 1914 losses were 49 per 1000 (cf Cole *et al.* 1962, pp. 18-19).

Marek's Disease

Marek's disease caused heavy losses of laying hens and broiler chickens in the United States during the 1935-1970 period. Research at the USDA Regional Poultry Laboratory at East Lansing, Michigan, determined that the disease is characterized by multiple tumors; i.e., it is a neoplastic disease.

B. R. Burmester and his colleagues at the Regional Poultry Laboratory developed a vaccine from a herpes virus of turkey origin. Its field use has been highly successful (Purchase *et al.* 1972). Lancaster *et al.* (1973) reported that chickens vaccinated against Marek's disease with turkey herpes virus laid 40 eggs more per layer housed than controls. The difference was largely due to lower mortality among vaccinated birds than among the controls. Hilleman of the Merck Institute for Therapeutic Research wrote, "The development of the Marek's disease vaccine has provided a new dimension in the prospects for eventual control of cancer in other animal species, including man. . ." (Hilleman 1972).

Foot-and-Mouth Disease

Foot-and-mouth disease (FMD) is a debilitating disease, endemic to many countries. Cattle, sheep, goats, hogs, deer and other cloven-footed animals are susceptible. The disease is caused by a virus which may be transmitted by contact or by virus-contaminated materials— meat, hides, forage, vehicles, feces, animal caretakers or their clothing. In the United States nine outbreaks, the latest in 1929, have been eliminated by quarantine, tests, and slaughter. Presence of the disease in other countries prevents or greatly restricts movement of livestock products from those countries into the United States (Shahan and Traum 1956).

USDA has conducted research on foot-and-mouth disease for more than 75 years. It is presently a major subject of research at the Department's Plum Island Animal Disease Laboratory (PIADL), an isolated, biologically secure facility. Research there has resulted in a very important recent achievement of substantial scientific importance and great economic potential. For the first time, PIADL scientists have produced an FMD vaccine from the protein coat of the virus rather than from the whole infective virus. Based on this research, future FMD vaccines may be made from fractions of a single virus protein and may even by synthesized. The new vaccine contains the purified subunit capsid protein VP_3 of the FMD virus. There are several immunological types of FMD virus, each of which requires a separate vaccine (Bachrach et al. 1975).

INSECT CONTROL

Cottony-Cushion Scale

Control by the Vedalia bettle (*Rodolia cardinalis*) of the cottony-cushion scale insect pest of citrus in California was achieved in 1888. The pest had threatened to destroy the infant California citrus industry.

This success was accomplished through the planning and direction of C. V. Riley, a USDA entomologist. He knew that the pest occurred in Australia where it was partially controlled by a parasitic fly (Clausen 1952). Funds were provided by the State Department to send another entomologist, Alber Koebele, to the Melbourn Exposition in Australia. Koebele had studied the cottony-cushion scale in California. In Australia he obtained and sent thousands of the parasitic flies to California where they were released.

But in addition, Koebele discovered in Australia that the Vedalia beetle and its larvae fed on the cottony-cushion scale and its eggs. Koebele sent Vedalia beetles to California, where they were released in cottony-cushion scale infested orchards by cooperating citrus growers. The Vedalia beetles thrived and within a year the pest was controlled; it has remained so ever since.

Following this early success, USDA has colonized at least 420 species of beneficial insects in the United States. About 110 of these species are exercising some degree of control over more than 60 insect and 10 weed pests (Klassen 1975).

Insecticidal Aerosols (The Bug Bomb)

In 1943, USDA chemists Goodhue and Sullivan invented the pressurized can method of aerosol dispersion—the "bug bomb." Its efficacy results from spraying a solution of an insecticide in a solvent of very high volatility which immediately changes to a gas at ambient temperatures, leaving very small particles of the insecticide suspended in the air (Sullivan *et al.* 1942).

The utility of the "bug bomb" for control of mosquitoes and other insect pests in homes, barracks and passenger planes was promptly established. Use of the bug bomb principle was soon extended to the production of many other products.

Pyrethroids Come Back

Pyrethrum, obtained from the powdered flowers of chrysanthemum-related *Pyrethrum spp.*, has been used as an insecticide for more than 100 years. USDA chemists determined the structures of the active principles, pyrethrolone and cinerolone, in 15 years of research (1932-1947). Schechter *et al.* (1949) prepared a stereo isomer of cinerolone; when esterified with chrysanthemumic acid it formed a pyrethrin-like product highly toxic to flies. An ester named allethrin was produced, announced, and application for public service patents was filed in 1949. Commercial production followed with substantial use—especially by the Armed Forces. After the Korean War its manufacture ceased. Schechter continued his attention to problems of disinfection and USDA research emphases turned to new fields, e.g., attractants and chemosterilants.

Research and development activity on synthetic pyrethroids continued in England, France, and Japan. The English workers synthesized a number of pyrethroids; one of them, "resmethrin," is now produced commercially in the United States.

There is currently a general resurgence of interest in pyrethroids. Some of the synthetics are more effective and may be safer than insecticides formerly available, particularly for quarantine control operations.

Screw Worm Fly

The concept of insect pest control through use of sterile males has gained worldwide attention. The concept was initiated by USDA entomologist E. F. Knipling in 1937. Propagation and release of cobalt-irradiated sterilized screw worm flies (*Callitroga americana*) eliminated this pest from Southeastern United States in a four-year program, from 1954-1958. The program was extended to Southwestern United States in 1962. In that area, suppression but not eradication of the screw worm fly has been achieved.

Insect Attractants

To combat the gypsy moth, a major exotic pest of oak, beech and some other species of shade, forest and orchard trees in the Northeastern United States, a sex attractant has been used in management programs for many years. The female gypsy moth is sessile and males come to her from miles around. The sex attractant segregated from the abdomens of these females was formerly used as bait for survey purposes. In 1970, USDA scientist M. Beroza and his colleagues (Bierl *et al.* 1970) synthesized this attractant, stimulating its use in insect pest management.

Extended use of the synthesized attractant has undergone field trial in "confusion" experiments. Infested areas have been aerially sprayed with minute amounts of disparlure. Males are indeed confused, their flights no longer directed by dispersed sex attractant from receptive females. Results have been variable; in some treated areas very few eggs become fertilized. This method has potential for use against other insect pests, e.g., the pink bollworm.

PLANT SCIENCE

Genetic Resistance to Plant Disease

William Orton was perhaps the first USDA scientist to score success in breeding for plant disease resistance. He was assigned responsibility to develop wilt-resistant cotton in 1899. He did. Sea Island Cotton had been devastated by Fusarium wilt (Orton 1900). The first resistant Sea Island variety, Rivers, was released in 1902. The first resistant upland variety, Dillon, was released in 1905.

Fusarium wilt of cotton occurs in all our cotton-growing States. Entrance of the fungus into cotton roots is facilitated by holes made by nematodes (Smith 1953). Within desirable strains of cotton, Orton selected individual plants that survived under heavy exposure to the pathogen. He bred these surviving individuals and subjected their progeny to further tests, a method still in wide general use.

With cowpeas, Orton found an existing resistant variety, "Iron," and simply exploited it. With watermelon, he introduced genes for resistance to cutworm and selected the progeny for both resistance to Fusarium and table quality. Orton was surely ingenious and versatile as well as innovative (Coons 1953).

Wheat Stem Rust

Wheat stem rust caused by the fungus *Puccinia graminis tritici* has caused catastrophic losses of spring bread wheat and Durum wheat. USDA has cooperated with individual states in cereal rust research since 1907 (Humphrey 1917). In 1916, losses of bread wheat in the United States were estimated at about 200 million bushels. This loss came during World War I, when export demand was trebled over pre- and post-war levels. Partial substitution of other cereal flours for wheat became necessary for the United States civilian population (Stakman 1918; Horsfall 1975).

The Durum wheat crop was almost totally destroyed in 1951–1952.

Wheat stem rust spreads in two ways: (1) Spores from rust-infested winter wheat fields in the Southern Great Plains are airborne to fields in more northern States as the spring advances. (2) In the Northern States, teliospores overwinter in straw or stubble and infest the common barberry in the spring. Development on barberry involves sexual reproduction of the fungus resulting in emergence of new strains. More than 240 strains are known (Martin and Salmon 1953).

Since 1917, USDA has participated in organized, cooperative research and surveillance with many States of the United States and with many other countries to identify wheat strains that are susceptible or that are resistant to new races of rust. The useful life of a resistant wheat variety is often only about five years when it must be replaced by new varieties resistant to the newly emerged virulent rust strains.

Evapotranspiration

Average annual precipitation in the United States amounts to 30 in. About 70% is lost by evaporation and plant transpiration. In areas where water is a scarce resource, efficiency of water use is a critical factor in plant productivity. Briggs and Shantz (1912, 1913) published the results of their measurements of transpiration by alfalfa, cereals, potatoes, other crop plants, and weeds. They reported wide variation in the amount of water transpired per unit of plant dry matter or grain produced. For example, the water required to produce 1 kg of alfalfa dry matter is about 3 times that required to produce 1 kg whole plant dry matter of corn or sorghum.

Briggs and Shantz demonstrated that drought begins when available soil moisture is so diminished that vegetation can no longer absorb water from the soil rapidly enough to replace the water lost to the air by transpiration.

Photoperiod

Garner and Allard (1920) demonstrated and then in 1920 announced that length of darkness and light in each 24 hours regulates the flowering of tobacco and many other plants. They discovered that some plants flower sooner when days are shortened and some when days are lengthened. Others are photoperiod insensitive.

This basic concept has been applied widely by plant breeders. For example, soybean breeders have produced varieties suited to seasonal differences in day length and growing period in nine zones from north to south in the United States. Consequently, soybeans flourish from North Dakota to Louisiana (Henson 1947).

Borlaug's work with semidwarf wheats for tropical areas included selective breeding for strains which were insensitive to photoperiod.

Florists use a method of dark period interruption to delay flowering of chrysanthemums and thus produce flowers of improved quality at a scheduled time.

Phytochrome

The response to light of many plants is mediated by a photo-reversible pigment which occurs in two forms with action maxima near 660 and 770 .mμ, respectively. The light stimulation involved in the germination of seed is one manifestation (Flint and McAlister 1935; Borthwick et al. 1952). The pigment was observed in living tissue by spectrophotometry; it was then separated from etiolated maize shoots and assayed by differential spectrophotometry (Butler et al. 1959).

Viroids

T. O. Diener, USDA Plant Virologist, independently discovered an infective particle which he named the "viroid" (Diener 1971). A viroid consists of naked nucleic acid much lower in molecular weight than any viral nucleic acid. When a tomato or potato host plant is inoculated with viroid particles, the infective particles multiply in the nucleus in association with DNA, causing "potato spindle tuber" or "tomato bunchy top" disease. The tomato disease occurs naturally in South Africa, and potato spindle tuber disease is widespread in potato-growing areas.

This initial discovery has opened the gate to discovery of several other viroids.

ANIMAL BREEDING AND GENETICS

Population Genetics and Animal Breeding

Conceptually, Sewall Wright was an outstanding contributor to science. His research with guinea pigs at Beltsville was the basis for prediction of, and quantitative methods of analyses for, the effects of inbreeding and crossbreeding (Wright 1921A). His paper on correlation and causation is a lucid account of his "path coefficient" method of quantifying effects with common, interrelated, or independent causes. The paper is also an important rationalization of the appropriate use of correlations to determine "the degree to which variation of a given effect is determined by each particular cause" (Wright 1921B). Wright was, with Fisher (1930), one of the principal founders of population genetics.

Meat-Type Hogs

Around 1934, Secretary of Agriculture Henry Wallace negotiated the importation of Danish Landrace swine for use in breeding research. The Landrace has contributed to genetic diversity and its concomitant heterosis in U.S. commerical swine production, which is now largely based on crossbred or other hybrid mating.

More important, the research at Beltsville under direction of John Zeller (1947), with that of the several States which cooperated in the Bankhead-Jones Regional Swine Breeding Laboratory, provided information indicating that all breeds and crosses of swine contain the genetic traits responsible for superior meat type. Application of this information by U.S. swine producers through the selective breeding of swine has been an important factor in the reduction of unwanted lard production and the concomitant increase in lean pork.

Small White Turkey

Beginning in 1934, M. A. Jull, then in charge of poultry husbandry research at Beltsville, began planning for the development of a turkey better suited to home use. A survey conducted for a chain store association indicated that a small turkey was needed. The proposed turkey was to be 8-10 lb in weight and white, so as to avoid the unattractive appearance of ready-to-cook turkeys with dark-colored pinfeathers. High egg production and hatchability of eggs as well as vigor were other objectives (Marsden and Martin 1944).

Breeding stocks of white, bronze, black, white and wild color patterns were obtained. A series of crosses were made among these stocks by S. J. Marsden in association with geneticist C. W. Knox. Breeding stock for successive generations was selected phenotypically for the desired traits. Within 15 years the objectives had been achieved. The Beltsville Small White Turkey was widely distributed. This turkey, with its derivatives and other small white turkeys subsequently developed, is currently produced at a rate of about 16 million per year (USDA 1974).

Parthenogenesis

During the period from 1952-1971, Olsen (1975) hatched more than 1100 parthenogenetic turkeys and chickens. Forty-six selected parthenogenetic males survived to an average age of four years. Adult homeothermic parthenogenetic amniotes are unusual, perhaps even unique. Olsen observed thousands of partially developed parthenogenetic embryos and extraembryonic structures. The parthenogenetic poults, chicks and embryos were all males and were automictic diploids. Olsen's findings have been confirmed by several other research workers.

Ovulation in the Domestic Fowl

R. M. Fraps and his colleagues (Fraps and Byerly 1942; Fraps and Drury 1943; Rothchild and Fraps 1949) were the first to induce ovulation in the domestic fowl. They contributed greatly to elucidation of the environmental endocrine and neurological parameters and their interaction which control ovulation and oviposition.

Artificial Insemination of Poultry

Burrows and Quinn (1935) reported a method for collecting sperm from roosters and inserting it into the oviducts of laying hens. Abdominal massage results in ejaculation, and collected semen is injected by syringe deep into the oviduct. The method has been adapted for use with turkeys and other avian species, and is now widely used.

,ANIMAL NUTRITION

Vitamin B-12, Soybeans and Chickens

Soybean meal is the principal protein supplement now used in livestock feeding in the United States. Its use is based on the research of many people who determined processing conditions and dietary supplements necessary to its efficient use in monogastric animals.

Beginning in 1929, research workers at Beltsville investigated soybeans and soybean meal use in poultry feeds (Byerly et al. 1933). These soybean products were compared with fish, milk, and meat by-products as sources of protein in chicken breeder diets. Hatchability was lower in the eggs of hens on soybean diets than in the eggs of those hens receiving animal protein supplements.

In research at the University of Maryland, H. R. Bird found that breeder hens receiving soy diets in cages produced eggs with very low hatchability while similar hens, maintained on litter, produced eggs of fair hatchability. Bird and his colleagues later showed that both cow manure and chicken manure contain a factor essential to hatchability of eggs from hens on soy protein diets. The same essential factor was supplied by fish, meat, and milk by-products. Potent concentrates of this factor from manure, at a level 0.004% of the breeder diet, supported normal hatchability (Bird 1947). In 1948, Lillie et al. (1948) reported that crystalline vitamin B-12 produced a comparable effect.

In 1951, Ellis and Bird (1951) wrote that the presence of Vitamin B-12 in poultry diets makes possible the successful rearing of chickens on diets in which all of the protein is of plant origin—most of it being supplied by soybean oil meal and grain. Calcium, phosophorus and riboflavin supplements are also necessary.

Energy Requirements for Milk Production

Rate of daily milk production per cow is limited by genetic capacity, energy intake, stage of lactation, and environment. Energy intake is limited by appetite, digestibility of ration, and rumen capacity. Digestibility tends to decrease as milk production and feed intake increase (Moe et al. 1971).

Moe and Flatt (1969) reported an equation for calculating net energy values which is now recommended for formulation of feeds for lactating cows in the United States (Moe et al. 1972; Loosli 1971). Moe et al. (1971) reported that cows with very high milk production, 45 kg or more per day, have net energy requirements during early lactation which exceed energy intake. Such cows may use body tissue for milk production with an efficiency of 80%, replenishing tissue reserves later in lactation, as rate of milk production diminishes.

SOIL EROSION BY RAIN

Erosion is a major cause of declining soil productivity. Conversely, soil erosion contributes the largest quantity of pollutants to our streams, lakes and reservoirs. The estimated amount of sediment entrained from the soil each year is more than 3.5 billion metric tons. Half this amount is deposited on the surface of upstream watersheds; ¼ is deposited in streams, lakes and reservoirs; and ¼ reaches the sea. Over long periods of time, soil erosion contributes to formation of new soil; in the short run, it is the major soil-degrading factor.

Predicting the amount of soil erosion likely to occur under local environmental, climatic, weather, and management conditions is essential to soil erosion control. A soil-loss prediction equation has been developed which is applicable "wherever locational values of the equation's individual factors are known or can be determined" (Wischmier and Smith 1965).

THE GREEN REVOLUTION

USDA scientists have contributed substantially to the package of technologies that is called the "Green Revolution." In 1946, S. C. Salmon brought back seed of a Japanese semidwarf wheat, Norin 10 (Reitz and Salmon 1968). This material was used by Vogel, a USDA wheat breeder stationed at Pullman, Washington, in the development of "Gaines" wheat. Gaines is a white wheat, adapted to growing conditions in Oregon, Washington and Idaho. It responds to fertilization with high yields, often exceeding 5 MT per ha, compared to the U.S. national average of about 2 MT per ha.

Borlaug used semidwarf wheat seed obtained from Vogel in developing semidwarfs insensitive to photoperiod and thus suitable for year-round growth in tropical countries.

USDA scientists have contributed to the supporting technologies needed to obtain high yields of semidwarf wheat and rice in developing countries. This technology consists of a package of practices systematically applied. It includes water management (timely irrigation and adequate drainage), fertilization, and pest control, as well as timely and appropriate preparation of seed bed, planting, and harvest.

This package of practices embodies the concept of interaction developed by C. E. Kellogg (1975) based on his own research and experience and that of his colleagues in USDA and in many other countries around the world as well as in the United States.

SUMMARY

USDA scientists have been and will continue to be productive. Their research information has been consolidated into a common body of technology. Application of that technology has been a major factor in the increased productivity and efficiency of farming in the United States.

Since I entered USDA service in 1929, meat and egg production in the United States have doubled. Milk production per cow has more than doubled, while the number of milk cows has decreased by half. Aggregate area planted in cereals and soybeans in 1929 and 1974 was 89 million hectares in each year. In 1929 the aggregate production of cereal and soybeans was 115 million metric tons; in 1974, 245 million metric tons.

At least half this increase in productivity must be due to application of research-based technology. There were a few years when weather or crop disease curtailed yield sharply but, generally, the past 45 years' weather has been more favorable for crop production than may be expected during the next 45 years.

BIBLIOGRAPHY

BACHRACH, H. L., MOORE, D. M., MCKERCHER, P. D., and POLATNICK, J. 1975. Immune and antibody responses to an isolated capsid protein of foot and mouth disease virus. J. Immunol. *115*, 1636.

BAKER, G. L., RASMUSSEN, W. D., WISER, V., and PORTER, J. M. 1963. Century of Service. U.S. Dep. Agric.

BIERL, B. A., BEROZA, M., and COLLIER, C. W. 1970. Potent sex attractant of the gypsy moth. Science *170*, 870.

BIRD, H. R. 1947. Feeding poultry. *In* Science in Farming: Yearbook of Agriculture. U.S. Dep. Agric.

BORTHWICK, H. A. *et al.* 1952. A reversible photoreaction governing seed germination. Proc. Nat. Acad. Sci. *38*, 688.

BRIGGS, L. J., and SHANTZ, H. L. 1912. The wilt coefficient of different plants and its indirect determination. Bur. Plant Ind. Bull. *230*.

BRIGGS, L. J., and SHANTZ, H. L. 1913. The water requirements of plants. Bur. Plant Ind. Bull. *285*.

BURROWS, W. H., and QUINN, J. P. 1935. A method of obtaining spermatozoa from the domestic fowl. Poult. Sci. *14*, 251.

BUTLER, W. L., NORRIS, K. H., SIEGELMAN, H. W., and HENDRICKS, S. B. 1959. Detection, assay and preliminary purification of the pigment controlling photo responsive development of plants. Proc. Nat. Acad. Sci. *45*, 1703.

BYERLY, T. C., TITUS, H. W., and ELLIS, N. R. 1933. Production and hatchability of eggs affected by different kinds and quantities of protein in the diet of laying hens. J. Agric. Res. *40*, 1.

CLAUSEN, C. P. 1952. Parasites and predators. *In* Insects: Yearbook of Agriculture, U.S. Dep. Agric.

COLE, C. G. *et al.* 1962. History of hog cholera research in the U.S. Department of Agriculture. USDA Agric. Info. Bull. *241.*

COONS, G. H. 1953. Breeding for resistance to disease. *In* Plant Diseases: Yearbook of Agriculture, U.S. Dep. Agric.

CRAFT, W. A. 1958. Fifty years progress in swine breeding. J. Animal Sci. *17*, 960.

DE SCHWEINITZ, E. H., and DORSET, M. 1903. A form of hog cholera not caused by the hog cholera bacillus. Bur. Animal Ind. Circ. *41.*

DIENER, T. O. 1971. Potato spindle tuber virus: a plant virus with properties of a free nucleic acid. Virology *43*, 75.

ELLIS, N. R., and BIRD, H. R. 1951. By-products as feed for livestock. *In* Crops in Peace and War: Yearbook of Agriculture, U.S. Dep. Agric.

FISHER, R. A. 1930. The Genetical Theory of Natural Selection. Clarendon Press, Oxford, England.

FLINT, L. H., and MCALISTER, E. D. 1935. Wave lengths of radiation in the visible spectrum inhibiting the germination of light-sensitive lettuce seed. Smithsonian Misc. Publ. *95*, No. 5, 1-11.

FRAPS, R. M. and BYERLY, T. C. 1942. The experimental induction of ovulation in the domestic fowl. Poult. Sci. *21*, 469.

FRAPS, R. M. and DRURY, A. 1943. Occurrence of premature ovulation in the domestic fowl following administration of progesterone. Proc. Soc. Expt. Biol. Med. *52*, 346.

GARNER, W. W., and ALLARD, H. A. 1920. Effect of the relative length of day and night and other factors of the environment on growth and reproduction in plants. J. Agric. Res. *18*, 553.

GOODHUE, L. D., and SULLIVAN, W. H. 1943. Methods of applying parasiticides. Pat. 2,321,023. June 8.

HENSON, P. R. 1947. Soybeans for the South. *In* Science in Farming: Yearbook of Agriculture, U.S. Dep. Agric.

HILLEMAN, M. R. 1972. Marek's disease vaccine: the implication in biology and medicine. Avian Dis. *16*, 191.

HORSFALL, J. G. 1975. The fire brigade stops a raging corn epidemic. *In* That We May Eat: Yearbook of Agriculture, U.S. Dep. Agric.

HUMPHREY, H. E. 1917. Cereal diseases and the national food supply. *In* Yearbook of Agriculture, U.S. Dep. Agric.

KELLOGG, C. E. 1975. Agricultural Development: Soil, Food, People, Work. Soil Science Society of America, Madison, Wisconsin.

KLASSEN, W. 1975. Examples of history of research by the U.S. Department of Agriculture on domestic and foreign crop pests. USDA. (Unpublished)

LANCASTER, J. E., BARR, W. K. and BARTLETT, H. R. 1973. A Marek's disease vaccination trial. Poult. Sci. *52*, 1450.

LILLIE, R. J., DENTON, C. R., and BIRD, H. R. 1948. Relationship of vitamin B$_{12}$ to the growth factor in cow manure. J. Biol. Chem. *176*, 1477.

LOOSLI, J. K. 1971. Nutrient Requirements of Dairy Cattle, 4th Edition. National Academy of Science, Washington, D.C.

MACKELLAR, W. M. 1942. Cattle tick fever. *In* Keeping Livestock Healthy: Yearbook of Agriculture, U.S. Dept. Agric.

MARSDEN, S. J., and MARTIN, J. H. 1944. Turkey Management. Interstate Printers & Publishers, Danville, Illinois.

MARTIN, J. H., and SALMON, S. C. 1953. Rusts of wheat, oats, barley, rye. *In* Plant Diseases: Yearbook of Agriculture, U.S. Dep. Agric.

MOE, P. W., and FLATT, W. R. 1969. Net energy of feed stuffs for lactation. J. Dairy Sci. *52*, 928.

MOE, P. W., FLATT, W. R. and TYRRELL, H. F. 1972. Effect of level of feed intake on digestibility of dietary energy of high producing cows. J. Dairy Sci. *55*, 945.

MOE, P. W., TYRRELL, H. F., and FLATT, W. R. 1971. Energetics. J. Dairy Sci. *54*, 548.

MOORE, L. A. IRVIN, H. M., and SHAW, J. C. 1953. Relationship between TDN and energy value of feeds. J. Dairy Sci. *36*, 93.

OLSEN, M. W. 1975. Avian parthenogenesis. UDSA Agric. Res. Serv., Northeast Regional Publ. *65*.

ORTON, W. A. 1900. The wilt disease of cotton and its control. Div. Veg. Physiol. Pathol. Bull. *27*.

PURCHASE, H. G., OKAZAKI, W., and BURMESTER, B. R. 1972. Long term field trials with the herpes virus of turkey vaccine against Marek's disease. Avian Dis. *16*, 57.

REITZ, L. P. and SALMON, S. C. 1968. Origin, history and use of Norm 10 wheat. Crop. Sci. *8*, 686.

ROTHCHILD, I., and FRAPS, R. M. 1949. The interval between normal release of ovulating hormone and ovulation in the domestic fowl. Endocrinology *44*, 134.

SCHECHTER, M. S., GREEN, N., and LaFORGE, F. B. 1949. The synthesis of cyclopentenolones of the type of cinerolone. J. Am. Chem. Soc. *71*, 1517.

SCHECHTER, M. S., and LaFORGE, F. B. 1952. Cyclopentenolone esters of cyclopropanecarboxylic acids. U.S. Pat. 2,603,652. July 15.

SHAHAN, M. S., and TRAUM, J. 1956. Foot-and-Mouth Disease. *In* Animal Diseases: Yearbook of Agriculutre, U.S. Dep. Agric.

SMITH, A. L. 1953. Fusarium and nematodes on cotton. *In* Plant Diseases: Yearbook of Agriculture, U.S. Dep. Agric.

SMITH, T., and KILBOURNE, F. L. 1893. Nature, causation and prevention of Texas Southern Cattle fever. Bur. Animal Ind. Bull. *1*.

STAKMAN, E. C. 1918. The black stem rust and the barberry. *In* Yearbook of Agriculture, U.S. Dep. Agric.

SULLIVAN, W. H., GOODHUE, L. D., and FALES, J. H. 1942. Toxicity to adult mosquitoes of aerosols produced by spraying solution of insecticides in liquefied gas. J. Econ. Entomol. *35*, 48.

USDA. 1941. Agricultural statistics. Agric. Res. Serv., U.S. Dep. Agric.

USDA. 1974. Agricultural statistics. Agric. Res. Serv., U.S. Dep. Agric.

WISCHMIER, W. H., and SMITH, D. D. 1965. Agricultural Handbook *282*, U.S. Dep. Agric.

WRIGHT, S. 1921A. The effects of inbreeding and crossbreeding on guinea pigs. U.S. Dep. Agric. Bull. *1090*.

WRIGHT, S. 1921B. Correlation and causation. J. Agric. Res. *20*, 557.

ZELLER, J. H. 1947. Progress in hog production. *In* Science in Farming: Yearbook of Agriculture, U.S. Dep. Agric.

4

Agricultural Utilization Research

Sam R. Hoover
Samuel B. Detwiler, Jr.

The use of chemistry in agricultural research did not start with the USDA Regional Laboratories.

On May 5, 1862, President Abraham Lincoln signed the Act to Establish the Department of Agriculture. One of the early actions of the first Commissioner, Isaac Newton, was to appoint the chemist Charles M. Wetherill, who set up a laboratory and started busily analyzing grapes for winemaking and sweet sorghum for sugar. Analyses of feeds, fertilizers, soils, and experimentally produced crops became more important over the years. In 1901 the Bureau of Chemistry was created, with the famous Dr. Harvey W. Wiley as Chief. His dynamic leadership and crusade for pure foods resulted in the Food and Drug Act of 1906, which significantly increased the Agricultural and Food Division staff.

By the 1930's there was an Industrial Farm Products Division, a Food Research Division, a Carbohydrate Division, a Fixed Nitrogen Laboratory, and several other research departments. In the heart of the Great Depression farm prices hit modern lows and the Chemurgic Movement was born. Cotton was meeting stiff competition from rayon; and industrial uses of corn and of the new crop, soybeans, were seen as possible saviours of the farm economy. Frozen foods were on the way, but the technological base was small.

Several most important developments in agricultural research were brought about by the Bankhead-Jones Act of 1935. It authorized basic research in agriculture: the Allergens Laboratory of Henry Stevens, E. Jack Coulson, and Joseph Spies was set up; the Enzyme Laboratory of A. K. Balls with Hans Lineweaver and Eugene E. Jansen was started (later to become part of the Western Regional Research Laboratory); research in plant physiology and pastures (at Penn State), in animal diseases, and in other areas of agricultural research.

Perhaps the most important of these new laboratories, for present purposes, was establishment of the Soybean Laboratory at Urbana. Concerned with the industrial utilization of soybeans, this laboratory opened in 1936 with the aggressive leadership of O. E. May and a young and enthusiastic staff, including Reid T. Milner, Klare Markley and Samuel B. Detwiler, Jr. The Soybean Laboratory became almost a blueprint for the later four Regional Research Laboratories.

In 1937–38, Senator Bilbo of Mississippi proposed the formation of a cotton laboratory to his colleagues on the Agricultural Committee. The Chairman and other members from the rest of the country could hardly support just a cotton laboratory; so the resulting Agricultural Adjustment Act of 1938 called for the Department ". . . to establish, equip and maintain four regional research laboratories, one in each major farm producing area, and, at such laboratories, to conduct research into and to develop new scientific, chemical, and technical uses and new and extended markets and outlets for farm commodities and products and by-products thereof . . .", as well as funds" . . . not to exceed $100,000—to conduct a survey to determine the location of said laboratories and the scope of the investigations to be made . . ."

Although the Act was signed February 16, 1938, the Department did not request the Bureau of Chemistry and Soils to carry out the required survey until June of that year. The survey was completed in about 6 months and the report (Senate Document No. 65, 76th Congress) was submitted to the Congress on April 6, 1939. In that time over 10,000 research projects in Federal and State agencies had been reviewed, over 1000 industrial groups consulted, as well as some 200 institutions and groups such as all the Land-Grant Colleges, the Rhode Island School of Design, the Oregon Grange, and the Yakima Valley Traffic and Credit Association. A set of ground rules for selection of locations had been set up, some 80 localities visited, and the 4 locations chosen. Plans for the buildings were under way, the Directors of the laboratories had been appointed and a basic commodity-oriented structure had been decided upon, with commodities assigned to each laboratory.

The Survey Report (337 pages) is followed in Senate Document No. 65 by a "Coordinated Research Program" of 65 pages. This proposed a relatively detailed plan of both basic and applied research on each commodity as well as cross-commodity basic work on proteins, carbohydrates, and vegetable oils. Food research was not in the original program, but Congress later directed its inclusion.

All of these things had been accomplished in 9 months! H. T. Herrick headed the planning committee; those in charge of the survey in each area, all of whom became the first Directors of their respective Laboratory: O. E. May, Northern; P. A. Wells, Eastern; T. L. Swenson, Western; and D. F. J. Lynch, Southern. All were experienced research leaders within the Bureau of Chemistry and Soils. Wells was a very youthful 32.

The appropriation for each laboratory was $1 million annually. The design called for a U-shaped building, but the first $1 million was expended on an L shape, a long laboratory wing with offices and services in the front or base of the L. The second year the contract was let for the chemical engineering pilot plant wing, completing the U, and fixed equipment was bought. A cadre of about 25 key staff members was recruited. The laboratories were occupied by a skeleton staff in the fall and winter of 1940, except for the Southern Laboratory, the completion of which was delayed until early 1942 by a fire in the half-completed attic.

Dr. J. J. Willaman, the great plant biochemist, says that moving into a laboratory without benches, equipment, or chemicals is one of his clearer memories of the early days at the Eastern Laboratory. There wasn't even toilet paper! He brought a roll which he kept in his desk and shared with his secretary.

Recruitment of staff in 1940–41 was rapid and quite successful. Because jobs were difficult to find, the Laboratories had their choice of new Ph.D's and young university staff members. M. J. Copley, (Univ. Illinois), C. H. Fisher (Illinois, Harvard, Bur. Mines), R. J. Dimler (Univ. Wisconsin), F. R. Senti (Johns Hopkins), and Ivan Wolff (Univ. Wisconsin) all moved up to become Directors of Laboratories. George Washington Irving, Jr., became Administrator of the Agricultural Research Service, and Aaron Altschul (Chicago) received a Rockefeller Public Service Award.

The planned program was disrupted by World War II (as it was later by the Korean conflict). Much of the research staff was put on so-called War Projects. The Western Laboratory mounted a large-scale project on dehydration of vegetables and fruit at the request of the Army and sent staff out to assist dehydrators. The Eastern and Southern Laboratories studied sources of natural rubber other than Hevea, (guayule, Kok-saghyz, and others in the east; Goldenrod in the south). Of course, the war ended long before a new agricultural industry could be established, and these natural rubber studies had no economic value at that time.

However, major contributions were made to the war effort that have continuing value. The Rubber Development Corp. had financed a crash program construction of large plants to produce GR-S rubber, going from a small pilot-plant scale in one bold step. The polymerization of butadiene was erratic, and the degree of polymerization was low. Waldo Ault, Brooks Brice and co-workers at Eastern Regional Research Laboratory showed that the commercial emulsifiers (soaps) used in these plants contained polyunsaturated fatty acids, which acted as chain breakers. They developed methods of hydrogenation to eliminate the polyunsaturates, and prepared purchase specifications and control tests to validate each lot of emulsifier. The results were excellent and the quality and quantity of GR-S improved dramatically.

The work at Northern Regional Research Laboratory on penicillin was classic! Florey and Heatley of England had convinced themselves that it was a tremendously valuable drug. At that time they recovered part of their penicillin from the urine of patients under treatment. In July of 1941 they came to the United States for help in its production. P. A. Wells (1975) tells a fascinating account of this: He sent the Englishmen to the Northern Regional Research Laboratory, where his fermentation group had been transferred. Using know-how developed in many successful fermentations (gluconic acid, citric acid, sorbose, etc.), a deep tank fermentation process for penicillin was rapidly established with a hundred-fold increase in yield. New strains of penicillia, use of lactose as a substrate, and corn steepliquor as a growth adjuvant made the commercial production of penicillin by deep tank fermentation a success. Later, other, more active strains were produced at Cold Spring Harbor, by X-ray irradiation, and the University of Wisconsin by ultraviolet radiation. In addition, lactose was replaced by glucose and many people in many laboratories contributed to the development of a group of penicillins with differing structures; these have often been called "miracle drugs."

This is a good place to make an important point. With many industrial developments it is very difficult to pinpoint the exact contributions of specific laboratories. Industrial research laboratories have the NIH syndrome—Not Invented Here! You can't expect anything else. What Director of Research will go to his management and say, "Here's a great product. We picked it up at the ACS meeting from a talk by Bill Tallent of the USDA Laboratory in Peoria." No; the usual thing for an industrial product developer to say is, "You had a clever idea, but it just didn't work out right; we had a devil of a time working it into a satisfactory process. We've applied for patents on the key steps, so I really can't talk about it."

In today's complex world, few product developments are the work of one person or one group. So there are fully justifiable claims from a number of groups for having made important contributions to developments that affect an entire industry. This latter point was previously demonstrated in the penicillin story.

Following are a number of cases in which the Regional Laboratories were in the forefront of solving major problems, although other laboratories also made valuable contributions.

The Southern Laboratory, for example, has fought an unceasing battle to improve cotton products and thus maintain its market, with limited success. Durable-press finishes for cotton have been under development since 1945, with strong supporting research on yarn and fabric structure, this "why" of wrinkling, and the organic chemistry of additives and their condensation with the fiber. Today, durable-press goods, many of them cotton blended with synthetics, dominate the apparel and household textile markets. Wilson A. Reeves, who recently retired after many years of handling the chemical side of these developments, has received some 9 awards for his work, including the Olney Medal of AATCC. J. David Reid of AATCC received the John Scott Award of the Franklin Institute in 1972.

The history of flame-retardant finishes is quite similar to that of durable-press treatments. The Southern Laboratory developed several finishes; all were effective in control of burning, but they had varying resistance to laundering. Much initial work was directed at bed sheets, since fires from smoking in bed cause many deaths and much property damage. Later emphasis was shifted to children's clothes because of public interest and the Flammable Fabrics Act. Today, fire retardants are most significantly used on carpets and rugs. The Southern Regional Research Laboratory treatments are now used by some eight firms. George L. Drake, Jr., who directs this research, has gained international recognition for his work on flame resistance.

Soybean oil has become the predominant food oil in the United States, largely because of the work at the Northern Regional Research Laboratory. Over a number of years, both basic and applied studies of off-flavors and their causes have led to the citric acid complexing of metals that catalyze oxidation, better use of antioxidants, and selective hydrogenation of linolenic acid. John C. Cowan and Herbert J. Dutton led this research. Both these men and their co-workers have received numerous awards from AOCS, as well as Canadian, British and French scientific societies. Here again, they worked closely with industry, which made many contributions to the overall success of the research on soybean oil.

The most important food processing development in recent years, frozen concentrated orange juice, came about in a curious way. The Florida Citrus Commission put three research workers—L. G. Mac-Dowell, E. L. Moore, and C. D. Adkins—in the USDA Winter Haven Laboratory in the early 1940's where they worked under the supervision of Matthew K. Veldhuis, and with the assistance of others on the staff. They developed the basic process of over-concentration and add-back of fresh juice, control of peel oil concentration, etc., and took out a USDA patent on it, as had been agreed. After World War II, the product took off; by 1970, Americans were buying $1 million worth of frozen concentrated orange juice a day.

There were some developments that were almost solely the result of Regional Laboratory Research. In many cases, the clear-cut impetus to a successful process or product came from one of our Laboratories.

Glutaraldehyde tannage of leather, an effective, stable, inexpensive tannage, was discovered in 1956 by E. H. Filachione and co-workers and commercialized in 1958 (a remarkably short time!). But the tanneries then did not have research on new tannages; nor did they have the NIH syndrome. Over 20% of U.S. tanning now uses this process. E. H. Filachione, W. Windus, E. F. Mellon, J. Naghski, A. L. Everett, S. H. Fairheller (all members of the Hides and Leather group at the Eastern Regional Research Laboratory) received the Alsop Award of the American Leather Chemists Association (ALCA). Everett also received another ALCA Award. (A number of these awards were for work not directly related to glutaraldehyde tannage.)

The Western Laboratory, now under the leadership of Arthur I. Morgan, Jr., set out in the late 1960's to alter vegetable processing so that there would be lesser amounts of waste. Of course, product quality had to be maintained. The peeling of potatoes contributes about 75% of the BOD of the effluent in the conventional lye peeling process. By innovation and hard work, R. P. Graham and his team developed the so-called dry caustic peeling method that essentially eliminates liquid effluents. A larger amount of product is thereby recovered and the peel, combined with trim, is converted to a useful cattle feed. The overall economics are, therefore, so favorable that dry caustic peeling has been rapidly adopted and, with modified conditions and equipment, extended to other root crops as well as tomatoes and peaches. The Western Laboratory received the Industrial Achievement Award of IFT for this work in 1972.

A relatively unique research group was established in the Southern Regional Research Laboratory to investigate and improve cotton mill machinery. True, there had been few basic changes in this age-old industry in 200 years, but many ingenious people had worked in the mills and there had been many step-by-step improvements in speed and efficiency. One invention was the fiber retriever, a simple device that takes out more trash and short fibers and reduces loss of spinnable fiber; over 20,000 are now in use. Another development was an Opener-Cleaner now installed in over 25 mills with the combined capacity to process 1/3 of all cotton processed in the United States. The Granular Card, using a new principle, was also invented and over 10,000 have been installed in mills.

This group also, with an operating mill in the Laboratory, established optimum conditions for the use of various cottons in different fiber and fabric structures. The results of this work have been a marked improvement in cotton mill speed and efficiency. Ralph Rusca directed this unusual engineering program.

Many groups have attempted the recovery of leaf protein for feed and food uses, most notably Pirie and associates of Rothamsted, England. The product has had serious drawbacks, such as dark greenish color, bitter flavor, insolubility, and poor digestibility. Dr. George O. Kohler of the Western Laboratory had a new idea. Alfalfa, as now produced, contains about 23% protein, far above the commercial level of 17% for dehydrated alfalfa. A high protein fraction can be obtained by pressing the fresh-cut alfalfa, thus partially drying it so that the dryer throughput is appreciably increased. The end-product meets all domestic and export specifications for quality. The press-juice, when dried, is an excellent feed ingredient, especially as a source of xanthophyll in poultry feed. It can be further treated to produce a bland, white protein of potential food use. Pilot-plant studies based on this work are under way in India and Yugoslavia.

The Fermentation Laboratory at the Northern Regional Research Laboratory did not rest on its penicillin work (Ward 1970). Later accomplishments were fungal amylase for production of ethyl alcohol, dextran for use as a blood plasma extender (discovered a new and better strain of the organism), a new sodium gluconate fermentation, and improvements in Vitamin B-2 and Vitamin B-12 production processes. The fermentation staff also set out to produce a gum-like product, a thickener, by fermentation. Xanthan gum (from *Xanthomonas campestris*) resulted. This gum has unique properties in that the viscosity of solutions is stable over wide ranges of temperature, salt concentration, and acidity. Crude xanthan gum products are used in oil-well drilling and flooding for secondary recovery of oil. Food uses of this gum developed rapidly after FDA clearance was obtained, based on toxicological studies by the Western Laboratory. This industrial development was by Kelco, of San Diego, California, a major producer of vegetable gums from kelp. Notably, the NIH syndrome did not operate in this case. Kelco gave full credit for the pioneering work of the Northern Laboratory, and the two organizations jointly received the Industrial Achievement Award of IFT in 1974.

The first IFT Industrial Achievement Award (1959) was received by the Eastern Laboratory that year for the development of potato flakes by Miles Willard and James Cording, Jr., who worked in the Chemical Engineering Laboratory under Roderick K. Eskew. This rather simple process is based on drum-drying under carefully controlled conditions so that the starch granules are not broken. Today over 120 million pounds of potato flakes are produced annually in 13 American plants, and the inventors have licensed the process in several foreign countries.

There are many other accomplishments. In 1966, the authors coordinated a review of accomplishments over the 25 years of the Regional Laboratories' existence (Agricultural Research Service 1966). There were 109 commercialized products and processes, 28 of potential applicability, and 20 major basic contributions. There also has been a lot of excellent basic organic, physical, analytical and biochemical research which has increased knowledge of our agricultural products and supported the applied research.

As part of the fermentation program at the Northern Laboratory, a culture collection was established under the leadership of Kenneth Raper, who later was elected to the National Academy of Sciences for his research in mycology. Bacteria, actinomycetes, fungi, and yeasts of known or potential usefulness are studied and maintained, as well as toxin-producing organisms. C. W. Hesseltine led this study with distinction; it is now led by T. J. Pridham. L. J. Wickerham, a world-recognized authority on yeasts, has had 4 species and 1 new genus named for him.

A. K. Balls, also a member of the National Academy, and his Enzyme Laboratory were transferred to the Western Regional Research Laboratory in the 1940's. Hans Lineweaver and Balls had previously crystallized papain, and Eugene F. Jansen crystallized chymopapain. Later, Lineweaver and Jansen worked out the pectin enzymes—polygalacturonase and methyl esterase—in a classical piece of work.

At the Eastern Laboratory, R. W. Jackson set up the Protein Division to study milk proteins, recruiting T. L. McMeekin from Harvard along with a number of others. Robert C. Warner of this group separated alpha and beta casein, and a steady stream of reports on the isolation of the milk proteins and their properties came from other members. The Borden Award in the Chemistry of Milk was won by T. L. McMeekin (1951), S. R. Hoover (1956), W. G. Gordon (1958), C. A. Zittle (1959), S. M. Timasheff (1963), and M. P. Thompson (1970). M. J. Pallansch, of the Dairy Products Laboratory, won the Borden Award in Dairy Manufacture in 1963. A number of the research workers of the Dairy Products Laboratory had won these awards before that laboratory became a part of the Eastern Laboratory in 1954.

The AOAC was not strongly supported by the Regional Laboratories for some time because some of its program leaders opposed the time and effort of collaborative work. There were exceptions among these leaders—three especially—whose jobs were basically in analytical chemistry. They were all recipients of the Harvey W. Wiley Award— Robert T. O'Connor of the Southern Laboratory for his work in spectroscopy (1967), and C. O. Willits (1969) and Clyde L. Ogg (1972), both of the Eastern Laboratory, for their work in putting the official methods on a micro basis.

One basic analytical technique, Thin-layer Chromatography (TLC), was discovered at the Pasadena Laboratory by J. G. Kirchner, G. J. Keller, and J. M. Miller. Originally developed for separating and identifying the volatiles of citrus, this technique has been used for analyzing and characterizing almost all classes of hydrophilic and hydrophobic organic compounds and it has broad applications in inorganic chemistry.

In addition, the innumerable contributions of Roy Teranishi and Ron Buttery of the Western Laboratory to Gas-Liquid Chromatography (GLC) over the past 20 years are clear in the minds of bench chemists throughout the world. The active participation of these two men in the Agricultural and Food Chemistry Division will continue to serve us all.

As stated earlier, the research and development studies cited are only examples of the types of work done. There are also the Western Laboratory's work on wool, the Northern Laboratory's basic and applied studies on starch, the excellent work of the Food Fermentation Laboratory (at North Carolina State) on pickling, cheese research at the Dairy Products Laboratory of the Eastern Laboratory, the successful development of new aerobic processes for treating dairy plant wastes at the Eastern Laboratory in the early 1950's, poultry research at the Western Laboratory now going forward at the Southeastern Laboratory, aflatoxin research at the Northern and Southern Laboratories, as well as other notable research and development studies and accomplishments that have benefited the country. First, there is the body of basic and applied research accomplishments such as those here mentioned. These benefit everyone in the long run. Second, there has been a detailed economic study on the report of research accomplishments by Harold B. Jones, Jr., of the Economic Research Service of USDA, in his Ph.D. thesis in the University of Georgia, and in two later journal papers (Jones 1971, 1973, 1975). Jones essentially concludes that 9% of the projects in the years 1939–1966 had an economic return. This compares favorably with the success rate of the food industry: 7.2% for all products entering the test stage (Buzzell and Nourse 1967). The cost of these projects was $43 million, or approximately 1/7 of the approximately $300 million cost of the total program over these years. The estimated total benefits were $16.6 billion. If the benefits were discounted at 6% (in other words, *not* doing the research and investing the money to return 6%, a favorite concept of economists), the resulting benefits would be $7 billion, a very satisfactory economic return.

These figures are, of course, suspect for they contain many judgments and estimates.

On the other hand, they relate solely to the dollar value of product and process developments. There is no way to evaluate the many contributions the UDSA Regional Laboratories have made to nutrition, to food safety and health, and to basic science. Unquestionably, one must conclude the USDA Regional Laboratories have proven their value in social, economic and scientific spheres of research activity.

BIBLIOGRAPHY

BUZZELL, R. D., and NOURSE, R. D. 1967. Product Innovation in Food Processing, 1954–64. Graduate School of Business Administration, Harvard University, Boston, Mass.

HOOVER, S. R. 1965. Research and purpose. Science *147*, 1532.

JONES, H. B., JR. 1971. Economic appraisal of the role of government in agricultural utilization research and product development. Ph.D. Thesis, Univ. Georgia.

JONES, H. B., JR. 1973. Evaluating cost and output levels for agricultural utilization research. Southern J. Agric. Econ. *5*, No. 2, 89–97.

JONES, H. B., JR. 1975. Empirical success ratios in USDA agricultural utilization research. Southern J. Agric. Econ. 7, No. 2, 123–128.

USDA. 1966. Achievements in agricultural utilization. Agric. Res. Serv., U.S. Dep. Agric.

WARD, G. E. 1970. Some contributions of the U.S. Department of Agriculture to the fermentation industry. Adv. Appl. Microbiol. *13*, 363–382.

WELLS, P. A. 1975. Some aspects of the early history of penicillin in the United States. J. Washington Acad. Sci. *65*, No. 3, 96–101.

5

Contributions of USDA in Nutrition and Food Research

Willis A. Gortner

The past century has been one of change. Food continues to be a part of this change. We have evolved new patterns in what we eat and where we eat. We have new technologies and new foods in the marketplace. We have a new awareness of the "hidden hunger" of improper nutrition in an era of plentiful food.

When we talk about food, we think of agriculture. Where has agriculture contributed? How has it responded to these challenges? A look at the record reveals some interesting data.

The federal government and the individual States began their efforts in nutrition nearly a century ago. They began with the nutrition of domestic animals, and studied the composition and digestibility of animal feed, not human nutrition. For a century there have been efforts to redirect priorities, so that human nutrition and food receive attention comparable to that given the nutrition and feeding of farm animals. The efforts continue, but the goal has yet to be attained.

The first major effort was in 1893, when Dr. W. O. Atwater (Fig. 5.1) urged recognition of the need for the Federal Government to be involved in human nutrition investigations. The result was a congressional appropriation of $10,000 to the Department of Agriculture to initiate the first nutrition investigations of the national government on "the nutritive value of the various articles and commodities used for human food" (Fig. 5.2). A Brookings Institution monograph states that "The nutrition work . . . was consistently expanded from year to year . . ." but cites figures showing that 7 years later, the appropriation had reached a mere $20,000. Indeed, only in the last decade has there been significant growth in support for human nutrition research by USDA.

Nevertheless, their productivity and significant pioneering re-

FIG. 5.1. DR. W. O. ATWATER

searches have provided us with many "bench marks" of the eight decades of USDA research in nutrition and food. (Reference here for the most part will be referring to research of the Agricultural Research (ARS) and its predecessors, and even then will be able to hit only a few highlights.) During these 8 decades the developing nutrition programs have been under the guidance of many outstanding men and women. Many of them have won distinguished places in the history of the biochemistry of metabolism, food chemistry, and nutrition. The Human Nutrition Research Coordinators for USDA

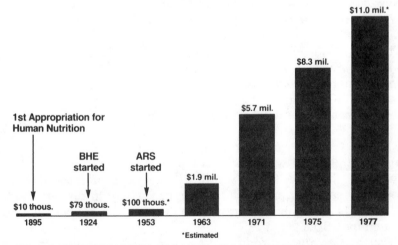

FIG. 5.2. APPROPRIATIONS FOR USDA FOOD/NUTRITION (ARS AND ITS PREDECESSORS)

from 1894 to 1976 were Wilbur O. Atwater, Charles F. Langworthy, Louise Stanley, Lela E. Booher, Esther L. Batchelder, Henry C. Sherman, Hazel K. Stiebeling, Callie Mae Coons, Ruth M. Leverton, and Willis A. Gortner.

Around the turn of the century, Wilbur O. Atwater made outstanding contributions to many facets of the broad spectrum we now encompass as "nutrition." As the first director of the Federal Program in Human Nutrition, he conducted pioneering studies on human energy expenditure under different stress conditions (Fig. 5.3). These metabolic

461

U. S. DEPARTMENT OF AGRICULTURE.

OFFICE OF EXPERIMENT STATIONS—BULLETIN NO. 109.

A. C. TRUE, Director.

EXPERIMENTS

ON THE

Metabolism of Matter and Energy in the Human Body,

1898-1900.

BY

W. O. ATWATER, PH. D., AND F. G. BENEDICT, PH. D.,

WITH THE COOPERATION OF

A. P. BRYANT, M. S., A. W. SMITH, M. S., AND J. F. SNELL, PH. D.

WASHINGTON:
GOVERNMENT PRINTING OFFICE.
1902.

FIG. 5.3. ATWATER'S BULLETIN
100 ON HUMAN ENERGY EXPENDI-
TURES

studies utilized the first respiration calorimeter suitable for human
subjects. These subjects could eat, sleep, and work in the test chamber for days and even weeks while their heat input and output were
measured. The Atwater-Rosa Human Calorimeter (Fig. 5.4) made
possible Atwater's calculation of energy conversion factors for fat,
protein, and carbohydrate. These factors allowed precise calculation
of energy values for individual foods or diets. The conversion factors
(9 calories per gram from fat, 4 calories from carbohydrate or protein)
are still in use today; they are cited in the latest edition of the Recommended Dietary Allowances of the National Academy of Sciences-National Research Council as the means for computing the energy
content of customary diets.

As a second facet of research during this first decade, USDA
under Atwater began a series of dietary studies to assess the food intake of persons with different cultural backgrounds and income
levels (Fig. 5.5). This research set the course for continuing and major
USDA involvement in dietary studies in this country. Indeed, such a
study is now being planned—a USDA nationwide survey of food consumption of households and of individuals, scheduled to begin early
in 1977.

A third pioneering activity in these early years of nutritional research was extensive analyses of foods, and the first compilation of
knowledge on the composition of foods (Fig. 5.6). Amazingly, the re-

FIG. 5.4. ORIGINAL ATWATER-ROSA HUMAN CALORIMETER

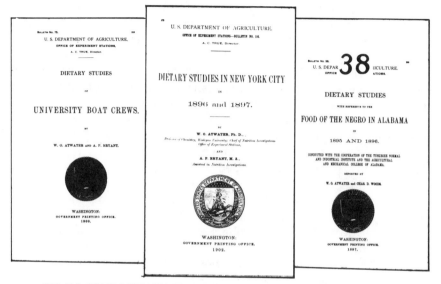

FIG. 5.5. SOME DIETARY STUDIES OF ATWATER'S GROUP, 1895-1900

BULLETIN No. 28.

U. S. DEPARTMENT OF AGRICULTURE.

OFFICE OF EXPERIMENT STATIONS.

THE CHEMICAL COMPOSITION

OF

AMERICAN FOOD MATERIALS.

BY

W. O. ATWATER, Ph. D.,

AND

CHAS. D. WOODS, B. S.

WASHINGTON:

GOVERNMENT PRINTING OFFICE.

1896

FIG. 5.6. THE FIRST FOOD NUTRI—
ENT TABLES, USDA BULLETIN 28

vised edition of this first bulletin in 1906 served as a reference standard for the next 34 years—after which a larger and more complete compilation finally emerged as USDA Circular 549, *Proximate Composition of American Food Materials*.

In the 1920's, reports appeared in print on some of the early USDA research studies concerning the biological availability of nutrients as they occur in foods. This included data on human utilization of the calcium in spinach. This calcium, surprisingly because of the known oxalate content of spinach, proved to be well utilized by the body toward meeting its need for this mineral.

During this decade, USDA issued the first comprehensive report on vitamin content of food (USDA Circular 84, *Vitamins in Food Materials*). During the 1920's research on vitamins was the exciting new frontier of an expanding interest in nutrition.

In the 1930's USDA researchers published the first quantitative work on the Vitamin A requirement of adults. This work demonstrated the possibility of obtaining quantitative data on human vitamin requirements. But it is regrettable that even now, some 37 years later, no definitive studies have been done to quantitate the Vitamin A requirement of teenage children, pregnant women, or elderly people.

The practical applications of nutrition research call for translating the findings into guidance material for use by nutrition educators or consumers. Early in the 1930's, the USDA developed flexible food plans to help limited income families select foods that would meet their nutrient needs. Over the years, these plans have been updated and expanded, the latest being the "Thrifty Food Plan" used to guide low-income families and to set the allotments for the Food Stamp Program.

The USDA Food and Nutrition Program was stepped up in the 1940's, despite the limited availability of funds. The first comprehensive survey of America's nutritional state, involving some 12,000 persons throughout the U.S., was undertaken in cooperation with numerous State Agricultural Experiment Stations. This culminated in a summary report by Agnes Fay Morgan, entitled *Nutritional Status—USA*.

A major contribution from USDA research came in 1950 with the first appearance of Handbook 8, *Composition of Foods—Raw, Processed, Prepared*. These new tables of food composition brought together knowledge not only of the proximate composition of foods, but also of 5 vitamins and 3 minerals, concerning some 750 food items. During the 1950's this information was supplemented by food composition tables on folic acid (Agriculture Handbook 29), pantothenic acid (Agriculture Handbook 97), amino acids in foods (Home

Economics Research Report 4), and fatty acids in food fats (Home Economics Research Report 7). In 1961, these were followed by publication of values for the vitamin B-12 content of foods (Home Economics Research Report 13). And in 1963, a revised and expanded Agriculture Handbook 8 appeared which covered nearly 2500 food items and some 20 nutrients, including 3 additional minerals and other food components of interest to the nutritionist, the dietician, and the food chemist.

USDA research contributions in the 1950's ranged from basic findings to applied programs of nutrition education. The very limited funds for contract and grant support of research led to the definition of the essential fatty acid requirement for infants and children. USDA extramural support also allowed nutrient metabolic studies at Nebraska, Wisconsin, and UCLA on over 50 young women, providing the first data quantitating their essential amino acid requirements. This series of studies established the requirements of women for threonine, valine, tryptophan, phenylalanine, leucine, methionine, isoleucine, and lysine.

During the 1950's it became evident that the dietary guidance device of the "Basic 7" food groups, which the USDA had helped devise in the war years, needed to be simplified if it was to be effective in nutrition education. Thus, USDA developed a new dietary guide based on knowledge emerging from research on nutritional needs, nutrients in foods, and food consumption habits. This new guide (Fig. 5.7) became known as the "Basic 4," and categorized

FIG. 5.7. THE "BASIC 4" FOOD GUIDE

foods into four main classes which could help assure nutritional adequacy of diets. It is still in use as a basic tool in dietary guidance.

The year 1955 saw USDA take the lead in a nationwide survey on food consumption and its implied effects on dietary levels. A decade later, the nationwide surveys were extended to include household food consumption in each of the 4 seasons, including individual food and nutrient intake in the Spring. For the first time, we had a picture of the diets of different sex and age groups of the U.S. population —their food needs and their nutrient needs.

Basic nutrition research has also maintained a prominent position in the USDA program. In the 1960's, research on chromium nutrition, under USDA leadership, established the essentiality and the role of the chromium-containing "glucose tolerance factor" in foods, demonstrating its key influence in working with insulin to regulate carbohydrate metabolism.

The small USDA extramural program allowed support of research on the effect of high protein levels on mineral metabolism (Table 5.1). Urinary losses of calcium proved to be so great that even a high dietary intake of calcium failed to keep the body in calcium balance. The body was unable to adjust to these calcium losses even after more than two months of being on a high protein diet. This raised some real concerns about the possibility that calcium losses induced by a high protein diet may, over the years, predispose one to osteoporosis, a common bone affliction of the elderly.

During the last decade, USDA has assembled an internationally recognized team of researchers on trace element nutrition. In-house studies in the 1970's have broken new ground in establishing the metabolic essentiality of two new trace mineral elements, nickel and vanadium. Each of these micro nutrients is needed by such species as rats and chickens, though in the extremely small range of parts-per-

TABLE 5.1

CALCIUM BALANCE 65-DAY PERIOD
820mg CALCIUM AND 142g PROTEIN DAILY
(INTAKE MINUS EXCRETION FOR FIVE MALES)

Days	Calcium Balance
1–5	–134
6–15	– 87
16–25	– 58
26–35	– 71
36–45	– 30
46–55	– 21
56–65	– 45

SOURCE: Data of Linkswiler, H., *Progress Report to ARS-USDA*, 1974.

billion in their diet. The Food and Nutrition Board cites these studies in the latest Recommended Dietary Allowances issued by the National Academy of Sciences.

Exciting new information has been coming from studies on other essential minerals, such as zinc, chromium, and iron. For example, rats which had been zinc deficient *in utero* but nutritionally rehabilitated after birth still showed marked behavioral deviation as adults (Fig. 5.8). What is noted here is the behavioral effects of such a diet history: Females that had been zinc deficient *in utero* became highly aggressive after a mild shock treatment; but where there had been no history of zinc deficiency, this aggressive behavior was not observed in the adult female. In other studies, learning was severly impaired following zinc deficiency, even when the deficiency occurred only *in utero*.

A study financed by a USDA grant showed that clinical deficiency of zinc is not uncommon in the United States. Many presumably healthy youngsters from middle class families showed a zinc deficency, as evidenced by poor taste acuity, poor growth, low zinc levels in hair or blood, and poor appetite. These symptoms were all reversed following zinc supplementation.

Our understanding of man's need for chromium in the diet is rapidly improving. USDA researchers in the last two or three years have shown that the essential and biologically-active form of trivalent chromium in foods is a complex organic molecule (Fig. 5.9). It contains two nicotinic acid molecules and several amino acid molecules as ligands on the coordination sites of the chromium. (This formula is a current guess as to how the amino acids may be coordinated around the chromium.) This essential food factor may prove to be a new vitamin. Adequate chromium nutrition is a requisite for

FIG. 5.8. AGGRESSIVE BEHAVIOR OF FEMALE RATS WHO WERE ZINC DEFICIENT *IN UTERO*

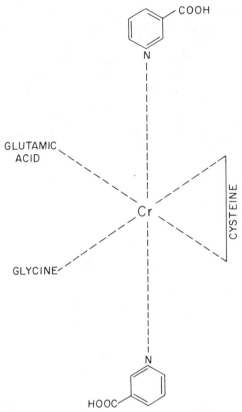

FIG. 5.9. PROVISIONAL FORMULA FOR
"GLUCOSE TOLERANCE FACTOR"

the full activity of insulin in regulating carbohydrate metabolism. Accordingly, the role of food in minimizing impaired glucose tolerance and regulating blood sugar is a research area of keen interest to food scientists, nutritionists, and health professionals.

Iron as a nutrient, and its role in anemia, is not at all new. For the past 30 years, wheat millers have been adding iron for enrichment of white flour. Only in the past few years have some of the puzzles about dietary iron been clarified. One basic truth emerged from USDA-supported grant research; another came from shattering an assumed "fact" as a result of in-house studies.

The first truth is that diet makeup is more important than the individual chemical forms of iron in the many American foods. Heme iron, such as occurs in red meats, is a good, biologically available source of iron; it is not affected by other components of the diet and is readily digested and absorbed. Iron occurring in foods in most other forms, however, seems to be absorbed as if the iron went into a common pool; for this pool, the makeup of the diet can be important in regulating whether considerable absorption or negligible absorption of the iron occurs. For example, if a food high in Vitamin C, such as citrus, is consumed along with the iron-containing food, good absorption of the iron is likely. If the two foods are consumed at different times, the synergistic effect is lost and iron absorption may well be poor.

The myth about iron that USDA studies dispelled within the past couple of years deals with the role of phytate. It was commonly held that food phytate depressed the biological usefulness of many metals, such as calcium, zinc, and iron. Yet USDA studies in the early 1970's proved that some high phytate foods, such as whole wheat, were quite good sources of some of these same minerals. Following this lead, one of our scientists took phytic acid out of the nutritionists' doghouse. He found that monoferric phytate is not only a major form in which iron occurs in the wheat kernel, but also that this chemical form has a biological value fully equivalent to the best iron salt. Perhaps even more surprising is the information given later at nutrition meetings: zinc phytate is also highly useful in meeting the zinc needs of the rat!

In summary, USDA research has sought the best use of our bountiful food supply for the nutritional health of all Americans. There are many exciting, provocative, significant contributions coming from the USDA Program in Nutrition and Food Research. We have come a long way from the pioneering beginnings of W. O. Atwater some 82 years ago, at a time when the American Chemical Society was a mere fledgling of 17 years. The future holds even greater promise, for the American Chemical Society and the USDA are now in their prime.

6

Food Science and Technology: Past, Present, and Future-The 1976 W.O. Atwater Memorial Lecture

Emil M. Mrak

.

The author has known about the prestigious Atwater Memorial Lecture since it was established in 1967 by the USDA Agricultural Research Service. The Lecture was to provide recognition in the fields of science that have contributed so much to the improvement of human nutrition, and the advancement of public understanding of problems involved in meeting world food needs. Needless to say, these concerns are now more important than ever. Chemistry, of course, has always played an important role in advances made in the fields of food and nutrition.

During the author's student days, many references were made to the nutrition work of Dr. W. O. Atwater, as well as his contributions to all of agriculture. Dr. J. George Harrar, in his Atwater Memorial Lecture, said quite aptly that Dr. Atwater ". . . embodied a fortunate combination of scientist and humanist; one who saw both the need and the opportunity to make science function effectively in behalf of the well-being of mankind. With a basic interest in agriculture and its role in human nutrition, Atwater effectively wedded his scientific and humanitarian concerns." Dr. William J. Darby, in his Atwater Memorial Lecture, cited as one of Atwater's great accomplishments, the Hatch Act of 1887, establishing in the Department of Agriculture an Office of Experiment Stations of which Atwater became Chief. With this Act, the seeds were sown for the evolution of the fields of nutrition and food science.

Later, Dr. Harvey W. Wiley, a chemist for the Department of Agriculture, and one of the first great pioneers in the field of food chemistry and toxicology, worked hard for the establishment of the Pure Food and Drug Act which became a reality in 1906. Wiley was the first to use a so-called "poison squad," with a number of volunteers from the Department of Agriculture serving as subjects for human toxicological research. An agricultural chemist in the Department of Agriculture, therefore, was the first U.S. scientist to be actively concerned with the safety of foods and food additives. This great man developed within the USDA an organization we now know as the Food and Drug Administration.

The U.S. Department of Agriculture did much to develop an early interest in foods and nutrition, and has continued to be active in these areas throughout the years. The early development of food science as a field of study parallelled the evolution of the food industries in the United States, particularly those concerned with canning and with dairy products. This is described quite well in a book by Bitting (1937) entitled, *Appertizing or the Art of Canning.*

Food science and technology first found their way into universities as a result of spoilage problems in the canning industry. As a matter of fact, the canning industry started to evolve in the United States, problems developed, and university scientists were called on for help. The first of these were H. L. Russell of the University of Wisconsin and Samuel C. Prescott of MIT. Both of these scientists studied the microbial spoilage of canned corn, the nature of the organisms involved, and the use of proper sterilization techniques. These early studies were done between 1895 and 1900.

Later, Andrew Macphail, a professor of history of medicine at McGill University, studied the discoloration of canned lobster. When he published his results, Bitting said, "This paper is all the more remarkable, since it presents a combined bacteriological and chemical study on the same product, the first of its kind ever reported." So here we have not only microbiology, but also chemistry involved in the slowly developing field of food science.

The Underwood Company in Massachusetts had experienced significant spoilage losses. Prescott studied the problem, determined the cause and its solution and, as a result of this experience, soon saw the need for teaching in this area. He then went on to develop courses of instruction in the Department of Biology at MIT.

Early in the twentieth century, Professor W. V. Cruess, a chemist at the University of California, became interested in enology, the art and science of wine making. He applied his chemical talents to the improvement of California wines and to the development of courses in the art and science of wine making. All this, however, terminated with prohibition. Rather than giving up in despair, Cruess turned his energies toward the development of courses in food chemistry, food preservation, food microbiology, utilization of surplus crops, and toward solving problems related to the canning and drying industries of California.

At the Oregon State University, Professor Ernest Weigand, a horticulturist, was faced, early in his career, with difficulties relating to the production of maraschino cherries and the freezing of berries. His attempts to solve these problems eventually resulted in the development of a teaching and research department concerned with horticultural products.

Teaching of dairy science actually preceded that of food science and technology as we know it today. There were many small creameries throughout the country and the need for research and teaching personnel for these establishments was great indeed. It is ironic that, as time went on, the need for such individuals decreased as a result of the growth and diversification of the dairy industry. Many colleges developed new teaching departments independent of those concerned with dairy science and dairy products, though eventually the trend went toward combining the two.

This new field did not evolve easily or gracefully. Scientists in other departments of agriculture looked down upon the new area of investigation. For some reason or other, it was perfectly scientific, respectable, and acceptable to work in the area of soil science and fertilization, including the use of manure, but not in a field concerned with the food we eat. As a matter of act, when the author first entered the field, he was termed by a soil scientist, concerned with fertilization, a "jam and jelly scientist." Be assured, this attitude has changed.

The pioneers in the field of food science and technology in USDA, in the universities, and in industry were remarkable individuals indeed. They had no orientation, they had no opportunity to develop a point of view, their training and experience were almost entirely limited to the basic sciences of chemistry and bacteriology, and, at times also, to one of the production fields. They realized there was a need; and their zeal, energy, and imagination enabled them to develop a new area of teaching and research. The breadth of their thinking, however, was quite narrow; the prevailing view was that food science and technology started at the time raw material reached the processing plant and ended when the processed product left the plant. This is certainly not the situation today.

As time went on there were other influences. World War I and especially World War II demonstrated clearly the need for food scientists and teaching in this field. It was difficult, for example, for the Quartermaster Food and Container Research Laboratory at Chicago to find scientists adequately trained in food science to fill its needs.

It was during this period that food science really came of age, for it was then that the importance of food science in the war effort became well recognized. In fact, it was at the Quartermaster Laboratory in Chicago that many new ideas evolved. For example, the concepts of acceptability, utility (convenience), stability, nutritive value, mobility and safety were conceived and pursued. These concepts are still followed today and are certainly given serious consideration by industry in both processing and, especially, in the development of new products.

The armed forces had great need for products that were convenient, safe, and easy to ship. They had great need for new, effective packaging—a field then quite strange to food scientists. In addition to the development of new products, there was pressure for new technologies, engineering applications, and so on. It became apparent that the food scientist had to expand his views and could no longer limit his area of interest to what took place in a processing plant. He had also to consider the product after it left the processing plant—the ease with which it could be transported and distributed, its stability during handling and storage, and the various factors relating to safety, nutritive value, convenience and, especially, acceptability. The latter, of course, involves color, texture, taste, and odor. It includes even the noise one makes when eating crisp foods such as potato chips or celery, and the sensory reaction one feels when eating a so-called "hot food." The interest of food scientists, therefore, expanded and ranged from the processing plant to the consumer, and included everything in between.

During the early movements of people to the cities some 50 years ago, those who moved were quite content to secure almost any type of processed food, and generally accepted it. Today, however, the situation is quite different; consumers no longer gratefully accept what is made available to them, despite the efforts of the food processor. On the contrary, unfortunately they often regard food products with mistrust and suspicion. The consumer is concerned about quality and all that quality means: color, texture, flavor, general appearance, convenience, stability, nutritive value, and above all safety. The food industry today is well aware that it cannot make a single mistake with respect to factors of quality, and especially with respect to factors of safety; for if it does, a whole organization or even an industry and its many products may be subject to severe criticism.

In considering safety, the food scientist today must be aware of the great and increasing concern about cancer. The publicity about carcinogenesis is fearful and alarming. So much of it seems to relate to the environment; and food and agriculture are certainly components of the environment. This situation persists even though the incidence of cancer (except perhaps lung cancer) seems to be leveling off and, in some instances, is even decreasing. The food scientist of today must maintain an intense interest in potential causes of cancer; this certainly includes all aspects of foods from production to consumption. He must be well informed about intentional and incidental additives, for he is the one responsible for the wholesomeness of a food.

The origin of raw materials and the use of pesticides have attracted a great deal of attention as potential carcinogens, teratogens, and even mutagens. Foods often do contain pesticides or other chemical residues, and even though they may be within the tolerances established by the government, these tolerances can and have been changed suddenly and frequently. Therefore, a food scientists must not only know what is in or on his product, but he must know and be aware of the many rules, regulations and judgements that continue to pour out of the many governing agencies and courts. He must, therefore, know what is applied to raw materials and why, and he must certainly be aware of what takes place on the farms, as far as the use of pesticides and other treatments is concerned.

He must have an understanding of the treatment of dairy animals with penicillin, for residues may occur in milk, whether or not they exceed the established tolerance; above all, he must be familiar with the required procedure for milk analysis, for even this may be and has been changed suddenly. In addition, he must know whether or not diethylstilbestrol has been used as a growth stimulant, even though this substance is now illegal. It is clear that the interest of the food scientist, therefore, has been greatly expanded to include the source and nature of raw materials, processing, distribution, consumer acceptance, safety, nutritive value, and even nutritional labeling.

With respect to safety, the author has had a great deal of exposure to the trials and tribulations of the food industry with its safety-related problems. It appears that, more often than not, industry scientists have been unaware of what is really expected of them with respect to the safety testing of foods, unaware of the proper protocols to be used, and unaware of what is meant by teratogenesis, mutagenesis, interaction, and even carcinogenesis. This is more the result of governmental uncertainties than neglect on the part of food scientists. The questions have been, and are, what to do about these matters and where to obtain information, for there is a frequent lack of agreement in government, often coupled with a lack of knowledge with respect to the protocols to follow in safety testing.

Lately, we hear much about allergies and especially hyperkinesis; and the question is what we should do about them. This is a matter of concern today. It has been claimed that certain food colors and flavors are responsible for hyperactivity in children, although this has yet to be substantiated by double blind tests.

Nutrition is a vital area that requires a thorough knowledge of new discoveries and actions, especially concerning fortification and labeling. W. B. Murphy, former president of the Campbell Soup Company, stated at a conference on nutrition held at Rutgers University in 1972, that one of the most perplexing and unsolved problems in the lives of people is the lack of sound knowledge of nutrition and sound eating habits. He then pointed out that there are more than 50 and probably more like 75 essential nutrients—possibly even 100. The minimum of 50 includes the eight essential amino acids, plus the arginine and histidine that are essential for infants; the 15 essential vitamins (others may yet be discovered); 19 essential minerals including copper, selenium, iron, nickel, tin, zinc, chromium; and others not yet known to be of vital importance. In addition, there are linoleic, linolenic and arachidonic fatty acids. We must include water, carbohydrates for calories and bulk, fats for energy and, of course, proteins. Murphy then added—as those in the school food program know full well—that there is also the importance of psychological and physiological factors involved in acceptance.

There is another area of increasing concern. Environmentalists have certainly made it clear that no issue is isolated and that ecology is a science of interdependence that overflows into the area of food science and technology. Today's food scientist must therefore gain an understanding of values and those who judge values. Richard Carpenter (Executive Director, Commission on Natural Resources, National Research Council) pointed out in a symposium held at Duke University in 1975 that some of the new laws involve many poorly understood words. Value words or phrases are found in environmental laws in such profusion that the adjudication of these statutes has centered on their definition. The reader will recognize these words or phrases: "acceptable," "best practicable," "achievable," "reasonable," "generally regarded as safe," "prevailing professional practice," "appropriate tests," and, perhaps the most value-laden of all phrases, "the public interest." These value words become "code words," signifying a whole set of assumptions. This means, of course, that the food scientist of today must be realistic and recognize, if he can, values or points of view with respect to a situation in which the government evaluator finds himself.

The modern food scientist must recognize this unscientific terminology as a source of some of the present-day confusion. Scientific data and facts are frequently overlooked; and science may not be considered as objectively as it is portrayed. This means that the food scientist of today, when considering environmental matters relating to foods, must not only consider them from a scientific and a legal point of view, but also in light of the implicit values.

A current environmental problem is the disposal of solid and liquid wastes. The food scientist today must realize that his organization may well face serious waste-disposal problems. For example, a vice president for research in a food organization, devoted a major part of his time in the past several years to the development of a means of disposal for tons of tomato and pear wastes. No sooner was this problem solved than another, relating to water pollution resulting from liquid waste disposal, confronted him. This is just one indication of the expanded interest and activity forced upon the food scientist as the result of an ever-changing legal environment.

Yes indeed, the food scientist of today must have a broad outlook. He is involved in a diversity of activities whether or not his training and orientation have been adequate. Teaching in this area must also change and indeed, it is changing. The field today embodies far more than was ever visualized early in this century. The food science field has come a long way but it will go further; already there are specializations, and as time goes on there will be more.

During the past few years, there has been more governmental activity than ever as a result of a number of new congressional acts, as well as increased activity on the part of agencies such as the Food and Drug Administration and the Federal Trade Commission. Then too, there is on the horizon the Hazardous Chemicals Act. All of these Acts, in one way or another, can and will influence the food industry and broaden the activities of food scientists more and more as time goes on.

Some of the developments resulting from Congressional action are the following:

National Environmental Policy Act of 1969
Council on Environmental Quality, established in 1970
Environmental Protection Agency, established 1970
Occupational Safety and Health Administration Act of 1970
Consumer Product Safety Act, 1972
The Federal Water Pollution Control Act, 1972
Clean Air Act, 1972
Energy Supply and Environmental Coordination Act of 1974
Safe Drinking Water Act, 1974

Federal Energy Administration Act of 1974
Federal Insecticide, Fungicide and Rodenticide Act of 1972, Amended 1975

Then there is the Hazardous Substance Act which was established a good many years ago.

The Toxic Substance Act is still in legislation.

These Acts, plus the many rules, regulations, standards, decisions, guidelines and judgments germinated by them, expand and change the whole outlook of a food scientist from one of scientific creativity to one of scientific defensiveness. Let's consider a few specifics: air pollution caused by smokestack emissions; odors escaping from such facilities as a catsup plant, a fermentation vat, or an onion drier; the pleasant smells of a coffee roaster or a hamburger stand (even these are demanding more attention).

In 1974 the Safe Drinking Water Act was passed. It directed the administrator of the Environmental Protection Agency to make, and report findings of a comprehensive study of water supplies. This study was to determine the nature, extent, sources, and means of control of contamination by chemicals and other substances suspected of being carcinogenic. At first sight, this appears to be a subject of but passing interest to those involved in the production of foods. Recently, however, a report appeared on the occurrence of 96 organic chemicals in New Orleans drinking water ranging in concentration from a few to several parts per billion. A later study was made of this situation in several cities throughout the United States and the number of chemicals found was increased to 411. Some of these substances are implicity carcinogenic, and of these chloroform, the result of chlorination, was specifically cited.

Dr. William Stewart, former U.S. Surgeon General, pointed out that a defensive posture has become more commonplace for public health officials in the last few decades because of our ability to detect foreign substances in the environment in increasingly smaller amounts. At the same time, our ability to understand the biological consequences of exposure to these small quantities over long periods of time has advanced slowly and with great uncertainty. This same defensive posture will become more commonplace for food scientists, as well.

A committee was appointed by the Science Advisory Board of the Environmental Protection Agency to study water safety. It made several comments of great interest to the food scientist. In the first place, the committee pointed out that it is likely that the majority of the drinking water purveyors and also most of the food research establishments do not have available sophisticated equipment and trained personnel to provide monitoring of individual contaminants on a routine basis.

One may wonder why water and its purity should be of such concern to the food scientist. Much of this same water may be used in food processing and by the consumer in home food preparation.

However, the committee pointed out that attention with respect to water purity has been focused largely on the concentration of contaminants in drinking water itself but a complete analysis of the problem would also require analytical and chemical data on exposure to these chemicals by ingestion of foods and beverages processed with contaminated water. Also, there are possible exposures resulting indirectly from environmental redistribution and biomagnification of the chemicals by food organisms which consume contaminated water. These problems will undoubtedly confront food scientists in the future.

There appear to be three great concerns with respect to water standards and requirements. One, of course, relates to the occurrence of infectious bacteria and viruses which is of great importance to those living in underdeveloped nations. In the United States, the present great concern appears to be about the presence of toxic organics. It is of particular interest to note that in certain parts of the world, particularly Finland, the importance of naturally occurring inorganic trace elements as water pollutants and their implications to man's health are of great concern.

In the United States, water softness has shown positive correlations with cardiovascular disease and arteriosclerotic situations. The question is what, if any, concern should this be to the food scientist. This remains to be seen.

Another factor bound to confront the food scientist in the future relates to packaging; we may be compelled to shift from the use of metals to plastic containers that are safe. Great progress has been made along these lines, but more work appears to be necessary. At present we import from other countries approximately 1/3 of our iron and all of the tin we use; one wonders how long this can go on.

Another area of concern is—and will be—the world food supply. We are becoming increasingly apprehensive about this problem even to the extent that the White House has asked USDA and the National Academy of Sciences to study the matter. There will be more pressure for creative thinking along these lines in the future for, as Lord Ritche-Calder pointed out, the world population only 25 years from now cannot be less than 7 billion. Even now, he says, 400 million go to bed each night with less food than they need. So there is much to do in addition to speculation about new products and formulations.

On the other hand, there seems to be more and more concern about safety, whether or not, in reality, it should be a matter of great concern. We are refining our methods of analyses so that we no longer speak only in terms of parts per million but in parts per billion and even parts per trillion. It appears that we may be going so far as to reduce our concerns about safety to the last molecule. This will indeed keep those of us worried about food safety busy for a long time—not to mention the food scientists trying to solve the problems relating to this last molecule.

Waste is one area that will require a great deal of attention, not only from the standpoint of pollution and its prevention, but also from the standpoint of utilization. Tomato and pear wastes have already been mentioned; and there are many others. For example, work at the U.S. Army Natick Research and Command is already under way on the conversion of cellulose wastes to sugars by the use of the organism *Trichoderma vitidans*.

Stress should be made here on what probably may become one of the most critical needs in the field of food science: that is leadership. Scientists, generally speaking, are not prepared to step up and do battle, although that is and will become necessary in the area of food science—as well as in other areas. Scientists normally take refuge in the need to do more research rather than in pointing out that there is a calculated risk. Advocates, on the other hand, too often speak with certainty, even in the absence of data to substantiate their statements.

Even though there are a number of methods for extrapolating risk from high doses to lower doses, or from animals to man, scientists are accustomed to working in areas of greater precision and thus do not want to stake their reputation on estimates which they consider unreliable.

Advocates and economic scientists, however, are often willing to come up with cost estimates of various control strategies designed for various specified levels of control. As a result, administrators of regulatory agencies frequently find themselves in situations where they have no quantitative risk estimation of health effects, but do have highly unreliable quantitative cost estimates from economists. Paradoxically, if scientists would use available risk-estimating methods, they might well come out with better estimates of risk than those provided by economists and advocates. Scientists should come out from behind their fortress of scientific certainty, and be willing to take the leadership in giving advice based on their expert knowledge.

Another quality of leadership is the ability to be willing to face that most uncomfortable of professional necessities: the ability to be self-critical. The scientist must be willing to run an experiment to disprove himself; if this is not done, his data should not be considered acceptable. The advocate must also be induced to do some testing of his ideas and results, a situation at best quite rare today. The business of the advocate is to win without violating the law; but the business of the scientist is to see that the whole truth is involved in a decision.

Unfortunately, responsible scientists often find it hard to present their views to the press or to political bodies, such as legislators or administrators. As Judge Jerome Framt pointed out, "Creative master minds seem to feel a personal aversion to the idea of unfolding before the public gaze the delicate threads of thought, out of which their productive hypotheses were woven, and the myriad of other threads which failed to be interwoven into any final pattern."

It appears that responsible scientists are well aware that perfection in science, especially in biological science, is a rarity, if at all existent. Realizing this limitation, they testify with uncertainty and always indicate the need for more research. On the other hand, those who may be advocates (biased, ill-informed, or even completely ignorant) testify with a positiveness that leaves no doubt in the mind of the politician or the average citizen. How then are the politicians or news media to know who is correct and responsible, and who is irresponsible?

In this connection Dr. William Darby, in his keynote address at the Annual Meeting of the Canadian Agricultural Chemical Association, commented, "Matters pertaining to environmental pollution, safety of the environment, safety and quality of foods, pesticides or additives are emotional subjects for many vocal members of our society. Statements pertaining to such subjects attract immediate attention. It is particularly important, therefore, that information provided *by scientists* to the public be accurate, balanced and objective and avoid creating a sense of alarm where there is no reason for disquiet." He also pointed out that the Food Protection Committee of the National Academy of Sciences—National Research Council had taken a position on this matter. This committee stated, "Standards of conduct of this sort are particularly necessary whenever a scientist, no matter how eminent, moves beyond his area of competence. In the field of toxicology, in particular, there is an increasing tendency for scientists working in a wide variety of disciplines to feel themselves capable of making pronouncements on hazards to the public. Evaluation of the safety of foods and food chemicals demands a specialized background of knowledge and experience without which scientific judgment falls short of what is desirable in the public interest. Public misinformation about food safety is an inevitable consequence of misplaced confidence of scientists in their ability and authority to pass opinions on questions of food safety. The scientific community as a whole has a duty to protect the public from the consequences of misinformation on such vital issues as food and nutrition in relation to health (and to preserve public confidence in science). . ."

This is certainly something the food scientist will have to keep in mind as time goes on.

We must realize that the image and nature of a person—a scientist or a food scientist—is critical. There is a real need for scientists who are articulate and willing to speak in terms of probabilities, but with certainty. The paranoic defensive behavior of the scientist, so common today, is just as deadly as taking refuge in the need for more research. It must be made clear that there are limits to perfection. Lawmakers, implementers, and enforcers must be made to understand that a scientist who says something is probable, but who refuses to speak in terms of absolutes, is more to be trusted than a biased advocate who speaks with specious certainty. We must make it clear that no one is or can be absolutely perfect, and that those who speak about perfection are anyting but perfect, themselves.

The food scientist of the future must be aware of the factors involved in making decisions, and these will certainly include science, economics, social factors, political factors, aesthetics, and emotions. Above all he must realize that, unfortunately, science may often play a minor role. It is the duty of the food scientist in the future to assume the leadership of the country in answering problems of food safety as well as in solving the problems of food quantity and quality.

Food science has come a long way from the days of its early development by Atwater, Russell, Prescott, Cruess, scientists in the Agricultural Research Service, and others. Times, too, have changed and the need for scientists who have the quality of leadership, and who possess great breadth of thought and comprehension is more apparent than ever. We will need those who will be willing to take responsibility and go to the lonely outposts of thought and action, who will persuade others to follow, and then act with positiveness, with persuasiveness, with constructiveness, and with certainty. These abilities, unfortunately, are the rarest of commodities; but truly, we must develop them for the future.

Section 2

Plant Production

7

Introduction

W. H. Gabelman

The evolution of man as a dominant species among the myriad organisms on earth mirrors an ability to survive in a hostile environment. Man's survival has been aided by his inventive genius. Many examples can be cited—fire, the wheel, clothes, medicine, language, etc. Unique among examples is his ability to domesticate competing species, both animal and plant.

Man as a nomad, gleaned his food from natural sources, often following migrating animal herds. At some point in history he discovered that the grains he ate could be deliberately planted and their productivity encouraged. Many of these same grains could help feed the animals he was domesticating. The domestication of plants initiated many beneficial biological changes; e.g., in the cereals, non-shattering grain was more apt to be retained than types that shattered readily; dormant seeds did not germinate so less dormant types were more apt to survive, etc. The change in culture of the crop provided selection pressure for traits adapted to domestication, and in a very simple way the art of plant breeding was launched. Genes came together by chance cross pollination, inbreeding occurred naturally, and selection pressure by man isolated desirable characteristics, such as nonshattering seed heads and nondormant seeds.

Another significant change occurred when man recognized the sexual processes in plants. Human curiosity, particularly among the educated clergy, compelled a few to find out the consequences of deliberately cross-pollinating and/or self-pollinating plants. The range of types derived from these matings provided an opportunity to select more desirable individuals; and if these superior individuals could be propagated asexually (cuttings, grafting, etc.) the superiority of the individual plant could be prepetuated indefinitely as a new cultivar. Thus the Bartlett pear and Russett Burbank potato came into being long before Mendel's laws of heredity were discovered.

The Luther Burbanks had a mystique about them which filled their contemporaries with awe.

When Mendel's papers were discovered early in the 20th Century, a solid basis for plant improvement was recognized and the "art" of plant breeding rapidly gave way to the "science" of plant breeding. Three topics have been chosen for this section to illustrate the evolution of this science as well as its contributions toward feeding a hungry world by eliminating the catastrophic consequences of plant parasites and improving the standard of living of all mankind. This is a story which demonstrates the potential creativity in one phase of the biological sciences. In the past century there are few success stories of greater importance to America.

8

One Hundred Years of Plant Breeding

G. F. Sprague

The food supply of man is provided directly or indirectly by plants. The efficiency of managed plant productivity is dependent upon the genetic potential of the species, environmental characteristics, the farming system employed, and the use to be made of the product. There is a strong interaction among these factors and we shall be concerned primarily with genetic modifications.

A detailing of accomplishments, crop by crop, would give little idea of the problems posed or solutions effected. Therefore, this chapter concentrates on a single crop—corn—in an integrated presentation illustrating the contributions from breeding developments which have permitted the changing agricultural production patterns. Major emphasis will be given to accomplishments, with only limited attention being given to the underlying genetic techniques and breeding systems employed.

Had another crop or series of crops been chosen for illustration, the developmental details would have differed, for each poses its own special problems. Regardless of the choice, there would be a strong common element; a substantial degree of progress in meeting a continually changing production and consumption pattern.

The details of plant breeding accomplishments are closely associated with changes in production patterns. This relationship justifies a brief consideration of some of the major changes that have occurred in American agriculture during the past 100 years. Developments attributable, wholly or in part, to plant breeding accomplishments will be covered in this chapter.

THE CHANGING PATTERN OF AGRICULTURE

In the 100-year period 1875-1975, we have moved from a predominantly rural to a predominantly urban society. Farming has changed from a labor intensive to a capital-energy intensive operation. Farm population has dropped drastically since 1950 and at present the number of persons supplied farm products per farm worker is approximately 50. This comparison underestimates the efficiency of production in two respects: (1) it ignores the current, very substantial exports of farm produce, and (2) it ignores the fact that over 3/4 of farm output comes from 20% of the farms under the present census classification.

In 1875, animals were almost exclusively the source of farm power. It has been estimated that approximately 10–20% of the total farm produce was required just to maintain this animal power. Tractors or steam power were of limited significance before 1920. Initially, the two power sources were complementary, tractors being used in peak seasons or for especially difficult tasks. The transition from animal to tractor power was very rapid during the 1920–1960 period and extensive use of animal power is now rare.

The transition from animal to tractor power freed substantial acreage for other crop use. Oats, corn, and timothy-clover hay were the preferred feeds for horses. Oats were widely grown for two reasons: directly as a feed crop and secondarily as a companion crop for the establishment of new legume seedings. The oat acreage has decreased rather steadily ever since. Breeding programs aimed at the improvement of timothy hay have been largely discontinued. This example illustrates that plant breeding goals and priorities are rapidly modified in response to changing economic conditions.

The machinery used in crop production has undergone tremendous developments. The one-bottom walking plow and one-row walking cultivator were replaced by riding plows and cultivators to be replaced, with the advent of the tractor, by even larger equipment. Harvesting of corn moved from a hand operation to the modern picker-sheller or corn combine. The net effect of these shifts has been a greater timeliness of all operations with attendant increases in yield. The man-hours required to produce 100 bushels of corn has dropped from 150+ in 1875 to less than 4 in 1975.

Another major change in production technology has been the increased use of fertilizers, particularly nitrogen. The level of crop productivity achieved is a function of water, nutrients and such various climatic characteristics as light, temperature, and length of growing season. Of this group, the component most susceptible to modification by man is nutrient supply. Most striking among the applied nutrients has been the change in use of nitrogen. In the 1940–1944 period a total of 29,000 tons of N were used in the Corn Belt states, with only a small fraction of this total being applied to corn. By

1972, total use had increased to nearly 4 million tons, of which most was applied to corn.

GENETICS AND PLANT BREEDING PRIOR TO 1900

The beginning date for the science of genetics is usually given as 1900. This is the date of the rediscovery and confirmation of principles reported by Mendel in 1866. Many findings prior to this date, however, contributed directly to subsequent developments in both plant breeding and genetics.

There is a long history, preceded by legend, of the desirability of inbreeding and outbreeding—procedures employed extensively in both theoretical and applied genetics. Greek and Norse mythology is replete with instances of the closest possible forms of inbreeding: parent-offspring and brother-sister matings. Close inbreeding was also a common practice in some early cultures. Planned inbreeding and crossbreeding were used extensively in the development of English breeds of cattle in the 18th Century.

Knowledge and understanding of sex in plants is a more recent development, despite the age-old practice of caprification of figs. One of the earliest records of plant hybrids was the report of Cotton Mather in 1716 in which he correctly interprets xenia involving aleurone color in corn. The next important development in plant hybridization was provided by Koelreuter in 1766. He reported on an extensive series of interspecific hybrids in Nicotiana, Datura and other genera. He observed hybrid vigor in certain of these crosses and also noted that the floral parts of many plants favored crossbreeding. This work on both floral morphology and on hybrid vigor was greatly extended by many investigators in the next 100 years.

EARLY WORK ON CORN

Darwin's report on inbreeding and hybridization in corn (1876) had a pronounced, if indirect, impact on corn research in the United States in the 1875-1900 period. In 1880, Beal reported yield increases of 50% from crosses between 2 varieties. The existence of heterosis in varietal hybrids was confirmed by Sanborn (1890), McCluer (1892), and Morrow and Gardner (1893). This series of investigations clearly established that appreciable heterosis was expressed in certain hybrids. No commercial use was made of varietal hybrids and, in the absence of a genetic understanding of heterosis, the impact of these studies on the subsequent course of corn breeding was limited.

A number of different types of corn were being grown by the Indians when the first settlers arrived. The two primary types were the New England Flints and the Southern Gourdseed. Agricultural history of this period is rather sketchy but there is a general consensus that the current dent varieties were derived from intercrosses of these two types, followed by selection for an intermediate type; the details of the selection and the plant and ear type sought varying with the individual farmer. None of these first varieties has persisted to the present time. The important varieties of the 19th Century, such as Reid (1846), Leaming (1856), and Boone County White (1876), were all of relatively recent development. Other important varieties of the prehybrid period, such as Krug and Lancaster Sure Crop, were developed in the early years of the 20th Century. All of these varietal developments were achieved by individual farmers. Experiment stations had little impact on corn breeding before 1900.

Experiments involving selection for oil and protein content of the corn kernel were initiated at the Illinois Experiment Station in 1896. These experiments have been continued to the present time with remarkable progress (Dudley *et al.* 1974). In the high series, oil content has been increased from 4.7 to 18.0% in the 75 years of the experiment. In the low series, oil content has decreased to approximately 0.5%. Comparable progress has been achieved in the high and low protein series.

During this same period comparable experiments were initiated at the Ohio Experiment Station for high and low yield. Similar experiments were conducted at several other stations for short periods of time. In each case, selection was deemed ineffective as practiced. As a result, the idea became prevalent that mass selection was ineffective in modifying corn yields.

In 1908, Shull presented his first report on inbreeding and hybridization of corn. He proposed that each variety was a complex mixture of genotypes. When these genotypes were isolated in pure form through inbreeding, there was a striking increase in uniformity of plant and ear type and a marked loss in general vigor. However, when certain inbred lines were recombined, yields greater than that of the parental variety were obtained. He also outlined in some detail the requirements for utilization of this increased yielding ability: hybrid vigor. Several stations initiated inbreeding experiments on the basis of Shull's report, but most of these were relatively short-lived. The experimenters were so impressed by the reduction in vigor from inbreeding that the potential for yield increases from hybridization appeared to hold very limited practical value.

THE HYBRID ERA

Connecticut was one of the few stations that continued an inbreeding program. Jones (1918) reported that certain single-cross parents could be combined to form double-crosses and to produce a substantial level of heterosis. This procedure overcame the limitations imposed by the reduced vigor of the component inbred parents. Double-cross hybrid development, therefore, opened the way for the commercial use of hybrid vigor. The earlier experiments with varietal hybrids had indicated that not all combinations were superior. Similarly with inbred lines, only a small percentage of those developed—certainly less than 1%—could be combined into hybrids having sufficient yield potential to justify commercial production.

Largely as a result of the double-cross development, new and extensive corn breeding programs were initiated in the early 1920's. The hybrids developed from these programs came into commercial use in the early 1930's. Two important developments during this period must be mentioned. Jenkins and Brunson (1932) developed the top-cross procedure for the preliminary evaluation of inbred lines, and Jenkins (1934) developed methods for predicting double-cross performance on the basis of either top-cross or single-cross data. These two developments undoubtedly contributed greatly to progress in this early developmental period.

Hybrids were available by the early 1930's and the replacement of older varieties by hybrids was quite rapid. In 1933, 0.7% of Iowa's corn acreage was planted in hybrids. Subsequent rates of adoption were a function of the importance of corn in the farming system of a region (Fig. 8.1) Replacement by hybrids was essentially complete in Iowa and Illinois by 1945, for the Corn Belt states as a group by 1950, and for the entire United States by 1960.

The impact of this substitution is illustrated in Fig. 8.2. Prior to the beginning of the hybrid era, U.S. corn production had exceeded 3 billion bushels on only two occasions, 1906 and 1920. This production was achieved with average per acre yields of 30.3 and 30.9 bushels, respectively. Since 1935 yields have increased at an almost linear rate; the per acre yield for 1975 reached 86.2 bushels.

Increases in per acre yield in the 1930–1945 period were directly and entirely due to the replacement of older varieties by hybrids.

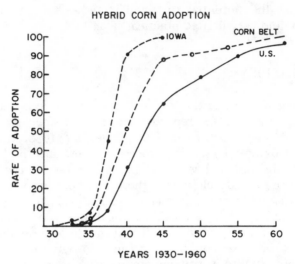

FIG. 8.1. THE ADOPTION OF HYBRIDS IN ILLINOIS,
THE CORN BELT, AND THE UNITED STATES

Hybrids were substituted in the production system with no other important changes in farm practices. Economists would designate this as a single-trait substitution. Subsequent to 1945, the role of hybrid substitution is less clear due to a number of concomitant changes in production practices. These include increases in planting

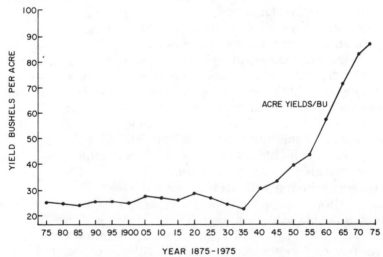

FIG. 8.2. AVERAGE YIELDS OF CORN (BUSHELS/ACRE) FOR THE UNIT-
ED STATES FOR THE PERIOD 1875-1975

densities; substantial increase in fertilizer use, particularly nitrogen; increasing use of herbicides; the shift from double- to single-crosses; improved soil preparation; and tillage, planting and harvesting machinery which permitted greater timeliness of all field operations. It has been assumed by many that the yield increases since 1945 have been due primarily to the above-mentioned changes in production practices. In effect, this assumption holds that the hybrids of the 1970's differ in no essential detail from those first introduced in the 1930's. This assumption is quite incorrect. A substantial number of the most widely used inbred lines of one era are replaced by newer and better lines in each following era. The gains resulting from such substitutions during the 1930–1970 period have been documented by comparative studies conducted in Iowa.

Russell (1974) has presented data involving one open-pollinated variety and a series of station hybrids representative of the 1930, 1940, 1950, 1960 and 1970 eras. All items were compared at 3 population densities, involving 11 environmental locations, and covering a 3-year period. The results for the best two hybrids from each era are illustrated in Fig. 8.3. Gains due to genetic improvement during this 40-year period have averaged 0.73 quintals/ha per year.

Gains in productivity arising from genetic improvements are closely related to changes in the entire production system. With changes in the genetic charactertistics of a species, new production practices become economically feasible. The total gain achieved in corn is illustrated in Fig. 8.2. We shall now explore how some of these gains were achieved.

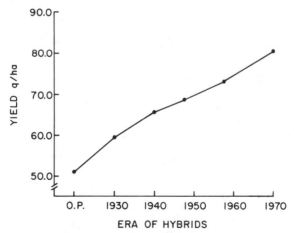

FIG. 8.3. GENETIC IMPROVEMENT IN CORN HYBRIDS REPRESENTATIVE OF THE 1930, 1940, 1950, 1960 AND 1970 ERAS

Plant Densities

Extensive data are available from the prehybrid era on optimum plant densities. These were shown to be related to soil fertility, amount and distribution of water, and genetic differences among varieties. In the Corn Belt region, plant densities in common use ranged from 12-16 thousand plants per acre. The inbred lines used in the first hybrids were developed and their hybrid potential evaluated at such stand levels. Early in the hybrid era it was noted that differences existed between open-pollinated varieties and hybrids in tolerance of increased plant densities. Even greater differences were noted among hybrids.

The data reported by Lang *et al.* (1956) are used to illustrate this point (Fig. 8.4). Single-crosses involving lines of the 1940-50 era were compared at plant densities ranging from 4-24 thousand per acre. Average yields of all entries increased with an increase in plant density within the 4-12 thousand range; 12 thousand was the opti-

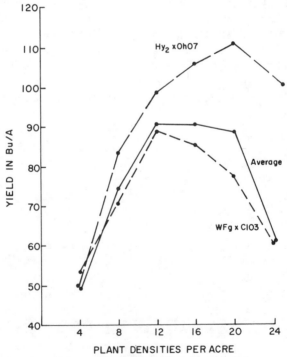

FIG. 8.4. YIELD RESPONSE OF A SERIES OF SINGLE-CROSS CORN HYBRIDS TO VARIATION IN PLANT DENSITIES

mum density for the single-cross WF9 X C103. One single-cross, Hy2 X Oh07, gave yield increases up to a maximum density of 20,000 plants. None of the lines used in this study are in current use, all having been replaced by new lines having greater response to high population densities. Plant densities in current use range from 20–26 thousand plants per acre in the central part of the Corn Belt.

Fertilization

All plants require several chemical elements for satisfactory growth. Dry matter production, grain or stover, is a function of the availability of these essential elements. If the elements removed in the harvested crop are not replaced, crop production becomes essentially a mining or soil-depleting operation.

The elements required in largest quantities are nitrogen, phosphorus and potassium. Prior to the introduction of hybrids, most Corn Belt soils contained adequate reserves of P and K to maintain the corn yield levels characteristic of that period. Nitrogen was more commonly limiting, and the addition of barnyard manure and plowdown of legumes that fixed atmospheric nitrogen were the common practices to minimize this shortage. Even with these practices, nitrogen remained a limiting factor at plant densities above 12–16 thousand plants per acre. Chemical forms of nitrogen were expensive or unavailable.

Nitrogen became available at reasonable prices after World War II. Numerous studies were undertaken to establish the effects of nitrogen applications on corn. Previous studies had established that substantial yield increases could be obtained with additions of barnyard manure that were, however, substantially higher than could be generally sustained. These early studies had established that hybrids had a greater capacity for fertilization response than did open-pollinated varieties. In later studies involving single-cross hybrids, a range of plant densities, and three fertility levels, a close relation between nitrogen response and plant densities became apparent. Typical results are illustrated in Fig. 8.5. These data show that hybrids differ in response to plant density and in the extent of their response to added quantities of applied nitrogen. Hybrids exhibiting the greatest response at the highest stand and fertility levels had also the highest yields at lower stand and fertility levels. This relationship, however, is far from perfect. There is no indication that selection for increased efficiency results in poorer than average performance when grown under less favorable circumstances.

The differential response among hybrids to both plant density and

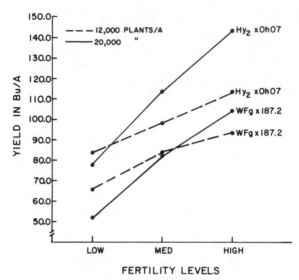

FIG. 8.5. YIELD RESPONSE OF A SERIES OF SINGLE-
CROSS HYBRIDS TO INCREASED LEVELS OF NITRO-
GEN FERTILIZATION

supplemental nitrogen indicates that if responsive hybrids had not
been developed we would still be using population levels of 16,000
plants or less per acre. Similarly, if supplemental nitrogen had not become readily available, the genetic differences among hybrids in efficiency of nitrogen utilization would not have been revealed. In the
absence of these two important changes—genetic substitution and
modification of cultural practices—it appears likely that yield levels
would not have increased materially above the 1950 level.

Disease and Insect Pests

In the prehybrid era the most damaging corn diseases were the
stalk and ear rots and seedling blights. The stalk rots were responsible
for stalk breaking, which added to the labor of hand-harvesting and
increased harvest losses when harvesting was done mechanically.
Techniques were developed for artificial inoculation with the more
important pathogens. The first hybrids were markedly more resistant
to stalk breaking than the open-pollinated varieties they replaced.
Each new generation of hybrids has exhibited increasing levels of
resistance. At present, stalk rot from *Diplodia zeae* is seldom a
problem.

Ear rots were responsible for two types of losses. One was a direct
loss caused by the reduced value and quality of the grain. The second

loss resulted from the use of healthy-appearing but infected seed, resulting in stand losses from seedling blights. The first loss has been minimized by the development of resistance lines and hybrids. The second has been further minimized by the almost universal use of chemical seed protectants.

As production practices have changed, so has the disease picture. Leaf blights such as *Helminthosporium turcicum* were known to infect corn long before the advent of hybrids. Possibly due to loss of genetic diversity this disease became a serious problem in some areas. The identification of sources of resistance and the incorporation of such resistance, both qualitative and quantitative, into lines and populations had made it possible to largely eliminate this leaf blight as a serious problem. Southern leaf blight (*H. maydis*) has been largely controlled in a similar manner.

Two additional illustrations serve to illustrate the rapidity and effectiveness of genetic control of new disease problems. In 1963, maize dwarf mosaic virus became a serious threat to corn production throughout the Ohio river valley and adjacent areas. Losses in Ohio alone were estimated at ten million dollars in 1963 and 1964. Within three years resistant lines were identified and used to produce commercial hybrids, thus averting further serious loss. A second illustration involves the leaf blight epidemic of 1970. A new race of *H. maydis* (race T) had occurred and become widely distributed throughout much of the corn growing areas of the United States. This race was strongly pathogenic to the Texas source (T) of cytoplasmic sterility, which was utilized in the production of approximately 80% of all seed. Losses in yield were estimated at 10–15%. An essentially complete return to normal cytoplasm provided a rapid solution to this problem.

Less effort has been expended in developing genetic resistance to the important insect pests attacking corn. Considerable progress has been achieved in developing improved levels of resistance to the first brood of the European corn borer (*Pyrausta nubilalis*), to the corn ear worm (*Heliothis armigera*) and to the corn leaf aphid (*Rhopalosiphum maidis*). Less progress has been made with root worms (*Diabrotica sp.*) and adequate control is still dependent upon the use of chemicals.

Weed Control

Yield reductions due to weed infestations in corn have been estimated at 10% (Anon. 1954). Chemicals were used for weed control only to a very limited extent before 1945; the primary control method was mechanical cultivation. The first widely-used herbicide was 2,4-D, which came into use in 1946. This herbicide was effective against most annual broadleaf weeds, but was much less effective against grasses. Atrazine was introduced in 1952, and under favorable conditions was effective against most broadleaf and grassy species. Since the introduction of atrazine, many additional chemicals, used singly and in combination, have come into general use.

The use of herbicides was delayed by two factors; effectiveness of the herbicides available and cost of application. Had materials such as 2,4-D or atrazine been available in the prehybrid period it is doubtful they would have been widely used. If one assumes a 10% increase in yield in the pre-1930 period from weed control, this increase, from a 25 to 30 bushel base at the prevailing corn prices of the time, would not more than pay for cost of materials and application—leaving little economic incentive. With the progressive increase in corn yields in the 1945–1975 period, gains from herbicide use have become economical. Today 80–85% of corn acreage is regularly treated with some herbicide.

Development of Hybrids with Special Industrial or Nutritional Value

Prior to 1940 we imported approximately 300 million pounds of tapioca flour. When milled, this starch was used extensively for such things as cloth and paper sizing and rewetting glues. Research at the Iowa Experiment Station (Sprague and Jenkins 1948) showed that the corn mutant "waxy" had starch properties similar to tapioca starch. Starch of the waxy mutant contains only the branched molecular form: amylopectin. Work was initiated in 1936 to introduce the waxy allele into a series of standard inbred lines of the period.

After the outbreak of World War II, tapioca imports from the Dutch East Indies were no longer possible. The hybrid Iowax 939 was rushed into production; the first commercial planting was made in 1943. For several years seed production of this and other hybrids was handled by the Iowa Committee for Agricultural Development. The commercial crop was grown under contract. Contractual arrangements and milling of the crop were handled by American Maize of Roby, Indiana.

Waxy starch was of uniform high quality and was, therefore, preferable to the variable tapioca product. Soon after World War II the Bear Seed Company, of Decatur, Illinois, undertook the development of waxy inbred lines and hybrid development. Still later, additional seed companies produced and marketed waxy hybrids.

Another mutant type (amylose extender, *ae*) conditions starch of a different molecular form. Starch from normal corn contains about 73% amylopectin and 27% amylose (the straight chain form). The gene *ae* conditions an increase in the straight chain fraction to approximately 45%. In combination with other modifying genes (*du*, *su2*, etc.), starches having 70–80% amylose have been developed. The largely straight chain form of starch has industrial value, particularly in the production of films and fibers. Exact figures on commercial use are not available, but the combined use of waxy and amylose hybrids probably amounts to less than 1% of the entire crop.

We have already mentioned the selection experiments involving oil and protein content, which were initiated at Illinois in 1896. These experiments were highly successful in modifying chemical composition, but the selected strains were too low in yield to be of direct commercial value. Several high-oil inbreds have been developed, and hybrids ranging from 6.5 to 8.0% oil show commercial promise. It has been established that oil quality, as well as quantity, is under genetic control. With sufficient demand, hybrids with high levels of linoleic acid (unsaturated) could be grown commercially (de la Roche *et al.* 1971).

High-oil hybrids probably have a promising future in U.S. agriculture. They produce more oil per acre than soybeans and are of growing interest to corn wet millers. Feeding trial data are not completely consistent, but in some trials with ruminants feed efficiency gains as great as 10% have been reported.

Corn protein has long been recognized as low quality, since it is deficient in two of the essential amino acids, tryptophan and lysine. In 1964, Mertz, Bates and Nelson reported that a mutant strain of corn, *opaque-2*, had substantially higher levels of both lysine and tryptophan. Work was immediately initiated at Purdue and in many other private breeding programs to transfer this allele into standard commercial types.

Feeding trials with swine indicated that this protein quality improvement, as indicated by chemical analysis, was reflected in weight gains. Protein supplementation was required for weanling and early growth stages, but during the finishing stages *opaque-2* corn required no supplementation. Striking results were obtained in Guatemala and Colombia in feeding *opaque-2* corn to protein-deprived children. Rates of recovery were almost miraculous. The protein efficiency ratio of diets in which *opaque-2* corn provided the protein source was approximately 90% that of casein. When generally available, *opaque-2* or high-lysine corn should largely alleviate protein malnutrition in those areas of the world where corn comprises a major portion of the diet.

SUMMARY

Plant breeding has been one of the major forces in increasing crop yields in the last 100 years. Additional improvements in both yield and quality should be possible through a more efficient utilization of techniques now at hand. New developments in cell biology may, in the longer term, provide even greater opportunities for plant improvement—if the required research support is provided.

BIBLIOGRAPHY

ANON. 1954. Losses in agriculture. U.S. Dep. Agric. Res. Serv. *ARS-20-1*.

BEAL, W. J. 1880. Indian corn. Mich. Board Agric. Rep.

DARWIN, C. 1876. The Effects of Cross and Self Fertilization in the Vegetable Kingdom. D. Appleton and Co., New York.

DUDLEY, J. W., LAMBERT, R. J., and ALEXANDER, D. E. 1974. Seventy Generations of Selection for Oil and Protein in Maize. Crop Science Society of America, Madison, Wisconsin.

JENKINS, M. T. 1934. Methods of estimating the performance of double crosses in corn. J. Am. Soc. Agron. *26*, 199–204.

JENKINS, M. T., and BRUNSON, A. M. 1932. Methods of testing inbred lines of maize in crossbred combinations. J. Am. Soc. Agron. *24*, 523–530.

JONES, D. F. 1918. The effects of inbreeding and crossbreeding on development. Conn. Agric. Exp. Stn. Bull. *207*.

KOELREUTER, J. G. 1766. Preliminary Information of Interspecific and Intergeneric Hybridization in Plants. Leipzig, Germany.

LANG, A. L., PENDLETON, J. W., DUNGAN, G. H. 1956. Influence of population and nitrogen levels on yield and protein and oil contents of nine corn hybrids. Agron. J. *48*, 284–289.

MCCLUER, G. W. 1892. Corn crossing. Illinois, Agric. Exp. Stn. Bull. *2*.

MERTZ, E. T., BATES, L. S., and NELSON, O. E. 1964. Mutant gene that changes protein compositions and increases lysine content of maize endosperm. Science *145*, 279–280.

MORROW, G. E., and GARDNER, F. D. 1893. Field experiments with corn. Illinois Agric. Exp. Stn. Bull. *25*.

DE LA ROCHE, I. A., WEBER, E. J., and ALEXANDER, D. E. 1971. Genetic aspects of triglyceride structure in maize. Crop Sci. *11*, 871–874.

RUSSELL, W. A. 1974. Comparative performance for maize hybrids representing different eras of breeding. Proc. 29th Annu. Corn Sorghum Res. Conf., 81–101.

SANBORN, J. W. 1890. Indian corn. Maine Dep. Agric. Rep. *33*, 54–121.

SPRAGUE, G. F., and JENKINS, M. T. 1948. The development of waxy corn for industrial use. Iowa State Coll. J. Sci. *22*, 205–213.

9

Utilizing Disease and Insect Resistance in North American Crop Plants

R. A. Fredricksen
G. L. Teetes
D. T. Rosenow
J. W. Johnson

The importance of agriculture in America, one of the few food exporting nations of the world, has vast political and humanitarian implications. Excluding agribusiness and forestry, our agricultural output for the past 5 years has been valued at over 50 billion dollars annually—principally from the production of corn, wheat, soybeans and grain sorghum (Table 9.1). Consequently, protecting these commodities is of vital concern to the nation's economic welfare.

American agricultural products have much in common with those of most other agricultural regions. Of the 200,000 known species of plants, only about 1% have been domesticated by man (Heiser 1973). The most important, by far, belong to the grass family (Graminae);

TABLE 9.1

PRODUCTION OF SOME MAJOR U.S. AGRICULTURAL PRODUCTS, 1974

Commodity	Production (1000 MT)
Corn	146,488
Wheat	58,073
Soybeans	41,405
Sorghum	19,266
Potatoes	14,318
Oats	9,534
Barley	8,338
Rice	5,789
Cotton and Cottonseed	3,569

SOURCE: USDA Statistical Reporting Service, 1975.

these cereal grains provide over 70% of the caloric intake of man (Edwin 1974). Wheat, barley, corn, rice, and sorghum, all members of the grass family, are grown in substantial quantities in the United States. In 1974 over 20 billion dollars worth of agricultural products were exported, principally wheat, corn, sorghum, rice, and the miracle legume, soybeans. Consequently, the resistance of these crops to disease is important to our gross national product (Blakesbee *et al.* 1973).

Host plant resistance provides perhaps the greatest single control of plant disease, and in certain crops it contributes greatly to pest control. Coons (1953) estimated the value of host resistance to disease, in developed varieties, at 10% of the value of the agricultural crops. In 1953, U.S. agricultural output was valued at 15 billion dollars; today that figure has more than tripled. Clearly, as the value of U.S. crops increases so does the value of plant resistance. Nelson (1973) estimated that more than 75% of current agricultural acreage in the United States is planted to varieties resistant to one or more plant diseases. More recently, Schalk and Ratcliffe (1976) argued that the value of host plant resistance could be measured in terms of an annual savings of 63 million pounds of insecticide that would be needed to maintain the present level of insect control.

EARLY DISCOVERIES OF HOST PLANT RESISTANCE

Most of our major agricultural crops were introduced from foreign countries and spread with colonization of new areas. As new crops such as small grains, sorghum, flax, soybeans, and even the native maize were being brought into greater cultivation, the brief and frequent "pest-free periods" disappeared.

During the colonial and expansion years, flax, *Linum usitatissimum*, had to be grown on new land because of a mysterious disease which destroyed the crop if grown on the same land for several years. By 1900, flax was grown as far west as rainfall allowed on the Great Plains and there remained no new lands for cultivation which received sufficient moisture to grow flax (Kommedahl *et al.* 1970). At the turn of the century H. L. Bolley discovered a way to help the crop fight back (Bolley 1901). He learned that the problem was flax wilt, caused by a seed- and soil-borne fungus (*Fusarium oxysporium lini*) and that certain varieties introduced from Budapest, Hungary survived in the presence of the disease. Seed from these plants was increased and in time these resistant plants were used as parents for new wilt-resistant varieties. Orton, working with cowpeas, melons and cotton during the same period, determined that varieties of these crops differed in their reaction to disease. He and E. L. Rivers selected disease-resistant plants and transferred resistance from one plant to another by hyridization.

During the 1920's, certain dwarf "milo" sorghums were selected, increased and distributed to farmers. A disease associated with this type of sorghum genotype was observed and by the 1930's it appeared that milo disease (*Periconia circinata*) would eliminate grain sorghum production. Milo disease, though not the first major or threatening disease of sorghum, was the first controlled by genetic host plant resistance. Almost as quickly as it appeared, it disappeared with the advent of host resistance (Quinby *et al.* 1951). Farmers, themselves, selected resistant plants and saved the seed. Fortunately, the progeny also were resistant. Thus, in a short time resistant milos were developed which solved the problem of milo disease. Had it not been for host resistance, milo disease would surely have then forced us to abandon grain sorghum. The resistance to this disease found in the 1920's is as good today as it was 50 years ago.

The Hessian fly, *Mayetiola destructor* (Say), was probably introduced into the United States in 1776. Larva of this pest feed in and kill culms of infested susceptible plants. Differences in resistance to the Hessian fly were first tested in 1785 (Painter 1951), but were not studied in detail until the 1880's. By the 1920's, the plant characteristics needed for the Hessian fly-resistant wheat varieties were known and resistance was incorporated into varieties grown in different regions of the country. By the mid-1930's breeding for resistance to Hessian fly was underway in both California and Kansas. This resistance is still available and effective.

The chinch bug, *Blissus leucopterus* (Say), attacks sorghum, often causing complete crop failure; particularly with early milos grown for grain. Differences among varieties and hybrids were noted and chinch bug-resistant sorghums ultimately became and remain the norm (Painter 1958).

CONCEPTS OF DISEASE AND INSECT RESISTANCE

Diseases

We will consider disease-resistant plants, in the broad sense, as any which possess characters that tend to reduce the incidence, severity or degree of damage by a pathogen. Hence, sorghum hybrids with 0–10% smut are more resistant than hybrids with 30% smut, and the hybrid yielding 98% of the disease-free control is more resistant (albeit tolerant) than the diseased hybrid yielding only 40% of its disease-free control.

Fundamentally, there are 6 methods of disease control: (1) Avoidance—avoiding conditions favorable to disease. (2) Exclusion—excluding the inoculum. (3) Eradication—eradicating the inoculum at the source. (4) Protection—rendering the inoculum ineffective in the site of infection. (5) Resistance—reducing the effectiveness of the causal factors in the host. (6) Therapy—curing infection in the diseased plants. Of these, the most useful in field crops has been resistance.

There are two recognized types of resistance to disease, generalized and specific. Generalized resistance, sometimes referred to as horizontal resistance, is effective against all physiologic races of a pathogen, whereas specific resistance is more effective against certain races of a pathogen than other races (Nelson 1973). Frequently, specific resistance is monogenic and generalized tends to be polygenic.

A noncompatible host-parasite interaction (resistance) results when the pathogen lacks the necessary virulence to establish itself in the host. A compatible host-parasite interaction (susceptibility) prevails when the pathogen has the necessary genes for virulence or the host lacks the necessary genes for resistance. Pathogens can be described in terms of their pathogenicity, that is, virulence and aggressiveness, as recently defined by Watson (1970). *Pathogenicity* is a general term used to describe the capacity of a pathogen to cause disease. Its rate of reproduction and host range directly affect its capability as a pathogen; whereas, *virulence* relates to a specific host-parasite interaction. *Aggressiveness* describes the ability of a pathogen strain(s) to reproduce more rapidly within one environment than within another.

Insects

Concepts regarding the host plant resistance to diseases and arthropod pests are quite similar. Descriptive terminology and definitions of host plant resistance to insects was recently defined by Teetes (1975) as modified from Painter (1958).

Resistance of plants to insects is relative and must be in comparison with plants which are more severely damaged under the same set of conditions. Since resistance is relative, degrees or levels can be qualified by use of such terms as highly, moderately, etc.

There are, in general, three types, bases or causes of resistance: (1) Nonpreference is used to denote the plant characters which lead pests away from its use for oviposition, food, shelter, or a combination of the three. (2) Tolerance is a type of resistance in which a plant is able to withstand damage in spite of supporting a pest population approximately equal to that damaging a susceptible host. (3) Antibiosis denotes some adverse effect of the plant on the pest such as reduced reproduction, decreased size, abnormal length of life or increased mortality.

Immunity.—An immune variety is one which a specific pest will never consume or injure under any known condition.

Susceptibility.—A susceptible variety is one which shows average or more than average damage by a pest.

NEED FOR RESISTANCE

During the past 6-12 years, researchers in the Texas Agricultural Experiment Station have developed programs to improve all kinds of host resistance in sorghums in the United States. Concomitant with expanded production of sorghum have been an intensification of disease, insect and mite pest problems, and a parallel increase in pesticide use. New pests have developed and/or old ones have become more serious. As has been the case with many of our other crops, insecticides have been relied upon as the major pest-controlling agent. However, their use for the most part, has been rather indiscriminate with little regard being given to pest population levels or adverse side effects that promiscuous use can cause. For a period of time this approach was successful and good yields were maintained. Unfortunately, prophylactic use of insecticides has resulted in problems which are destined to become even more serious and could threaten the sorghum industry.

What then is to be the alternative? The current philosophy is that pests should be "managed." This simply means that a strategy is developed using all available means of suppressing pest population levels (culturally, biologically and chemically) below economically damaging levels in an orderly, compatible system. Insecticides may remain a part of the system, but their use is based on need as related to the damage tolerance level of the crop (Maxwell *et al.* 1972; Sprague and Dahms 1972).

Insect-resistant varieties provide an ideal way to control or suppress insect damage to crops. In some cases where resistance is at a relatively high level, it may be sufficient in itself to effect control. The greatest use, however, of resistant varieties in the future will undoubtedly be one component part of a pest management system. In this type of management system, the value of low levels or resistance is magnified because the resistance works as just one of possibly many suppressant factors integrated to prevent the target species from exceeding the economic threshold.

Disease-resistant varieties often represent the only effective control available for many plant pathogens. Widespread use of chemicals in field crops has not been accepted. At first this was because of the relatively low return per acre from cost of application and more recently because of nonavailability of effective labeled products. Today, there are extremely few general chemicals labeled for use on field crops; none for virus diseases, none for mycoplasms and none for bacterial diseases. Examples of biological control are few and extremely fickle. Consequently, host plant resistance for the past 30–40 years has become an ever-increasingly more important form of disease control.

Disadvantages

A major problem in host resistance has been the genetic variation in species of pathogens or insect pests resulting in strains, races or biotypes which cause damage. In 1955, Stakman reviewed the history of the development of rust-resistant wheats in the infamous Puccinia path. Marquis wheat was resistant to rust from 1909 until 1916. Stakman (1955) reports that in 1916 "a terrible epidemic swept over most of the spring wheat area and destroyed about 300 million bushels of wheat in the United States and Canada. For a distance of more than 600 miles from eastern Minnesota to eastern Montana I did not see a single normal wheat field and most fields were not worth harvesting." This epidemic resulted in the discovery of the existence of physiologic races of pathogens. By the 1930's, a new race of the rust pathogen appeared that destroyed the varieties resistant to races that appeared in 1916. Again in 1950, the rust race cycle repeated itself with an epidemic caused by the race 15B.

Over the next 40 years concepts were developed on a highly sophisticated host parasite interaction regarding coexistence of hosts and their parasites based on the genetic interaction of races of pathogens and the genetics of resistance in the host (Flor 1956). Today, we know that most pathogen populations are composed of many forms of the same species. Some pathogens differentiate among host varieties (virulent races) and others differ in the degree of damage they cause to their host (aggressive races). Some pathogens, such as race T of *Helminthosporium maydis*, had more virulence and aggressiveness than its predecessor race 0.

Parallel examples of variation in the pest population have occurred. Frequencies of many biotypes (races) of both the Hessian fly and greenbug have resulted from the introduction of host resistance in cultivated varieties.

CURRENT PROGRAMS

During the past 40 years, an era of modern agricultural research, host plant resistance as a component of disease and pest control became a commonly-accepted procedure. Teams of workers, whether they be in land grant institutions, federal workers or employees of private firms, quickly learn of hazards which cause, or may cause, devastating losses. Losses can be local, developing to a limited extent each year (endemic) or they may be epidemic. Some pathogens can be carried or disseminated from field to field or location to location and cause widespread damage or epidemics. Epidemics may cause a major portion of the crop to be lost over large geographic areas, affecting the national economy. This situation is referred to as a pandemic. Fortunately for mankind, pandemics are rare. Modern agriculture in the United States has, however, created potential for pandemics as great as any in our history (Harlan 1972). Southern corn leaf blight in 1970 which resulted in over a 1 billion dollar loss convinced many Americans of how vulnerable the U.S. agriculture is to diseases.

Diseases all too frequently result in serious losses or have the potential to do so in any given year (Klinkowski 1970). Downy mildew is a good example where yields of sorghum are reduced when the disease incidence becomes high (Fig. 9.1).

At the turn of this decade, grain sorghum represented nearly 1/5 of our nation's feed grains, most of which is grown in a concentrated area on the western regions of the Great Plains in Texas, Kansas, Nebraska, Colorado, Oklahoma and New Mexico. After a relatively long and peaceful expansion of hybrid sorghum acreage from the 1950's, diseases and insect problems increased dramatically (Frederiksen and Rosenow 1971; Johnson 1974).

Head smut, a fungus disease, appeared with the first hybrids in the late 1950's but the problem was "solved" with resistance. In 1961, downy mildew was observed and by 1967 became epidemic. In 1966, maize dwarf mosaic virus appeared in Texas and in 1967 it forced the abandonment of a major group of sorghum hybrids which were genetically known as Redlan X Caprock hybrids (Edmunds and Zummo 1975).

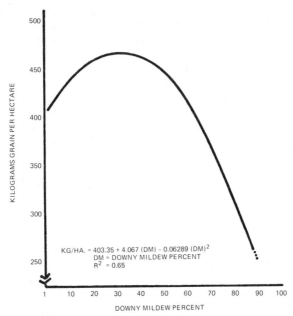

KG/HA. = 403.35 + 4.067 (DM) - 0.06289 (DM)^2
DM = DOWNY MILDEW PERCENT
R^2 = 0.65

FIG. 9.1. EFFECT OF DOWNY MILDEW ON YIELD OF
GRAIN SORGHUM—BERCLAIR, TEXAS, 1971

By 1968, a new biotype of greenbug, *Schizaphis graminum* (Rondani), appeared that adapted to sorghum and increased on individual plants by the thousands. Damage losses in the millions were estimated (Harvey and Hockerott 1971). To this day, grain sorghum growers on the Great Plains remember the year of the first greenbug outbreak.

In 1968, anthracnose raged along the South Texas coast line. Late rains and warm weather favored the spread and development of the pathogen. The disease plus delayed harvesting commonly resulted in losses up to 50% (Reyes 1972). In 1971, a new root rot and lodging complex appeared on the Plains and in 1972 fusarium head blight was serious. In each of these years, damaging losses were caused by midge (*Contarinia sorghiola*) and various mite (*Oligonychus sp.*) species.

Observations such as these can be listed for all of our major agricultural commodities. Such perils and problems are commonplace and can be expected (Schafer 1974).

How does one go about providing American agriculture with host plant resistance? We begin by studying the real and potential enemies.

In sorghum there are 47 insect pests and about 42 diseases. Of these, only about 10 or 12 each cause substantial losses and generally about 5 or 6 have a potential of causing severe losses annually. The six major disease groups in sorghum include head smut, downy mildew, maize dwarf mosaic, anthracnose, stalk rots and seed molds. These can be characterized by the nature of host parasite interaction and the potential usefulness of resistance to each disease (Table 9.2). Host resistance to each of these disease problems is evaluated annually in sorghum areas in Texas, the United States, and internationally. In this way, we are continually monitoring for changes that may occur in pathogen populations and effectiveness of host resistance in many different environments. The monitoring process includes international testing through cooperative agencies such as the International Centers, FAO, USAID and with mutually cooperating countries. Some disease and insect problems are of international concern. Southern corn leaf blight appeared to have spread to Africa, South America and Asia all within a year of its appearance in the United States in 1969 (Ullstrup 1972). Soybean rust, Phillipine downy mildew of corn, sugary disease of sorghum, green ear (downy mildew) of Pennisetum millets, the shoot fly and sorghum stalk bore are diseases and insects not found in the United States but each has the potential of causing severe damage. Consequently, we must also cautiously guard against introduction of these pests and develop sound control for those diseases and insects that have a potential of being introduced. For example, as recently as 1975, we learned of a race of *Colletotrichum graminicola* causing anthracnose in Africa

TABLE 9.2

CHARACTERISTICS OF CERTAIN KEY SORGHUM DISEASE PROBLEMS

Major Disease Problems in Sorghum	Genetic Nature of Host Parasite Interaction	Degree of Genetic Vulnerability
Downy Mildew	General	Low
Head Smut	Specific	High
Maize Dwarf Mosaic	General	High
Stalk Rots:		
Charcoal	General	High
Fusarium	Intermediate	Intermediate
Red Rot	Intermediate	Intermediate
Foliage Diseases:		
Anthracnose	Intermediate	Low
Bacterial Stripe	General	Low
Cercospora Leafspot	General	Low
Helminthosporium Blight	Intermediate	Low
Rust	Intermediate	Low
Zonate Leafspot	General	Low
Root Rot:		
Pythium	Unknown	Low

that differed from that appearing in the United States (King and Frederiksen 1976).

In addition to direct evaluation and study of host resistance, efforts to reduce genetic vulnerability include the introduction of new germ plasm. Traditionally, scientists have found that the greatest diversity in types of pathogens, pathogenic races and variation in the host occurs at their centers of origin (Leppik 1970; Browning 1974). Whereas, Harris (1975) argues that sources of resistance to insect attack have evolved in plants in the absence of the insect pest to which they are resistant or susceptible. In either case, hosts and their diseases and pests are ultimately brought together where they must interact. In today's research or breeding programs as much concern is given to the variability of the pathogen or pest as to the resistance factor. Short-lived resistance has little value in a program of crop production. Recent treatises have stressed approaches to host resistance (Roane 1973; Van der Plank 1968) and conclude that generalized or horizontal resistance tends to reduce hazards associated with pathogen variations more than the traditional specific resistance.

To increase the genetic diversity in sorghum, tropical sorghums from the world collection are being introduced through a research program developed cooperatively by the Texas Agricultural Experiment Station and USDA (Stephens et al. 1967; Quinby 1974). In this program, exotic tropical sorghums from the world collection are being converted to dwarf, photoperiod insensitive, early maturing varieties adapted to the temperature zone. This new germ plasm can be used directly in private and public breeding programs for improving sorghums. To date, about 200 tropical sorghums have been converted, many of which have better and different sources of disease and insect resistance than previously available in U.S. breeding programs. In the past 5 years, effective, stable resistance to downy mildew, anthracnose and maize dwarf mosaic have greatly lessened the potential of damage by diseases. Head smut resistance has been found from a variety of sources; however, it has not proven to be as stable as needed. Resistance to seed mold and stalk rot is available only in experimental varieties. Similarly, progress has been made with the identification of spider mite, sorghum midge and greenbug resistant sorghums.

In 1976, about 2 million hectare of grain sorghum was planted to greenbug, *Schizaphis gramium* (Rondani), resistant sorghum. Cost of controlling greenbug with insecticides has averaged about $20 per hectare. Sorghum resistance to one pest means an estimated savings of 400 million dollars in 1 year. Equally as important if not more so, will be the indirect benefits of less pollution, less interference with biological control and higher yields because host resistance is present every day and not dependent on timing of insecticide application for maximum benefit.

THE FUTURE OF HOST PLANT RESISTANCE

Presumably, we have established the traditional framework for our needs in host plant resistance. As you might suspect, not all agree that host resistance is a panacea. In the past two decades pleas from eminent plant pathologists have been made for chemotherapeutic and systemic fungicides, like those used to control insect pests, to augment host resistance. Their arguments usually include a phrase such as: had as much time and effort been spent on developing chemical and other disease controls, as resistant varieties, then they would be more feasible. However, Martin (1975) states that the fallacy of depending solely on any one form of disease or pest control is clearly recognized by those of us in crop research.

Currently, historically and probably in the forseeable future, we expect yield capacity or production efficiency to prevail as the single most important aspect of plant or crop improvement. Disease and pest control become significant considerations when they materially affect production and marketing economics.

Host plant resistance in American agriculture became an important pest control tactic out of necessity. Early development of plants with resistance by Bolley, Orton and others paved the way for the next 75 years of agricultural progress in this increasingly important discipline.

U.S. technology contributed to the southern corn leaf blight of 1970 (Nat. Acad. Sci. 1972). The pathogen, a new race of relatively minor fungus that attacks corn, spread rapidly over genetically and cytoplasmically uniform hosts. In 1971, switching to corn with resistant cytoplasm and less favorable environment virtually eliminated the importance of southern corn leaf blight.

Host plant resistance has been firmly established as a viable tool in the cultivation of crops. Host resistance to diseases and pests will be increasingly more important as the world's population grows and the need for food increases. A major responsibility of American agricultural scientists will be to efficiently and effectively utilize host resistance. Historically, the investment in agricultural research has paid rich dividends in production. With the increasing demand for agricultural products, it is imperative that sound programs for study of host resistance be continued. More efficient production of U.S. agricultural products represents a savings in food costs to all Americans.

BIBLIOGRAPHY

BLAKESBEE, L. L. HEADY, E. O., and FROMINGTON, C. F. 1973. World Food Production, Demand and Trade. Iowa State Univ. Press, Ames.

BOLLEY, H. L. 1901. Flaxwilt and flax sick soil. N.D. Agric. Exp. Stn. Bull. *50*.

BROWNING, J. A. 1974. Relevance of knowledge about nature of ecosystems to development of pest management programs for agro ecosystems. Proc. Am. Phytopathol. Soc. Monogr. *1*, 191-199.

COONS, G. H. 1953. Breeding for resistance to disease. *In* Plant Disease, the Yearbook of Agriculture, U.S. Dep. Agric.

EDMUNDS, L. K., and ZUMMO, N. 1975. Sorghum diseases in the United States and their control. Agriculture Handbook 468. ARS, U.S. Dep. Agric.

EDWIN, E. 1974. Feast of Famin. Carter House, New York.

FLOR, H. H. 1956. The complementary genic systems in flax and flax rust. Adv. in Genet. *8*, 29-54.

FREDERIKSEN, R. A. and ROSENOW, D. T. 1971. Disease resistance in sorghum. 26th Annu. Corn and Sorghum Res. Conf. Am. Seed Trade Assoc. *26*, 71-82.

HARLAN, J. R. 1972. Genetics of disaster. J. Environ. Qual. *1*, 212-215.

HARRIS, M. K. 1975. Allopatric resistance: searching for sources of insect resistance for use in agriculture. Environ. Entomol. *4*. 661-669.

HARVEY, T. L., and HOCKEROTT, H. L. 1971. Research on greenbug and resistance in sorghums. Proc. Seventh Bienn. Program. Grain Sorghum Res. & Utilization Conf. *7*, 84-86.

HEISER, C. B. 1973. Seed to Civilization, a Story of Man's Food. W. H. Freeman and Company, San Francisco.

JOHNSON, J. W. 1974. Breeding for insect resistance in sorghum. 29th Annu. Corn and Sorghum Res. Conf. of the Am. Seed Trade Assoc. *29*, 1-13.

KING, S. B., and FREDERIKSEN, R. A. 1976. Unpublished Data.

KLINKOWSKI, M. 1970. Catastrophic plant diseases. Annu. Rev. Phytopathol. *8*, 37-60.

KOMMEDAHL, T., CHRISTENSEN, J. J., and FREDERIKSEN, R. A. 1970. A half century on research in Minnesota on flax wilt caused by Fusarium oxysporium. Minn. Agric. Exp. Stn. Tech. Bull. *273*.

LEPPIK, E. E. 1970. Gene centers of plants as sources of disease resistance. Annu. Rev. Phytopathol. *8*, 323-344.

MARTIN, W. P. (Editor). 1975. All-out food production: strategy and resource implications. Am. Agron. Soc. Spec. Publ. *23*.

MAXWELL, F. G., JENKINS, J. N., and PARROTT, W. L. 1972. Resistance of plants to insects. Adv. Agron. *24*, 187-265.

NATIONAL ACADEMY OF SCIENCE. 1972. Genetic Vulnerability of Major Crops. Nat. Acad. Sci. USA. Washington, D.C.

NELSON, R. R. 1973. Breeding Plants for Resistance, Concepts and Ideas. Pennsylvania State University Press, University Park, PA.

PAINTER, R. H. 1951. Insect Resistance in Crop Plants. Univ. Press of Kansas, Lawrence.

PAINTER, R. H. 1958. Resistance of plants to insects. Annu. Rev. Entomol. *3*, 267-290.

QUINBY, J. R. 1974. Sorghum Improvement and Genetics of Growth. Texas A&M Press, College Station.

QUINBY, J. R. *et al.* 1951. Grain sorghum production in Texas. Texas Agric. Exp. Stn. Bull. *912*.

REYES, L. 1972. Personal communication, Texas Agricultural Research and Extension Center, Corpus Christi, Texas.

ROANE, C. W. 1973. Trends in breeding for disease resistance in crops. Annu. Rev. Phytopathol. *11*, 463–486.

SCHAFER, J. F. 1974. Host plant resistance to plant pathogens and insects: history, current status and future outlook. *In* Proceedings of the Summer Institute on Biological Control of Plants, Insects and Diseases. Univ. Press of Mississippi, Jackson.

SCHALK, J. M., and RATCLIFFE, R. H. 1976. Evaluation of ARS program on alternative methods of insect control: host resistance to insects. Bull. Entomol. Soc. Am. *22*, 7–10.

SPRAGUE, G. F., and DAHMS, R. G. 1972. Development of crop resistance to insects. J. Environ. Qual. *1*, 28–34.

STAKMAN, E. C. 1955. Progress and problems in plant pathology. Ann. Appl. Biol. *42*, 22–23.

STEPHENS, J.C., MILLER, F. R., and ROSENOW, D. T. 1967. Conversion of alien sorghums to early combine genotypes. Crop Sci. *7*, 396.

TEETES, G. L. 1975. Insect resistance and breeding strategies in sorghum. 13th Annu. Corn and Sorghum Res. Conf. Am. Seed Trade Assoc. Chicago, Illinois.

ULLSTRUP, A. J. 1972. The impacts of the southern corn leaf blight epidemic of 1970–1971. Annu. Rev. Phytopathol. *16*, 37–50.

VAN DER PLANK, J. E. 1968. Disease Resistance in Plants. Academic Press, New York.

WATSON, I. A. 1970. Changes in virulence and population shifts in plant pathogens. Annu. Rev. Phytopathol. *8*, 209–230.

10

Plant Breeding Contributions to Food Production in Developing Countries

Peter R. Jennings

EARLY PLANT BREEDING IN DEVELOPING COUNTRIES

In the present less-developed areas of the world, plant breeding has a long history of contributing to food production. We should remember that all our important food crops evolved and were domesticated there. The art of breeding was practiced from the time Neolithic man, or more likely woman, shifted from hunting to food gathering and finally to crop cultivation. Early peoples were adept at observing variations, selecting seed of better plants and replanting the outstanding selections. This constant selection pressure over thousands of years did not result in sudden, immense progress, but the total, long-term process rivals modern contributions by professional plant breeders. Crops were not only domesticated, but the native variability was introduced to and selected in innumerable environments. This resulted in the creation of thousands of varieties, each having a certain value in narrowly restricted areas. Over 30,000 varieties of rice, for example, have been collected recently; and the collection process continues. Early peoples in the tropics are thus largely responsible for the selection of the multiplicity of varieties that constitutes modern plant breeding's greatest resource: abundant crop germ plasm.

MODERN PLANT BREEDING IN DEVELOPED NATIONS

Modern breeding commenced at the turn of the 20th Century with three scientific phenomena: the rediscovery and application of Mendel's laws; the understanding of limitations to genetic advance through pure line selection of mixed populations; and the appreciation of deliberate varietal building through hybridization of distinct varieties.

In the developed countries generous public support of agricultural research and broad educational programs resulted in enormous increases in food production. Plant breeding based upon scientific principles became a highly developed and respected profession. Its many successful practitioners were production-oriented men of keen observation and intuition developed by years of handling plants in the field. These plant breeders typically worked in disciplinary isolation; close collaboration with scientists in other agricultural disciplines was haphazard. Breeders for each major crop were located in many states and each worker concentrated on the problems of his area. This approach resulted in the development of narrowly adapted varieties highly suitable for limited geographical areas.

By mid-century, the developed countries had large commodity surpluses, cheap food and a growing feeling that future bounty was inevitably assured. Perhaps as a consequence, support for production-oriented programs languished; false sophistication began to replace field work; and in some institutions plant breeders became second class researchers. Today, public supported plant breeders, as opposed to geneticists, are becoming rare; excellence in graduate training in applied breeding exists in only a few universities. Some of the new laboratory research now in vogue eventually will contribute to varietal development and higher farm productivity. Yet there is danger that a rapid shift to new pioneering research at the expense of traditional methodology will lead temperate agriculture into a period of stagnancy and abstruse specialization at a time when world population growth threatens to exceed food production.

MODERN PLANT BREEDING IN DEVELOPING COUNTRIES

Plant breeding in the developing countries lagged far behind the decades of advance made in the temperate nations. Even so, until recent years many developing nations were net food exporters. But as populations expanded more rapidly than production increased, they gradually became food importers. The availability of cheap food importations from surplus areas stifled national investments in research. It took local research managers many years to accept that temperate varieties, agronomy and farming systems are not directly transferable to the tropics. As demand for food increased, some nations laid the groundwork for sustained progress in food production by strengthening research, education and extension. Financial support increased somewhat as did national recognition of agricultural research as a tool for development.

Nevertheless, in most developing countries plant breeders and other agriculturists do not hold the status comparable to that enjoyed by lawyers, physicians or the military. Salaries and research funds are uniformly inadequate, and few bright young poeple are attracted to careers in agriculture. Many start in research but after acquiring experience, seek the better salaries of the commercial sector. This results in a continual training process and weakens research leadership and continuity. The growing need for trained plant breeders in developing countries unhappily coincides with a decline in the practicality in plant breeding instruction offered in developed nations. It is disheartening to see young graduates, expert in the mysteries of genetics and computer science but inexperienced in old-fashioned field work, trying to cope with yield restraints.

The delayed start and the erratic performance of plant breeders and other agriculturalists make it difficult to evaluate the contribution of agricultural research in recent decades to food production in the developing countries. In one such analysis, Cummings[1] surveyed the present state of food crops technology in low income countries and compared the average annual gain in production from 1961–1965 to 1971–1973 for 15 basic commodities. The data roughly indicate the contribution to food production of research following the Second World War. Two salient conclusions appear from the study:

1. The present research status for all major food crops was classified by a group of authorities as seriously-to-critically inadequate. Only the specialized sectors of irrigated rice and irrigated spring wheat were considered to receive adequate research attention.

2. Annual gains in production of seven crops (cassava, white potato, soybeans, cowpeas, broad beans, maize and wheat) came to more than 2.5% thereby keeping up with population growth; but most of the gains came from increase in the acreages planted. Although there were some exceptions for particular countries, it was generally concluded that research contributed significantly to rising yields only in irrigated rice and irrigated wheat. Other crops, including sorghum, pigeon

[1] "The State of Present and Expected Technology and Agricultural Research on Food Crops for the Low-Income Countries" written by Ralph W. Cummings, Jr., of the Rockefeller Foundation which appeared in Agricultural Economist, May 1976.

peas, dry beans, chick peas, groundnuts, sweet potatoes and millets had production and yield growth below the rate of population increase.

These are discouraging data and indicate the inadequacy of earlier research in developing countries. Tropical agriculture suffers such intense disease, insect and weed pressures that new crop varieties for the tropics remain useful only about half as long as in temperate areas of agriculture. This means that plant breeders must work harder to keep pace, and indicates the difficulty of improving productivity. The Cummings' analysis shows that the record in developing countries is spotty at best and clearly indicates that plant breeding and other research prior to about 1965 was inadequate to achieve a sustained overall advance in food productivity.

EXAMPLES OF PROGRESS

The question today is whether any recent progress has been made in agricultural research that would lead to a more optimistic outlook. I believe there has. My guarded optimism stems largely from the new network of international agricultural institutes and their interplay with national research programs. The international institute system began with the International Rice Research Institute in the Philippines in 1960. Six institutes concentrating on crop research are functioning, and another is in the planning stage (Table 10.1). Along with two others working exclusively with cattle, they are funded largely through the Consultative Group on International Agricultural Research (CGIAR), an international consortium of donors estab-

TABLE 10.1

CROPS EMPHASIZED AT THE CGIAR-FUNDED INTERNATIONAL INSTITUTES

Institute	Crops	First Funded	Location
IRRI	Rice	1960	Philippines
CIMMYT	Wheat, maize, barley, triticale	1966	Mexico
IITA	Corn, rice, cowpeas, soybeans, lima beans, root and tuber crops	1968	Nigeria
CIAT	Field beans, cassava, rice, corn, forages	1969	Colombia
CIP	Potatoes	1972	Peru
ICRISAT	Sorghum, millets, chick peas, pigeon peas, groundnuts	1972	India
ICARDA	Wheat, barley, lentils, broad beans, oilseeds, cotton	Planning stage	Lebanon

lished in 1971, with support in 1976 reaching about $65 million. The system covers most of the major food crops and extends to most of the agricultural areas of the developing world.

The research programs of the international institutes follow a now-familiar pattern. The challenge is to:

1. Assemble teams of young career scientists representing several disciplines and to provide them with superior facilities and program support.
2. Adopt a hard-nosed push for increased production as the goal of the programs.
3. Identify and attack yield-limiting restraints on a systematic, international and interdisciplinary basis.
4. Train large numbers of scientists from national programs in practical research methods and materials, and forge international linkages.
5. Evaluate new technology on farms in many countries.

In my opinion one of the most important contributions of the international institutes has been the formation of teams of plant breeders and other specialists who interact to contribute collectively to farm productivity. The need for plant breeding teams is especially critical in the humid tropics where the biological restraints to farm productivity are most varied and severe. The integrated team approach to agricultural research is so widely accepted and successful within the international institute system that today it is inadequate to single out plant breeding in discussions of food production; although as a plant breeder I emphasize the genetic component in production-oriented investigation.

Career staffs of the international institutes have not only made direct massive contributions to food production, but have helped to revitalize national research programs. Young scientists from developing nations now receive practical training in team-oriented research on yield-limiting problems in their own environment. Hundreds of researchers, working at their experiment stations after 6–12 months of training in the institutes, constitute a corps of production-oriented researchers around whom national programs are growing. These people evaluate the research findings of the institutes in their own countries, modify them as needed in their national programs and form a direct link from international institutes to farmers. This growing corps of agriculturists is having a positive effect on food production for some crops in the developing countries. The contributions of IRRI in the Philippines and of CIMMYT in Mexico to food production over the past ten years have been widely cited in the world press as the basic elements of the popularly named Green Revolution. Two new examples of progress in Latin America are reported here: one on rice and the other on cassava.

Rice in Colombia

Spectacular progress in rice breeding and agronomy in the early years of the past decade culminated in the release and wide acceptance in tropical Asia of the new dwarf varieties. In an effort to extend the impact of this new technology to Latin America, an IRRI rice breeder was transferred in 1967 to CIAT, a sister international institute, in Colombia. A comprehensive regional breeding effort was established with the Colombian national program, several of whose workers were sent to the Philippines and elsewhere for academic and practical training in specialized areas of production research. The breeding team having a competent extension arm and experience in seed multiplication and sales, worked closely with the Colombian Rice Federation, an association of rice growers. Linkages to national programs throughout the Americas were developed through the training of more than 90 nationals in Colombia who returned to their countries to evaluate the latest breeding materials and cultural practices under their conditions.

This association of an international institute, a national program and an organization representing producers generated technology that was tested directly on farms. Several new dwarf varieties were released regionally and others were named locally from promising lines distributed from Colombia.

The following data summarize some of the results from this breeding program. Table 10.2 presents the changes in rice varieties in Colombia since 1964. The first plantings of high yielding varieties were made in 1968. By 1974 these modern dwarf varieties accounted for more than 99% of the irrigated rice area. The impact of this new technology is measured by the changes in yield shown in Fig. 10.1. The upland rice sector has neither benefited from the new technology nor changed in productivity. Upland rice continues to yield about 1.5 tons/ha and is declining rapidly in area. Irrigated rice, with the total adoption of modern varieties, increased in national average yield from about 3.0 tons/ha to 5.4 tons/ha in 1975. Thus, each hectare using the new technology produces nearly an additional 2.5 tons of grain. This added productivity and the rapid expansion of area in irrigated rice have dramatically influenced national production. Figure 10.2 shows an increase in annual pro-

TABLE 10.2

PERCENTAGE OF AREA IN DISTINCT RICE VARIETIES IN IRRIGATED SECTOR OF
COLOMBIA, 1964–1974

Year	Bluebonnet (%)	Traditional Varieties Napal (%)	Tapuripa (%)	ICA 10 (%)	Others (%)	Modern Dwarf Varieties[1] (%)
1964	87.0	5.1			7.9	
1965	86.6	5.0			8.4	
1966	89.9		0.2		9.9	
1967	80.2		6.7	0.2	12.9	
1968	52.8		42.1	0.1	4.6	0.4
1969	50.1		36.2	0.9	7.3	5.5
1970	36.0		26.2		9.2	28.6
1971	34.6		13.5		8.2	43.7
1972	12.4		0.2		0.2	87.2
1973	2.2					97.8
1974	0.8					99.2

[1] Including IR8, IR22, CICA 4, CICA 6.

duction from 680,000 tons in 1966, the year before the program began, to 1,632,000 tons in 1975. Improvement in productivity accounted for this increase as there was little change in the area planted during this period. The value of the additional rice production through 1974 was about $450 million.

On a regional basis the Colombia-based program contributed to the

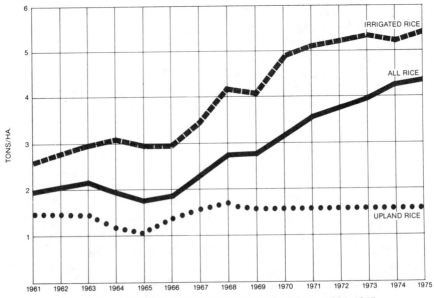

FIG. 10.1. AVERAGE YIELD OF RICE IN COLOMBIA, 1961–1975

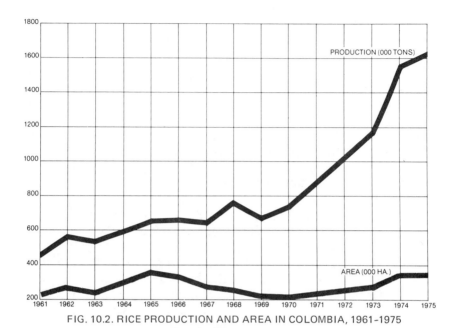

FIG. 10.2. RICE PRODUCTION AND AREA IN COLOMBIA, 1961–1975

harvest of an additional 1.5 million tons of grain in 1974 with substantial progress made in Mexico, Central America, Ecuador and
Venezuela.

Cassava in Latin America and Africa

Cassava, an important root crop in the developing countries, is an
excellent example of a crop for which the technology is being rapidly
accumulated for large increases in farm productivity in the near
future. Cassava research was long neglected until CIAT in Colombia
and IITA in Nigeria recently undertook its improvement. Despite its
enormous yield potential for relatively poor soils, most national
yields still average less than 10 tons/ha. The breeding teams of the
international centers in collaboration with national programs have
progressed toward the formation of a viable technological package
of variety, cultural practices and crop protection. Yields of 30–40
tons/ha on farms are being obtained over a wide range of environments. A pragmatic rule of thumb suggests that when farm trials
yield 2–3 times more than national averages, the potential is available
for rapid increases in regional production. Once a technique is found
to minimize storage losses of freshly harvested roots, we can expect
cassava production to increase.

Other crops for which it appears that plant breeding teams are generating the technology to effect productivity jumps in the developing countries are durum wheat, potato, maize and barley.

SUMMARY

Despite these examples of progress, the overall record remains too uneven to claim a broad front of progress on all major food crops. The several species of grain legumes still present a dismal picture. Productivity remains low and stagnant, and the essential progress in plant breeding is not being generated. In recognition of this situation, the international centers, as shown in Table 10.1, are now systematically attacking the problem of low yields in the world's most important grain legume crops.

There appears to be a new danger in these elements of progress in most of our important food crops. A growing number of people concerned with the allocation of resources in developing nations has become overly impressed with the recent surge in research capability stemming from the international centers and their ties to national research programs. Many have been led to believe that there is a surplus of useful technology and that the bottleneck now is in the delivery systems used to convey new varieties and cultural practices to farmers. Some even argue for a shift in financial support from research to extension. This, unfortunately, is quite unrealistic. In some cases, there are enormous amounts of written information, but the data and observations are not unified into farm-evaluated technology capable of increasing crop productivity. It is simple to generate bits of information, but excruciatingly difficult to unite and apply them to improve farming systems.

Even with the contributions of international research centers and reviving some national programs, the overall status of research for most crops continues to be seriously inadequate, largely because of lack of governmental support of agricultural research and development. In the rush for industrialization, short-term politicians are reluctant to make long-term investment in the human and institutional resources necessary to produce a strong agricultural base.

Agricultural research aimed at increasing productivity of food and other crops is the best investment the international community can make in developing countries. Since 1967, the total annual expenditure on rice research and related activities by CIAT, The Colombian government and the Rice Federation never reached $1 million. The production increase coming from new technology in 1974 in Colombia alone was valued at about $230 million, giving a splendid return on investment. More examples, such as these, should be brought repeatedly and energetically to the attention of world leaders to persuade them to invest more in research—of climatic, soil, water and human resources—so the developing world might realize its potential in food production.

Section 3

Animal Production

11

Introduction

Clair E. Terrill

We are here to present a series of success stories. Agricultural scientists, with chemistry intimately involved, have produced technology that along with the innovation and application by the American farmer has produced an abundance of food that is the envy of the world. Of course, much of our success has been due to the wonderful combination of soil and climate in this country that few other countries are blessed with. In spite of this tremendous success in food production the farmer has often failed to obtain his fair share of the returns. For many years subsidies were required to maintain production in spite of prices below the cost of production. But still the farmer worked under a very narrow margin of returns over costs.

For many years the cost of food in relation to other costs has been declining. Now we have passed over the top of the curve of food production relative to need and food costs will probably take an increasing part of our incomes from now on. The downhill road will probably be quite bumpy as drouths and other natural disasters influence production. The world population is likely increasing at a more rapid rate than food production. It increased from 3 to 4 billion in just 16 years but I doubt that the capacity for food production went up at such a rapid rate. A large part of the world resources for food production that can be practically used at present economical levels is already in use. Our ability to increase food production will rest largely on our ability to improve technology and to obtain more efficient production.

Unfortunately, there are some restraints on food production that tend to depress the farm price more than the retail price. Thus, they discourage increased production and sometimes they even discourage more efficient production that might more likely hold down retail prices. Retail prices often tend to keep in pace with inflation better than farm prices. Restraints on export of farm products, designed to hold down domestic food prices, will probably be short lived as we need to steadily increase exports to pay for increasing imports of fossil fuels and scarce raw materials. The best hope for slowing the upward trend of retail food prices is increased efficiency of production. There are large opportunities to accomplish this through research, especially with livestock products.

Present relatively low research efforts on livestock production may be inadequate to permit potential increases in efficiency of livestock products to slow the probable upward trend in prices. For example, there are unlimited opportunities to increase efficiency of meat production through genetics. This generally involves application of genetic principles by a research agency rather than by the farmer. The farmer usually cannot afford to improve efficiency of the larger farm animals through breeding because of the slow rate of improvement and because the public, not the farmer, generally obtains the benefits from any improvements.

Still, we cannot afford to be pessimistic about livestock production. People have demanded animal products in their diet for millions of years. We can ask the questions: Could primates have evolved without animal products in their diet? Could civilization have developed without the domestication of animals? Could we exist without the ruminant to convert feedstuffs people cannot digest? Could we maintain our intelligence on purely vegetarian diets? You know, carnivores tend to be more intelligent than herbivores. Anyway, let us hope that the information presented here will lead to even greater production of animal products at even lower cost.

12

Cattle and Chemistry: An Effective Partnership

J. E. Oldfield

Chemists and cattlemen are not as improbable partners as one might think, at first. Both are characterized by independent thinking, resourcefulness and inherent ability to cope with complex and often frustrating situations. Although most often they have worked separately, they have nevertheless constituted an effective research and development team that has helped move beef cattle production from the art of husbandry to the technology of science and contributed to this country's having one of the most plentiful meat supplies in the world. Much of this has taken place during the past 100 years, which serves as the basis for this Symposium.

Cattle production was a tough, but relatively simple business a century ago, characterized by bringing together the animals and the forages they used as feed. When there was plenty of grass, the cattle usually did well; when the soil became "worn-out," as much of it did in the East midway through the last century, the crops failed and so did the animals, and man when he could, pushed westward and pre-empted new lands.

The turning point in agricultural production in this country is usually identified with the understanding of what constituted soil fertility and the development of chemical fertilizers to maintain it. Samuel W. Johnson, at Yale, introduced much of the developing European concept of soil composition and plant growth to this country in his book, *How Crops Feed*, published in 1870, and the new agriculture was underway. The next step was to match composition of this increased crop production to the nutritional requirements of animals, like cattle, and it was accomplished in a similar way—through chemistry. Analysis of animal and plant tissues showed that, in some respects, they were composed of the same things: protein and fats, or oils; and in others, a constituent of one was a useful substrate for production of a constituent of the other, e.g., plant carbohydrate could contribute to body or milk fat.

While both the scientific basis and much of the early inspiration for these analytical explorations were provided by European laboratories, especially those of Liebig, at Giessen, and Lawes and Gilbert, at Rothamsted, added impetus came rapidly in this country. One of the early contributors was W. O. Atwater, who was significantly involved in the establishment of the nation's first agricultural experiment station in Connecticut, just 100 years ago, (Rossiter, 1975). Many people contributed to the knowledge of plant composition and its significance in animal nutrition including Armsby at Pennsylvania State College who exhaustively reviewed and summarized earlier studies in addition to his own in his classic text, *The Nutrition of Farm Animals*, (Armsby 1917) and Morrison at Cornell, who provided the first extensive tables of feed composition in his book, *Feeds and Feeding*, which became virtually the bible of American animal agriculture (Morrison 1956). Early analyses showed not only the reasons for the different nutritional values of different types of feeds (e.g., hays, grains, straws, oil meals) but also showed that considerable variation existed within feeds of the same type. Thus, hay and grain and other feeds were evaluated, not as constant entities, but as changeable mixtures of nutrients.

The analytical approach to scientific feeding of cattle began with the determination of major components of the feeds used—those that contributed the energy and protein required—and progressed to the lesser items, the micronutrients. At first, the methods used were crude, as befitted the state of the art at the time, and for many years the accepted practice was the "proximate analysis" scheme developed by Henneberg and Stohmann, at Weende, over a century ago. This scheme did not deal with chemical entities specifically, but by a series of chemical treatments broke feedstuffs down into recognizable fractions that had certain nutritional implications for the animal (Henneberg and Stohmann 1860). Despite its obvious shortcomings, proximate analysis was the start of an orderly assessment of feed values and it survived almost unmodified until portions of it were challenged and superceded by van Soest only about a decade ago (van Soest 1967). As far as grazing animals were concerned, it was shown to be important to analyze what the animal actually ate, rather than what it was exposed to, and various techniques for assessing consumption were developed (Hardison and Reid 1953). The early methods were successful enough that many held the belief up to the beginning of the 20th Century that the specific functions of various nutrients for animal life could be determined by ever-more-detailed chemical analysis alone. Although this belief was naive, it did give moral support to a number of excellent chemical studies, as Synge (1963) has observed.

The "purified diet" concept was the next major step forward. This was the practice of feeding experimental animals diets of defined composition under carefully controlled conditions. This greatly simplified the task of determining which nutrient did what when it was ingested by the animal. Some of the beginning efforts are humorous in retrospect but must have been trying at the time. When Babcock, at Wisconsin, first compared performance of cattle fed diets composed entirely from single plant sources, he was asked to cease and desist by his Experiment Station Director who valued the animals lost more highly than the knowledge gained (Hart et al. 1911). Much of the credit for the development of the purified diet technique goes to McCollum, one of Babcock's younger colleagues, who carried the definition of diet composition to the point of chemical description. This ushered in the biochemically-frantic decades of the 20's and 30's that marked the discovery of the vitamins and of the essential functions of many of the micronutrients, both of which had important implications in cattle production.

The last essential nutrient to be identified for beef cattle was selenium, and its recognition was so long delayed because the needed amounts were so minute that they were undetected by laboratory procedures available. It was shown in Oregon some 20 years ago, using sheep as experimental animals, that a deficiency of Se causes a myopathy called "white muscle disease" to develop (Muth el al. 1958); and the same holds true for cattle. This observation has eliminated annual losses of hundreds of thousands of young animals for whom the difference between life and death is now known to be less than 0.1 ppm of Se in the diet. It has also been shown, with purified diets, that Se is essential for normal growth. Selenium has another side; in quantity it is one of the most toxic substances known; but it has changed its image greatly in the last few years.

A final reference to analytical applications concerns the interfering effects that may come from presence of non-nutrient substances in feeds. One of the most widely publicized examples was a problem called "X disease" that affected thousands of beef cattle in this country in the early 1950's. This was a situation in which advances in feed-mill engineering had unexpected effects on the lives and productivity of beef cattle, and it provided a fascinating study in cause-and-effect relationships. Outward symptoms of hyperkeratosis and dermatitis accompanied inward stomatitis and a significant drop in plasma vitamin A. After very extensive investigations, including analysis of involved feedstuffs for a wide range of nutrients, it was concluded that the culprit was a chlorinated napthalene which was contained in certain mill lubricants, and contaminated the feed (Olson, 1969). Subsequently, different lubricants have been used and the problem has not reappeared.

These few examples may illustrate the era of analytical contributions to modern cattle production, during which some hundred-odd chemical entities were found capable of constituting an adequate diet. They also anticipate the future, in which we will surely be concerned about a whole host of chemical materials that may be present in feeds by accident or design. These are substances, not necessarily nutrients themselves, which influence the effects of the diet for better or for worse. They include such things as antibiotics, hormones, anthelmintics, antioxidants, crop residues from pesticide and herbicide treatments, gelling agents or binders and many other things that complicate modern cattle nutrition.

ADVANCES IN PHYSIOLOGY

Knowing what was present in cattle feeds was one thing, but knowing what cattle could do with those feeds often proved to be something quite different and this kind of knowledge was the result of research in physiology, or perhaps more correctly, physiological chemistry. One of the first attempts to quantitate the growth process in cattle by physiological and chemical analysis was provided by the studies of Waters (1908) at the University of Missouri, who showed that the form of cattle changed with age, and that growth priorities existed. The length of bones in the head or legs of steers increased even on maintenance or submaintenance diets, while width at chest or hips decreased at the expense of other dimensions when nutrients were restricted. This work was important to the understanding of the influences of time factors in nutrition. It was continued at Missouri for many years and culminated in the monumental text, *Bioenergetics and Growth*, by Brody (1945). Dissection techniques, by which various organs and tissues were studied in relation to such parameters as age, sex, fat, protein and mineral composition have also yielded a great deal of fundamentally-useful data. These have been reviewed by Palsson (1955).

Another dimension of physiologic investigation that has had spectacular effects on beef cattle production in this country is the concept that growth and productive processes are hormone-controlled, and thus are capable of manipulation. As early as 1905, Starling commented on the possibilities of control of different functions of the body by the production of definite chemical substances within the body (Starling 1905).

Cattlemen, of course, had been "practising endocrinologists" for many years, and recognized beneficial effects of castration on animal tractability and meat quality. In the 1930's and early '40's, physiologists tried to duplicate these effects by artificial administration of hormones or hormone-like substances to animals. Some of the early studies were carried out on laboratory animals and showed that dosing animals with either natural estrogens or synthetics like diethylstilbestrol tended to reduce true growth, since they hastened closure of the growth cartilage in the epiphysis and thus hastened skeletal aging. These demonstrations did not deter many physiologists, however, and after Andrews, Beeson and others showed that subcutaneous implantation of diethylstilbestrol pellets increased both rates of gain and efficiency of feed conversion in beef steers (Andrews 1958), the race was on. Much of the impetus for the commercial application of diethlystilbestrol in cattle diets came from the work of Burroughs and his group (1954) at Iowa State, who recognized that estrogenic materials caused a true anabolic response in ruminants that accelerated their weight gains. This was different from that found in many other species, including laboratory animals (Burgess and Lamming 1960) and explained some of the early negative results. Soon, a significant percentage of the cattle finished in feedlots in the United States were receiving diethylstilbestrol (DES), either orally or implant form, and the time period that it took to bring beef to marketable condition was measurably reduced. An added dimension has now appeared on the scene: that of testing product safety. This, too, has involved chemistry, specifically the application of improved analytical technology.

The dramatic departure of DES from the cattle feeding scene was due to the "zero-tolerance" concept which prohibits the use of certain materials in livestock feeding if such use results in measurable residues of the materials in the carcasses of the animals fed. There is no question that safety must be assured; but continued use of DES under the law has become not so much a matter of biological safety as it has one of analytical sensitivity since DES and other hormone-like compounds can now be detected at levels far lower than were possible when the regulations were first proposed (Zimbleman 1974).

The principle of chemical growth stimulation in animals is still a valid one; in fact, an exciting one in view of concern for the future of man's food supply. There are a number of alternatives open for its continued application. An obvious possibility is that sources of hormonal activity be found that do not involve carcinogenic chemicals and that such sources be natural or synthetic. It has been found, for example, that significant estrogenic activity is exhibited by certain plants that may serve as forage for animals (Bickoff 1963) and there is some evidence that these can induce anabolic activity in animals (Oldfield et al. 1966) although such knowledge has not been applied commercially. Alternative compounds like melengestrol acetate (MGA) have been synthesized by the chemical industry which appear to have significant growth-promoting properties for cattle (Zimbleman et al. 1970).

But growth stimulation has been only one facet of the hormonal control concept of improving production efficiency of cattle. The animal industries (beef production is certainly no exception) have been plagued for years by the uncertainties of the reproductive process. Not only was reproductive performance often lower than it should be, but it often happened at the wrong time, economically speaking. During the 1930's physiologists made significant progress in identifying and isolating progesterone from the corpus luteum, which was known to delay estrus and ovulation in various farm animal species even if the corpus luteum regressed (Hansel 1959). The implication was that cattle could be made dependent on exogenous progesterone, rather than the natural secretion, and synchronized estrus followed. This work naturally attracted the attention of the pharmaceutical industry, which responded by the synthesis of a number of progestogens including medroxy progesterone acetate (MAP) which has been used commercially to synchronize estrus in cattle (Kleckner and Zimbleman 1966). Although these compounds effectively control onset of estrus, the fertility rate following their use has been somewhat disappointing. This approach to reproductive improvement is continuing now with the involvement of prostaglandins (Hafs et al. 1974).

THE CONCEPT OF FEED ADDITIVES

As knowledge of the physiology and metabolic processes of cattle increased, it led to intense efforts to manipulate and control them by the addition of a variety of substances to their feed. The era of chemical feed additives has been one of significant change and improvement in efficiency of the American cattle industry.

Some of the earlier feed-additive efforts were directed towards overcoming nutrient deficiencies; one of the commonest of these was protein. It has been recognized since about 1890 that nonprotein nitrogen can be converted to protein by the microorganisms in the ruminant stomach (Hagemann 1891) and a tremendous amount of literature confirms the practicality of using simple compounds like urea for this purpose (Reid 1953).

Although the initial research on urea as a protein replacement was done in Germany, it is fair to say that most of the development to the point of application was conducted in this country. Naturally, there were some problems. It was found, for example, that urea was unpalatable which meant that its use in feed mixes for cattle could depress both protein and energy intakes. Moreover, the conversion of urea-nitrogen to protein could not take place without a carbon skeleton and a source of energy. These difficulties have been surmounted by using urea mixed with sources of easily digested carbohydrate, such as molasses (Mills *et al.* 1944) or starch. Much of the landmark work on this was done at Kansas State University (Bartley 1969). Urea can also be toxic to cattle if the rate of absorption of ammonia freed from it by bacterial urease exceeds the capacity of the liver to remove it from the portal blood (McDonald 1958). Chemists provided one answer to the problem by synthesizing biuret, which is a condensation product of urea. The rate of ammonia formation from biuret was significantly slower than it was from urea, and dangers of toxicity were reduced (Berry *et al.* 1956). This, and other similar research has led to the widespread use of non-protein-nitrogen (NPN) compounds as protein extenders for cattle and other ruminants. It has been dramatically demonstrated by Virtanen (1966) in Finland, and Oltjen and others (1969) in this country that cattle can be maintained in a healthy and productive state on diets which are supplemented with NPN and contain no preformed protein at all. In the light of present concern about the future world-population:food problem, it is attractive to consider the use of urea or other NPN compounds as supplements for low quality roughage, which might make cattle virtually independent of feeds that could be consumed directly by man. Some experiments conducted in Oklahoma have been discouraging in that they showed mixtures of urea and molasses or ammoniated molasses when sprayed on thick stands of native grass resulted in slight depressions of cattle weight gains (Pope *et al.* 1955). Nevertheless, total usage of urea, in particular, for feed purposes, has increased greatly (Hodges 1965) and is now estimated at over a quarter-million metric tons per year.

Along with inadequacies of nutrition, infectious disease ranks as a serious impediment to the productivity of cattle. The search for a "magic bullet" to control such diseases in humans has been well publicized; it had its counterpart in domestic animals in the discovery that antibiotics when added to feed produced a positive growth effect. Here the magic bullet had hit an unexpected target. As a matter of fact, research on the effects of antibiotic feed supplements for cattle proceeded slowly at first, on the assumption that they could interfere with the very processes of microbial digestion on which ruminants depend. Early investigations of additions of chlortetracycline at high levels—up to 600 mg per animal daily to beef cattle rations—did, indeed, indicate depressed appetite and cellulose digestion (Bell *et al.* 1950; Horn *et al.* 1955). Later, as levels of the antibiotics fed were reduced, the effects on the animals improved. Cattle fed about 10 mg per 100 lb body weight generally grew faster than unsupplemented controls and were noticeably freer from scours and diarrhea (Matsushima *et al.* 1954; Perry *et al.* 1954). The cattle also showed lowered incidence of liver abscesses. A number of antibiotics have been used in rations for cattle; much of the earlier work was done with chlortetracycline. Again, it is significant in terms of current interest in lowering grain consumption by animals that the best responses were usually obtained on high-roughage rations (Jukes 1955). Recently, concern has been expressed that continued feeding of low levels of antibiotics would result in the emergence of resistant strains of microorganisms. Episomal transfer of antibiotic resistance can, indeed, take place (Moorehouse and McKay 1968) but the implications of this revelation for human health are still uncertain (Walton 1971). It is reassuring that after some 16 years of widespread use, the beneficial effects of antibiotics in animal production continue to occur (Jukes 1969). But this matter should obviously be kept under close surveillance.

MODIFYING THE ANIMAL AND ITS FEED

After using chemistry to learn the basic makeup of both cattle and their principal feed materials, man logically next directed it to the business of modifying or changing both to his advantage. The use of NPN compounds as extenders of natural protein in low-quality roughages has already been mentioned. There are vast quantities of such feeds in this country much of which is inefficiently used. Such feeds have another major limitation in their low digestibility and this, too, has yielded to the chemical approach. The basic problem is one of loosening the bonds between the highly indigestible lignin and the more easily digested cellulose. It has been attacked by many workers in a number of different ways. One of the earliest efforts was that of Kellner and Kohler in Germany, who boiled rye straw under pressure with sodium hydroxide and other alkalies and were able to increase its digestibility from about 50 to 88% (Woodman and Evans 1947). Later modification of this treatment became the Beckmann process which was described in the literature in 1921 and has been frequently copied or modified since (Beckmann 1921). Essentially, this involves soaking chopped straw in 1.5% NaOH in proportions of 1 part solution to 8 parts straw for a minimum of 4 hr, after which the pulp is drained, washed, and fed wet. The Beckmann process is milder than Kellner's and does not increase digestibility of the roughage as much; but it does improve it significantly and it has important advantages of economy and simplicity. The volumes of liquid used in these processes are unwieldly and in order to meet mounting concerns of their potential danger as pollutants, Canadian workers developed a "dry" process using greatly reduced volumes of a concentrated solution of NaOH (Wilson and Pigden 1964). Ralston and Anderson (1970) at Oregon State University have been able to increase digestibility of ryegrass straw significantly by soaking it in 8% NaOH, and have found that no neutralization of the product is necessary before feeding.

Other feeds have caused other problems for cattle. Many legumes, desirable for their high content of protein and other essential nutrients, are troublesome because of their tendency to cause bloat. This is a particularly difficult problem if the potential of legumes as nitrogen-fixers is to be fully realized in future. Early efforts to control bloat were heroic and, though frequently effective, they were dangerous—often consisting of tapping the rumen with a trochar and cannula inserted through the body wall, to release the entrapped gases. Biochemical studies in this and other countries during the last couple of decades have revealed information that offers encouragement for the ultimate control of the bloat problem in cattle. They have shown that much of the bloat that occurs in cattle on pasture is the result of formation in the rumen of durable bubbles having a a rigid skin, composed of protein and lipid components that will not burst normally and release the gases they contain. Australian chemists have experimented with alcohol ethoxylates which are used in household detergents, and have found them effective antibloating agents (CSIRO 1972). In this country, poloxalene, a polyoxypropylenepolyonyethylene polymer, has been similarly effective as a nonionic surfactant (Bartley et al. 1965). The topic has been interestingly reviewed by Reid (1959).

A matter of great current concern to the consumers of beef and other animal products is the effect that they may have in creating a nutritional environment conducive to coronary heart disease (CHB).

This topic has deservedly received a great deal of attention and extensive reviews have been written on it (see, for example, Royal Society of New Zealand 1971). It is generally agreed CHD is a complex condition, the onset of which is contributed to by a number of factors, only some of which are nutritional. High serum cholesterol levels have been generally associated with CHD incidence (Epstein 1968; Keys 1970), although a direct cause-effect relationship has not been proven. It has been observed, further, that reductions in human blood cholesterol levels have followed substitution of unsaturated for saturated fat in the diet (Peifer and Holman 1959). This had led to conjecture whether the degree of saturation of the fats of beef and other meats can be reduced.

The process that has to be circumvented is the ability of cattle and other ruminant animals, through their symbiotic rumen microorganisms, to hydrogenate dietary fats (Reiser 1951). A series of interesting experiments has shown this to be, indeed, possible.

Australian workers reported in 1961 that the composition of depot fats of ruminants could be altered by infusing triglycerides intraruminally, thus by-passing the rumen (Ogilvie et al. 1961). Practical application of the principle has been established by feeding "protected" fats (coated with hardened protein) to resist rumen digestion. By this procedure, the polyenoic fatty acid content of the meat has been increased from a normal level of 2-4% up to about -20-30% (Cook et al. 1970).

Chemistry has concerned itself also with the esthetic aspects of meat as food. Again, some notable progress has been made particularly where color and tenderness are concerned. Normally, when a beef carcass is cut, the color of the muscle is purplish, but it brightens on exposure to air to a cherry red. Biochemically, the color change is caused by the oxidation of myoglobin in the muscle to oxymyoglobin, and it has come to be associated with quality meat. In some cases, especially where cattle are subject to stress before slaughter, this oxygenation does not take place, or else it occurs at the greatly reduced rate; and esthetically-objectionable, dark-cutting carcass results (Hedrick and Stringer 1964). It has been shown that a contributory cause of this problem has been lowered muscle glycogen, which, in turn, limits both the formation of lactic acid and the penetration of oxygen into the muscle tissues (Munns and Burrell 1966). Through understanding the nature of the chemical changes that take place both anti- and post-mortem, it has been possible to formulate procedures in cattle handling that have significantly reduced the incidence of "dark-cutters" (Hedrick 1958).

Most people will agree that the single most important factor governing acceptability of beef is its tenderness. This, too, has yielded to the chemical approach. It has been shown, for example, that enzymes from exotic plants, like papain, bromelin and ficin, can cause degradation of the collagen fibers and thereby increase tenderness (American Meat Institute Foundation 1960). Application of tenderizing enzymes to beef cattle immediately preslaughter has been practiced with some success by at least one commercial beef processor in this country.

Such procedures carry some implications for the future if beef cattle are to be raised on larger proportions of roughage feed and lower proportions of the more digestible grains.

THOUGHTS FOR THE FUTURE

It is easy to yield to the feeling that most of the improvement in beef cattle production achievable by chemistry has already been done, and certainly the progress during the last hundred years has been remarkable. There is real reason, however—made urgent by the pending problems of world population and food supply—to seek further avenues of improvement and better applications of some already existing. It is encouraging that thought and discussion have been organized to anticipate further research needs in this topic area. Several prestigious reports have already appeared. One of these, entitled *Research to Meet U.S. and World Food Needs*, prepared by a working conference of the Agricultural Research Policy Advisory Committee (ARPAC), may serve as an example (Bentley and Long 1975). This conference brought together scientists from the Agricultural Research Service of the U.S. Department of Agriculture and the National Association of State Universities and Land-Grant Colleges in a common effort to focus attention on priorities for continuing effective research in support of food production. Several of the recommendations for research with beef cattle have important chemical implications.

They suggest, for instance, that attention be directed toward increased use of "noncompetitive" feeds including crop residues, animal manures and various industrial wastes. It seems likely that chemical or biochemical treatment may be heavily involved in this. There are newly emerging problems in this context that need a chemical research input also, notably the removal or neutralization of potentially harmful residual materials that may exist in such products.

They propose also that research continue on the improvement of reproductive performance. This would surely involve continued investigation of the biochemistry and physiological function of reproductive hormones and perhaps synthesis of appropriate analogs.

Prominence is also given to the protection of beef cattle against disease—a matter that becomes increasingly important with the increasing trend toward concentration of large numbers of cattle in confinement during at least a part of their growth cycle. Here again, chemistry must contribute to the knowledge and development of protective agents that will function effectively with no potential danger to the consumers of beef.

Two cautions are pertinent: (1) We should ensure that, as better ways of manipulating the performance of cattle are identified, we do not lose the inherent abilities of the animals. (2) Current and future developments should be evaluated completely, to the point of determining ultimate safety of meat and milk to the consumer.

These few examples may serve to show that the partnership of chemist and cattleman has been productive and is far from ending its usefulness. Indeed, as Winston Churchill remarked, "We have not reached the beginning of the end, but perhaps we have reached the end of the beginning."

BIBLIOGRAPHY

AMERICAN MEAT INSTITUTE FOUNDATION. 1960. The Science of Meat and Meat Products. W. H. Freeman & Co., San Francisco.

ANDREWS, F. N. 1958. Fifty years of progress in animal physiology. J. Anim. Sci. 17, 1064–1078.

ARMSBY, H. P. 1917. The Nutrition of Farm Animals. Macmillan Co., New York.

BARTLEY, E. E. 1969. Non-protein nitrogen supplements for ruminants. Feedstuffs 41, 24–25.

BARTLEY, E. E. et al. 1965. Bloat in cattle. Efficacy of poloxalene in controlling alfalfa bloat in dairy steers and in lactating cows in commercial dairy herds. J. Dairy Sci. 48, 1657–1662.

BECKMAN, E. 1921. Conversion of grain straw and lupins into feeds of high nutritive value. Festchr. Kaiser Wilhelm Ges. Forderung Wiss. Zenjährigen jubiläum., 18–26. (German)

BELL, M. C., WHITEHAIR, C. K., and GALLUP, W. D. 1950. The effect of aureomycin on digestion in steers. J. Anim. Sci. 9, 647–648.

BENTLEY, O. G., and LONG, R. W. 1975. Research to meet U.S. and world food needs. Report of a working conference. Agric. Res. Policy Advisory Committee, Kansas City, 1, 217–222.

BERRY, W. T., Jr., RIGGS, J. K., and KUNKEL, H. O. 1956. The lack of toxicity of biuret to animals. J. Amin. Sci. 15, 225–233.

BICKOFF, E. M. 1963. Estrogen-like substances in plants. In Physiology of Reproduction. F. L. Hisaw, Jr. (Editor). Oregon State University Press, Corvallis.

BRODY, S. 1945. Bioenergetics and Growth. Van Nostrand Reinhold Co., New York.

BURGESS, T. D., and LAMMING, G. E. 1960. The effect of diethystilbestrol, hexoestrol and testosterone on the growth rate and carcass quality of fattening beef steers. Anim. Prod. 2, 93-103.

BURROUGHS, W. et al. 1954. The effects of trace amounts of diethystilbestrol. Science 120, 66-67.

COOK, L. J., SCOTT, T. W., FERGUSON, K. A., and MCDONALD, I. W. 1970. Production of polyunsaturated ruminant body fats. Nature 190, 725-726.

CSIRO. 1972. Bursting the bloat bubble. In Rural Research, CSIRO Quart. Dickson, A.C.T. Australia, Dec. 2-6.

EPSTEIN, F. H. 1968. Multiple risk factors and prediction of coronary heart disease. N.Y. Acad. Sci. Bull. 44, 916-936.

HAFS, H. D., LOUIS, T. M., NODEN, P A., OXENDER, W. D. 1974. Control of the estrous cycle with prostaglandin F_2 in cattle and horses. J. Anim. Sci. 38 (Suppl. 1), 10-21.

HAGEMANN, D. 1891. Contribution to the knowledge of protein metabolic products (urea) in the animal organism. Landwirtsch. Jahrb. 20, 261-291. (German)

HANSEL, W. 1959. The estrus of the cow. In Reproduction in Domestic Animals. H. H. Cole and P. T. Cupps (Editors). Academic Press, New York.

HARDISON, W. A., and REID, J. T. 1953. Use of indicators in the measurement of the dry matter intake of grazing animals. J. Nutr. 51, 35-43.

HART, E. B. et al. 1911. Physiological effect on growth and reproduction of rations balanced from restricted sources. Wis. Agric. Exp. Stn. Res. Bull. 17.

HEDRICK, H. B. 1958. Etiology and possible preventive measures in the dark cutter syndrome. Vet. Med. 53, 466.

HEDRICK, H. B., and STRINGER, W. C. 1964. Dark-cutting beef—its cause and prevention. Univ. Missouri, Folder 134. Ext. Div.

HENNEBERG, W., and STOHMANN, F. 1860. Experiments on maintenance feed for mature cattle. In Contributions Basic to a Rational Feeding System for Ruminants. Schwetschke, Braunschweig. (German)

HODGES, E. F. 1965. Feed use of urea in the United State. U.S. Dep. Agric. Econ. Res. Serv. Admin. Rep.

HORN, L. H., Jr., SNAPP, R. R., and GALL, L. S. 1955. The effects of antibiotics upon the digestion of feed nutrients by yearling steers, with bacteriologic data. J. Anim. Sci. 14, 243-248.

JOHNSON, S. W. 1870. How Crops Feed: A Treatise on the Atmosphere and the Soil as Related to the Nutrition of Agricultural Plants. Orange Judd & Co., New York.

JUKES, T. H. 1955. Antibiotics in Nutrition. Medical Encyclopedia, New York.

JUKES, T. H. 1969. The use of drugs in animal feeds. Natl. Acad. Sci. Publ. 1679. (cf. pp. 56-62.)

KEYS, A. 1970. Coronary heart disease in seven countries. Am. Heart Assoc. Monogr. 29.

KLECKNER, M. D., and ZIMBLEMAN, R. G. 1966. Estrus synchronization: key to better beef production. Mod. Vet. Pract. 47, No. 2, 53-55.

LAWES, J. B., and GILBERT, J. H. 1859. Experimental enquiry into the composition of the animals fed and slaughtered as human food. Trans. R. Soc. (London), 149, 493-680.

MATSUSHIMA, J., DOWE, T. W., and ADAMS, C. H. 1954. Effect of aureomycin in preventing liver abscess in cattle. Proc. Soc. Exp. Biol. Med. 85, 18-20.

McDONALD, I. W. 1958. The utilization of ammonia nitrogen by the sheep. Proc. Aust. Soc. Anim. Prod. 2, 46-51.

MILLS, R. C., LARDINOIS, C. C., RUPEL, I. W., and HART, E. B. 1944. Utilization of urea and growth of heifer calves with corn molasses or cane molasses as the only available carbohydrate in the ration. J. Dairy Sci. 27, 571-578.

MOORHOUSE, E. C., and McKay, L. 1968. Hospital studies of transferable drug resistance. Br. Med. J. 2, 741-742.

MORRISON, F. B. 1956. Feeds and Feeding, 22nd Edition. Morrison Publishing Co. Ithaca, N.Y. (Note that this is a successor to earlier versions by Henry, and later Henry & Morrison.)

MUNNS, W. O., and BURRELL, D. E. 1966. The incidence of dark-cutting beef. Food Technol. (Chicago) 20, 1601-1603.

MUTH, O. H., OLDFIELD, J. E., REMMERT, L. F., and SCHUBERT, J. R. 1958. Effects of selenium and vitamin E on white muscle disease. Science 128, 1090.

OGILVIE, B. M., McCLYMOAT, E. L., and SHORLAND, F. B. 1961. Effect of duodebal administration of highly unsaturated fatty acids on composition of ruminant depot fat. Nature 190, 725-726.

OLDFIELD, J. E., et al. 1966. Coumestrol in alfalfa as a factor in growth and carcass quality in lambs. J. Anim. Sci. 25, 167-174.

OLSON, C. 1969. Bovine hyperkeratosis. Adv. Vet. Sci. Comp. Med. 13, 101-151.

OLTJEN, R. R., BOND, J. and RICHARDSON, G. V. 1969. Growth and reproductive performance of bulls and heifers fed purified and natural diets. I. Performance from 14 to 189 days of age. J. Anim. Sci. 28, 717-722.

PALSSON, H. 1955. Conformation and body composition. In Progress in the Physiology of Farm Animals, Vol 2. J. Hammond (Editor). Butterworths, London, England.

PEIFER, J. J., and HOLMAN, R. T. 1959. Effect of saturated fat upon essential fatty acid metabolism of the rat. J. Nutr. 68, 155-168.

PERRY, T. W., BEESON, W. M., HORNBACK, E. C., and MOHLER, M. T. 1954. Aureomycin for growing and fattening beef animals. J. Anim. Sci. 13, 3-9.

POPE, L. S., HUMPHREY, D. D., and MacVICAR, R. W. 1955. Urea-molasses and ammoniated molasses as supplements for beef cattle on native grass. Okla. Agric. Exp. Stn. Misc. Publ. 43, 51-54.

RALSTON, A. T., and ANDERSON, D. C. 1970. Utilization of grass straw. Proc. 5th Annu. Pac. Northwest Anim. Nutr. Conf. Richland, Wash., 19-24.

REID, C. S. W. 1959. The treatment and prevention of bloat with antifoaming agents. Proc. Nutr. Soc. 18, 127-130.

REID, J. T. 1953. Urea as a protein replacement for ruminants: A review. J. Dairy Sci. 36, 955-996.

REISER, R. F. 1951. Hydrogenation of polyunsaturated fatty acids by the ruminant (abstract). Fed. Proc. 10, 236.

REXEN, F., and MOLLER, M. 1974. Use of chemical methods to improve the nutritional value of straw crops. Feedstuffs 46, 46.

RIGGS, J. K. 1958. Fifty years of progress in beef cattle nutrition. J. Anim. Sci. 17, 981-1006.

ROSSITER, M. W. 1975. The Emergence of Agricultural Science. Yale University Press, New Haven, Conn.

ROYAL SOCIETY OF NEW ZEALAND. 1971. Coronary Heart Disease. Rep. of a committee of the R. Soc. N. Z.

STARLING, E. H. 1905. The Lancet, cited by A. L. Maisel. In The Hormone Quest. Random House, New York (1965).

SYNGE, R. L. M. 1963. Some nutritional aspects of the "non-protein nitrogen" fraction of plants. In Progress in Nutrition and Allied Sciences. D. P. Cuthbertson (Editor). Oliver & Boyd, London, England.

UNDERWOOD, E. J. 1952. Trace Elements in Human and Animal Nutrition, 1st Edition. (See also 3rd Edition, 1971.) Academic Press, New York.

VAN SOEST, P. J. 1967. Development of a comprehensive system of feed analyses and its application to forages. J. Anim. Sci. 26, 119-128.

VIRTANEN, A. I. 1966. Milk production of cows on protein-free feed. Science 153, 1603-1614.

WALTON, J. R. 1971. The public health implications of drug resistant bacteria in farm animals. Ann. N.Y. Acad. Sci. 182, 358-361.

WATERS, H. J. 1908. The capacity of animals to grow under adverse conditions. Proc. Soc. Prom. Agric. Sci., 71-96.

WILSON, R. K. and PIGDEN, W. J. 1964. Effect of a sodium hydroxide treatment on the utilization of wheat straw and poplar wood by rumen microorganisms. Can. J. Anim. Sci. *44*, 122-123.

WOODMAN, H. E., and EVANS, R. E. 1947. The nutritive value of fodder cellulose from wheat straw. I. Its digestibility and feeding value when fed to ruminants and pigs. J. Agric. Sci. *37*, 202-210.

ZIMBLEMAN, R. G. 1974. Approach to hormones as drugs for animals—relationship to the Delaney Amendment. J. Anim. Sci. (Suppl. 1) *38*, 68-76.

ZIMBLEMAN, R. G., LAUDERDALE, J. W., SOKOLOWSKI, J. H., and SCHALK, T. G. 1970. Safety and pharmacologic evaluations of melengestrol acetate in cattle and other animals: a review. J. Am. Vet. Med. Assoc. *157*, 1528-1536.

13

A Century of Progress in Dairy Cattle Production in the United States: 1876 to 1976

J. T. Reid
Ottilie D. White

This chapter outlines a few significant landmarks in dairy cattle production during the centennial period 1876-1976. Doubtlessly, many equally noteworthy advances may also have been included here.

NATURE OF DAIRY CATTLE POPULATION AND MILK PRODUCTION INDUSTRY IN 1876

Origin and Evolution of Dairy Cattle in the United States

First cattle in North America were of Spanish origin, probably descending from those which Columbus brought to the West Indies during his second voyage in 1493. These reached Mexico in 1521 and eventually some were taken to Florida, the Coastal Plains and northward up the Mississippi Valley and some reached the English Colonies as well as California via Mexico.

In 1611, the English brought cattle to the Jamestown Colony and to other English colonies along the Atlantic seaboard. In 1625, the Dutch imported cattle to New York. The French brought cattle, especially during Cartier's third voyage in 1541, to the settlements along the St. Lawrence River. Later these were taken westward as far as Minnesota via waterways and then southward to Wisconsin,

Illinois, Indiana, Michigan, and Ohio. During westward migrations, the English, Irish and Dutch cattle, brought previously to New England and other English colonies along the Atlantic seaboard, became interbred with cattle from the Swedish settlements along the Delaware River, Danish large yellow cattle from New Hampshire, Flemish cattle from Jamestown, and Devon cattle from Massachusetts.

These animals were mixed indiscriminately, and their descendants became the "native" cattle of the next 200 years as the initial mass importations ceased from about 1640 to 1830.

The following are importations of purebred cattle from the Channel Islands, Britain and Europe:

(a) *Holsteins:* First imported as early as 1795, but the first ones to be kept pure came in 1861, and only a few were imported before 1875; from 1875 to 1905, almost 8000 Holsteins were imported.

Holstein Herd Book Association was organized in 1871 and became the Holstein-Friesian Association of America in 1885.

(b) *Jerseys:* A few cows were imported into Connecticut in 1850, but many were brought to the United States between 1868 and 1890.

The American Jersey Cattle Club, a registration agency, was established in 1868.

(c) *Guernseys:* First importation in 1830 to Massachusetts; others in 1840 and later; but about 1240 were imported between 1880 and 1895.

The American Guernsey Cattle Club was organized in 1877.

(d) *Ayrshires:* Of the first importation of Ayrshires in 1822, none was kept pure. Frequent importations into the United States came between 1830 and 1860.

The American Ayrshire Breeders Association was organized in 1875.

(e) *Brown Swiss:* The first cattle of this breed were imported into Massachusetts in 1869 and others came to the United States in 1882; however, through 1931 only 185 animals had been imported.

The Brown Swiss Breeders Association was formed in 1880, but it did not publish a herd book until 1908.

(f) *Milking Shorthorn:* First imported in Virginia in 1783 and into New York State in 1817; many were imported during the remainder of the century.

(g) *Red Polled:* First importation into New York State in 1873 and several other importations were made before 1890.

The Red Polled Cattle Club of America was organized in 1883.

The first step towards the improvement of dairy cattle in the United States was the importation of purebred animals. This was well begun at the beginning of this centennial period; and it continued at a considerable rate during the next 25 years.

Three of the breed associations had been established by 1876 and others soon followed.

General Conditions Related to Dairy Cattle Production in 1876

In colonial days, it was customary for cows to calve in the spring, produce milk while pasture herbage was available, dry off in the fall, and barely survive the winter. Cows sometimes were used for draft; many cows served as "family" cows; and a herd consisted of relatively few cows occasionally with a bull which served cows from miles around as well.

These conditions changed only slowly and not by much until about 1850.

Some examples of notable events occurring mainly between 1850 and 1876 that establish the state of development of the dairy industry in 1876, are:

(a) In 1841, Fuchs discovered bacteria in milk. In 1857, Pasteur's observation of the formation of lactic acid during the souring of milk later led to the pasteurization process.

Thirty years later, Pasteur turned his attention to anthrax, chicken cholera and rabies: this research led to methods of immunization.

About 1880, refrigeration was introduced.

(b) In 1850, the first milk train was run.

(c) In 1851, the first cheese factory was established at Oneida, N.Y.; in 1871, the first creamery was operated in Iowa; in 1856, Gail Borden invented condensed milk; in 1878, the centrifugal separator was invented.

(d) The region known as the Corn Belt had stabilized in its present location between 1860 and 1870.

(e) The first vertical silo in the country was built in Howard County, Maryland in 1876; in 1881, Roberts at Cornell and Henry at the University of Wisconsin conducted the first research concerned with ensiling forage crops.

(f) In 1862, the USDA was established, but without cabinet status (this granted in 1889); also in 1862, the Morrill Land Grant College Act was passed by the Congress.

(g) Beginning about 1870, a few State colleges initiated experimental work in agriculture.

Estimates of the dairy cattle population and milk yield in 1876: (a) The average cow produced *ca.* 650 kg of milk/yr. (b) The United States had over 11 million milking cows; there were about 4 people/

cow. (c) The country's total milk output was *ca.* 7.4 billion kg/yr. (d) In 1876, 45.6% of the people who were gainfully employed worked in agriculture.

Thus, in 1876, the dairy industry was primitive and poorly developed; milk yield per cow was low; and the knowledge of the sciences related to milk production and dairying was essentially nonexistent.

Nevertheless, fragments of knowledge resulting from research conducted prior to 1876 in Britain and Europe formed the beginning of some of the sciences on which dairying was built during the centennial period, 1876-1976.

For example, Darwin's *Origin of the Species by Natural Selection* in 1859, and Mendel's studies in 1866 of the mechanism of inheritance, in retrospect, represented the beginning of the research that led to dairy cattle improvement by genetic means which reached a pragmatic level in about 1950.

TRENDS IN THE DAIRY COW POPULATION AND MILK YIELD IN THE UNTED STATES FROM 1876-1976

Cow Population

Taking a census of the cattle in the United States was not developed in detail until 1920, the estimates made prior to that time indicate that there were about 11×10^6 dairy cows in 1876. As shown in Figure 13.1, the cow population increased from that number to 25.2×10^6 in 1933, declined somewhat during a drouth in the late 1930's, reached a maximum of 25.6×10^6 in 1944, and declined precipitously thereafter to reach the level of 11.04×10^6 cows in 1976. Thus, the same number of cows existed in 1876 as in 1976.

Milk Yield per Cow per Year

Extrapolation of yields estimated in the early 1900's indicates that the average milk output/cow/yr was about 650 kg in 1876. As shown in Fig. 13.1, the milk yield/cow/yr increased relatively slowly from that year, reaching 2410 kg in 1950, and then increased rapidly to the level of 4763 kg in 1976.

Thus, of the total improvement in milk yield/cow during the centennial period, 57.5% occurred since 1950, i.e., during the past quarter of the centennial period.

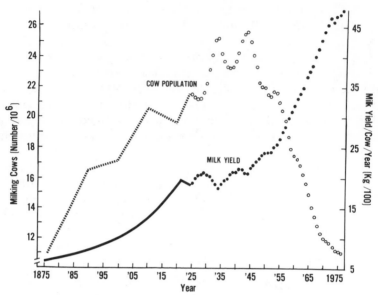

FIG. 13.1. TRENDS IN THE DAIRY COW POPULATION AND IN MILK YIELD PER COW FROM 1876 TO 1976

Annual Total Milk Yield

The total milk output by U.S. cows increased from about 7.4×10^9 kg/yr in 1876 to about 52×10^9 kg/yr in 1942; from 1942 through 1976, the total yield of milk has varied between 52×10^9 and 57×10^9 kg/yr. This is shown in Fig. 13.2.

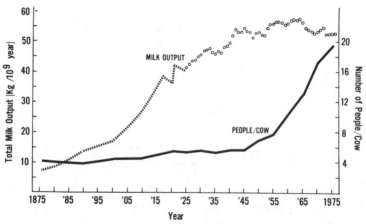

FIG. 13.2. TRENDS IN ANNUAL TOTAL MILK OUTPUT AND IN NUMBER OF PEOPLE PER COW IN THE UNITED STATES

It is noteworthy that in 1976, only 50.3% as many cows existed as in 1950, but each cow produced 2 times as much milk; therefore, the total volume of milk produced in the United States was the same in 1976 as it was in 1950.

Figure 13.2 shows also the number of people/cow in the United States.

From 1876 to the end of World War II, 1 cow served 4–5.8 people. After 1945, the number of people served by 1 cow increased markedly from 5.8 to 19.6 by 1976.

EVENTS DURING THE PERIOD, 1876–1976, THAT BENEFITED DAIRYING IN A GENERAL WAY

Legislative Acts and Establishment of Organizations

The Hatch Act of 1887 established State Agricultural Experiment Stations. At that time, only 17 experiment stations existed in 15 States; by 1893, there were 49 experiment stations.

The Dairy Cattle Branch of USDA was established in 1895.

The American Dairy Science Association was founded in 1906 and published the first Journal of Dairy Science in 1917.

The Smith-Lever Act was passed in 1914, establishing the State-Federal Cooperative Extension Agency and thereby a vehicle to extend research findings and technology to practice.

In 1922, the Capper-Volstead Act provided legal status for agricultural cooperatives. By 1928, there were 12,000 organizations in the United States. Eventually, dairying was benefited by feed, machinery, milk-marketing, cow-testing, and artificial insemination cooperatives.

ADVANCES IN DAIRYING RESULTING FROM RESEARCH AND EXTENSION DURING THE PERIOD, 1876–1976

Genetic Improvement of the Dairy Cow

Improvement in the genetic composition of dairy cattle required coordinated effort in production testing and recording, reproductive physiology and artificial insemination, and population genetic methods.

Milk Production, Testing and Recording.—In 1890, Babock invented a simple determination for the fat concentration in milk.

In 1902, the University of Illinois demonstrated the significance of fat testing and of measuring the milk volume with 221 cows in 18 herds; 1/3 of the cows were found to be unprofitable.

The first Dairy Herd Improvement Association (DHIA) in the United States was organized in Michigan in 1905; in 1906, it began operating with 31 herds enrolled. Four more associations were formed in Michigan in 1907; and the first association was formed in 1908 in each of Maine and New York. Most States soon followed, with the last of the 48 States forming a DHIA organization in 1929.

The DHIA replaced Cow Testing Associations and became a milk testing and measuring organization which eventually provided data on herd health, reproductive characteristics and feeding recommendations. Some of the data were a source of practical (e.g., economic) information, and other data constituted sources for genetic research.

After the adoption of the Smith-Lever Act (1914), the DHIA became a State and Federal Cooperative Extension instrument.

Special aspects of DHIA data are the provisions for identification of animals and milk production records both of which are essential for sire testing and genetic improvement by selection. (The U.S. dairy cattle breed associations, organized between 1868 and 1883, provided a means of identification of those animals which were registered; also, some of these associations had testing programs.)

The DHIA instituted a national program for sire proving in 1936; between 1937 and 1955, 50,000 sires were proved.

Noteworthy are the development and widespread use of the computer since 1945 that made it feasible to process massive volumes of DHIA data and to conduct population genetics research.

Reproductive Physiology and Artificial Insemination (AI) Technology.—Prior to the centennial period, Spallanzani (1785) had artificially inseminated animals; and as early as 1897, mares had been bred artificially in the United States.

In 1902, Ivanoff artificially bred cows in Russia, but the practice did not reach a practical scale until the work of Milanov in 1931.

Perry organized the first AI Cooperative in the United States in New Jersey in 1938; he had studied the Danish AI Cooperative which had been established in 1936 as the first one in the world.

In the 1940's, a yolk-citrate diluent for bull's semen, developed by Phillips at the University of Wisconsin and modified by others, greatly improved the success of artificial insemination, bringing the 60–90-day non-return-to-service rate to 55% of the cows served. In 1940, the fertility rate was only 40%.

Beginning in 1949, the addition of antibiotics to semen extenders improved the non-return-to-service rate to 70-75% by 1960, as the result of reducing the embryonic mortality rate.

Nevertheless, in 1976, the percentage of cows conceiving at the first service was only 50-55 on the average; this figure, as compared with the 60-90-day non-return rate, infers that some estrus periods go undetected and that embryonic mortality still might be a factor contributing to inefficiency.

After 1965, frozen semen, rather than liquid semen, was employed. This made it possible to establish a semen bank to be stored until the genetic merit of a young sire could be determined or to be used years after a high-transmitting sire is dead. The first successful freezing of semen was accomplished in England in 1949 by the use of glycerol.

During the 1950's and 1960's, it was found that a high plane of nutrition induces early sexual maturity and maintains a high level of spermatozoa production by a young bull (Bratton) and that sexual stimulation increases the number of spermatozoa produced per unit of time. These findings combined with those concerning the number of spermatozoa to allocate to a service, and with frozen semen storage, make it commonplace in 1976 for 1 bull to provide the semen for 50,000 inseminations in 1 year.

Thus, at the end of the centennial period, a bull transmitting high-milk production could sire 500,000 or more offspring in his lifetime, even after his death.

The milk-production level of AI-produced daughters was about 500 kg greater in 1976 than that of daughters resulting from natural service; as recently as the early 1950's, there was no difference.

Unfortunately, only about 53% of the nation's cows were bred artificially even in 1976.

Endocrinology perhaps began in 1888 when Brown-Sequard at the age of 72 years injected an aqueous extract of dogs testes into himself. His claims of sexual rejuvenation, certainly imagined, though reported in intimate detail, attracted much attention and gave impetus to the formation of a new branch of physiology.

By 1910, in his *Physiology of Reproduction*, Marshall pointed-to the ovaries and testes as organs of internal secretions (hormones) which influence the growth and development of other generative organs and the secondary sexual characteristics, and that the influence is chemical rather than nervous.

In his work between 1917 and 1922, Marshall demonstrated that an estrus-producing hormone is elaborated by the follicle.

Despite the discovery since 1900 of many hormones and some of their roles, the prominent regulatory role of hormones in reproduction was still being studied intensively and gives great promise of future progress in animal production.

In the 1960's and 1970's, Hansel at Cornell and others were able, by the administration of sex hormones to cattle, to synchronize the estrous cycle of cows with good success (30% conception in poorest, to 75% in best treatments). This development offers the possibility that cows singly or in a group soon can be programmed for insemination at an appointed time. This also would alleviate the need to detect estrus in cows which was a major problem under the confinement, group-handling management systems employed in 1976.

In 1953, Wisconsin scientists reported the first calves born as the result of embryo transfer from donor to recipient cows; this possibility in combination with hormonally-induced superovulation offers a tool for genetic improvement which could exploit the female side.

Up to 1976, most of the improvement in dairy cattle has been effected by sire selection based on the milk yield of daughters. Except in very restricted situations, very few dairy cattle by 1976 had resulted from superovulation and the transfer of superior embryos to donor animals.

Genetics and Breeding.—After Mendel (1866) described the mechanism of inheritance, his discoveries went unused until 1900 when three botanists (de Vries, Correns and Von Tschermak) rediscovered independently the bases of inheritance.

Between 1890 and 1900, Galton applied statistical methods to heredity and variation. His methods concerned measuring the degree of resemblance between the offspring and parents; and he tried to determine the contributions of each of the ancestors to individual offspring.

Galton and Pearson developed theories of selection and statistical methods which in the 1920's and 1930's were fused with genetic principles proposed by Wright, Fisher and Haldane to establish the basis of quantitative genetics.

During the same period and later, Lush explored the mating structure of farm animals and developed theories of genetic selection; also, he developed many disciples of the quantitative genetic approach.

In the 1940's and 1950's, artificial insemination was developed, reaching a degree that enabled one bull to mate with as many as 50,000 cows and DHIA thrived to produce milk production records on thousands of progeny.

Thus, in 1948 the machinery was in hand to carry out genetic research on large populations of cows; in that year, Henderson and his colleagues undertook such research at Cornell.

This work led to the following: (a) A herdmate method to evaluate sires on the basis of limited numbers of daughters in herds of various management levels. (b) A sire comparison method which took into account genetic improvement. (c) In 1952, he proposed a young sire program in which young bulls were selected from the top 1% of the cows and evaluated on the basis of 50 daughters' records and then only the very best of these were put into general AI service. (d) By 1970, these methods were generally accepted.

Although the total research and extension efforts expended in recording milk production, processing DHIA data, reproductive physiology, AI technology and sire selection based on progeny testing have accounted by 1976 for about 30% of the improvement in milk yield per cow made since 1950, there remains much opportunity for additional improvement, e.g., only 53% of the cows are bred artificially.

Improvements in Nutrition and Feeding

State of Knowledge in 1876.—Knowledge underlying nutrition and feeding was considerably more advanced than that related to reproduction, lactation and genetics of cattle.

For example: Chemistry was fairly well developed. Fiber digestion had been studied considerably by Sprengel, Haubner, Sussdorf, and Henneberg and Stohmann between 1832 and 1860. Volatile fatty acids, as digestive products in the rumen, had been recognized by Tiedmann and Gmelin (1826) and Grouven (1870), but their role as an energy source for ruminants was found after 1876. Lignin was known by Konig (1871) to be indigestible. Ca, P, and Mg balances in cattle had been determined by Lehmann in 1859 and by others, and Ca PO_4 had been used as a dietary supplement. Deetz, in 1873, had observed a relationship between the nutritive value and growth stage of forage; feed values had been proposed by Einhof in 1800 and by Thaer in 1809. Data on the digestibility of feedstuffs began to accumulate in 1860; the first feeding standard had been proposed by Grouven in 1859 and was based on the crude nutrients proposed by Liebig in 1842. Wolff, in 1864, published a feeding standard based on individual digestible nutrients, but it was not until 1897 that Kuhn proposed that rations should be formulated on the basis of one portion for maintenance and another for milk production. The knowledge of energy metabolism began with Lavoisier in 1770; heat production was related to animal surface area by Sarrus and Rameaux in 1839, and to the diet ingested and work performed by Regnault and Reiset (1849) using the first closed-circuit respiration apparatus which later was improved and used extensively by Pettenkofer and Voit in 1862 to study farm animals. From that time until

1876, many German scientists, including Kellner, contributed data that eventually culminated in Kellner's feeding standards in 1905 based on "starch-equivalent" values and digestible "true" protein. None of the vitamins had been discovered; the essentiality of a few mineral elements was recognized; but remedies to certain syndromes associated later with specific dietary defects were devised without knowledge of the specific cause. The digestible protein requirements of 0.26 kg (Henneberg and Stohmann in 1860) and of 0.32 kg (Wolff in 1864) and the digestible dry matter allowance of 3.62 kg (Henneberg and Stohmann in 1860) for maintenance of a 450-kg cow are not much different from those cited in the feeding standards employed in 1976.

In 1876, many of the feedstuffs and practices were the same as those in 1976. Some exceptions are that more by-product feeds existed, more concentrates per cow were fed, reliance on the grazing of pastures was less, and the dependency on silages was greater in 1976. (The first upright silo was built in 1876.)

A few examples of major developments in the nutrition and feeding of dairy cattle between 1876 and 1976:

Note: Many noteworthy advances will be neglected here such as the discovery of vitamins; or of the causes of mineral-deficiency diseases which occur (either under impractical situations, or occasionally, or not at all) but which (under most practical situations) have either no or only a small impact on dairy cattle production.

Ruminal Digestion, Synthesis and Absorption.—Between 1876 and 1885, Zuntz, Tappeiner and others demonstrated that some of the end-products of digestion in the rumen are CO_2, CH_4, and acetic, propionic and butyric acids. Tappeiner, in 1881, suggested that these products resulted from the digestion of cellulose by microorganisms in the rumen. In 1879, Zuntz pointed out that the digestion of fiber, different from that of starch, requires a considerable expenditure of energy. Much attention was given to microbial digestion both *in vivo* and *in vitro* from 1880 through the 1950's by Baker, Becker, Mangold, Elsden, Hungate, Doetsch and others. Eventually, in 1944 and later, Barcroft, Phillipson, Elsden *et al.* showed that the volatile fatty acids are absorbed directly from the rumen and that they comprise a major energy source for ruminants. After 1951, McClymont, Popjak, Folley, Cowie and others showed that the volatile fatty acids are the metabolites from which the mammary gland synthesizes especially the short-chain fatty acids which are secreted into milk. During the 1950's and 1960's, Armstrong and Blaxter drew attention to the difference in the energetic efficiency between the volatile fatty acids and much effort was devoted from 1955 to the present time to manipulate ruminal fermentation by dietary means.

In 1879, Weiske *et al.* observed that the nitrogen of asparagine is utilized by ruminants. In 1891, Hagemann was the first to suggest that protein formed in the bodies of microorganisms from nonprotein nitrogen is digested in the small intestines to avail the host of amino acids. Despite considerable study in Germany, the utilization of nonprotein nitrogen by ruminants was not generally recognized until 1937 when Fingerling employed nitrogen balance experiments to demonstrate the phenomenon, and others demonstrated the protein-sparing value of nonprotein nitrogen for the growth of cattle (Bartlett and Cotton in 1938) and for milk production by cattle (Owen *et al.*, Rupel *et al.*, and Archibald in 1943).

In 1976, the maximum extent is not known to which nonprotein nitrogen can replace preformed protein in the diet and how to employ nonprotein nitrogen in high-cellulose diets.

That the B-complex vitamins are synthesized as the result of microbial action in the rumen has been known since 1926; in that year, Bechdel, Eckles and Palmer observed that, a diet sufficiently low in the B-complex vitamins to cause death within a few weeks in rats to which it had been fed, would support normal growth in calves which, when mature, reproduced and performed normally. In 1928, rats fed a B-complex-deficient diet supplemented with ruminal ingesta taken from a cow consuming a B-complex-deficient diet grew much more rapidly than their controls ingesting the unsupplemented diet. In the 1940's, Goss *et al.* demonstrated that vitamin K is also synthesized in the rumen and Link showed that dicumarol present in spoiled sweet clover is antagonistic to vitamin K to cause massive hemorrhage in cattle, a condition that is overcome by the supplementation of the diet with vitamin K.

Nutrient Requirements and Feeding Standards.—In 1976, most of the nutrient requirements of dairy cattle were known, at least in practical terms; but in many situations it remains for certain interactions among nutrients to be determined in an integrated way. However in 1976, the most frequently occurring and severe dietary defect remains to be inadequate energy.

Early studies of feed input-milk output of dairy cows by Ezekial (1927-1940), Woodward (1933), Headley (1935), Cannon (1940), Graves (1940), and Jensen (1942) led to studies of the causes of diminishing returns. In 1955, an analysis by Reid suggested that the efficiency of energy utilization by milking cows varies with the deposition and utilization of body fat and the level of energy input. Work at Cornell between 1956 and 1965 demonstrated a decided decline in the digestibility of energy as the level of energy ingested increases; this observation coupled with the results of the detailed energy balances conducted by Moe, Tyrrell and Flatt led to improved feeding standards beginning in 1966.

Feeding standards changed abruptly at intervals during the centennial period, building on the early German guidelines. In 1903, Haecker proposed that the amount of nutrients allowed for milk production should depend not only on the size of the cow and the quantity of milk she is producing, but also upon the fat concentration of the milk produced. Other standards with increased protein allowances were proposed by Savage (1912), and Henry and Morrison (1915). Later, other features were proposed in the feeding standards of Morrison (1937-1959), and of the National Research Council, published at intervals since 1945. Over the century, the energy rationing unit has changed from digestible matter to total digestible nutrients to metabolizable energy to net energy for milk production, and in some instances feeding standards record two or more energy units.

Nutritive Evaluation of Feeds.—Since 1876, improvements in the evaluation of feedstuffs resulted in tests which provided estimates of the biological values of feeds and their replacement values in the diets of dairy cattle. Noteworthy advances were: The advent of *in vitro* digestion methods. The separation of cell-wall and intracellular substances by Van Soest (1959-1976). The development of indicator methods to estimate the digestibility and intake of pasture forage by freely grazing animals. The development of the relationship between the growth stage and nutritive value of forages (Reid in 1948-1969; and others). Detailed examination of energy metabolism by cows (Moe, Tyrrell, Flatt in 1957-1976). Recognition of the reduced milk-production value per unit of intake of a diet ingested at a high level as compared with that of the same diet at a lower intake (Reid, Moe, Tyrrell and others in 1956 to 1976). The consequent feeding of a higher proportion of concentrates in the diet with higher quality forage than was fed previously.

IMPROVEMENT IN MILK YIELD PER
COW DURING THE LAST QUARTER
CENTURY (1950-1976)

Partition of the Extent of Improvement
Attributable to Various Causes

The average milk yield per cow per year in the United States was 650 kg in 1876; 2410 kg in 1950; and 4763 kg in 1976. Thus, 57.2%

of the improvement achieved during the past century occurred since 1950.

It is estimated that during the period, 1950 to 1976, the proportions of the total improvement attributable to research and extension in various subjects are as follows: genetics and breeding, 30%; nutrition and feeding, 60%; and management, including disease control and manipulation of the environment, 10%.

The major changes in dairy cattle feeding that occurred between 1950 and 1976 are: increasing proportions of concentrates fed; increasing forage quality, thereby increasing the concentration of available energy; provision of more energy per unit of time; and a more effective distribution of energy relative to needs during the lactation period.

Of the total improvement in milk yield, 40% is attributable to the increased rate of concentrate feeding, and 20% is the result of drastic changes in the energy value and acceptability of forages to cows.

The following improvements in management were made between 1950 and 1976: increasing trends toward free-stall and confinement housing; bulk handling of materials; linear programming of ration formulation, provision of "complete" rations, silage feeding, mechanical feeding, group handling of cows, general mechanization of the dairying process, rapid milking methods, and labor-conserving changes in milking arrangements; improved sanitation and disease control; and less grazing of pastures and reduced rates of feeding of hay.

Coincidentally to, or the direct result of, with decided changes in management methods invoked since 1960, a new set of problems evolved, e.g., increased calfhood mortality; reproductive inefficiencies; udder, feet and leg ailments.

Economic Implications of
Improvements in U.S. Dairying
During the Last Quarter Century

A comparison of milk-production statistics for New York State in 1950 and 1976 are examples of U.S. changes:
(a) Number of dairy farms: 60,715 in 1950 and 18,000 in 1976.
(b) Number of dairy cows/farm: 21.4 in 1950 and 51.0 in 1976.
(c) Milk yield kg/cow/yr: 3091 in 1950 and 4953 in 1976.
(d) Concentrates fed kg/cow/yr: 903 in 1950 and 1905 in 1976.
(e) Labor hr/cow/yr: 125.3 in 1950 and 53.0 in 1976.

A comparison of the economics of dairying in New York State in 1950 and 1976, using 1976 prices and costs:
(a) To produce the same volume of milk (252,619 kg/yr) produced by 51 cows/herd in 1976, a herd of 81.7 cows of the 1950 version would be required.

(b) Because of the greater maintenance cost of the 30.7 additional cows needed in 1950, the feed costs would have been $5181 more in 1950 than in 1976.

(c) Because of management differences as well as the 30.7 additional cows needed to produce the same volume of milk in 1950, the labor costs would have been $17,335 more in 1950 than in 1976.

(d) The income above the costs of feed and labor would have been only $889 for the 81.7-cow herd in 1950 and $23,405 for the 51-cow herd in 1976; thus, dairying under 1950 conditions could not survive under the 1976 price-cost structure.

(e) The milk outputs kg/1 kg of total digestible nutrients ingested were 1.28 in 1950 and 1.63 in 1976; this 27% increase in efficiency in 26 years reflects chiefly the reduction in the proportion of the total energy intake that is required for body maintenance, i.e., with increasing energy input and milk output, a higher proportion of the total energy intake is put into milk production.

(f) As a consequence of the massive improvements made in dairying since 1950, the cost of producing milk in 1976 is approximately $0.11/kg less than it was in 1950.

SUMMARY

The 1976 U.S. dairy cow was supreme among man's food-producing animals.

The amounts of protein (g) produced in food products per Mcal of digestible energy ingested by various animals under average U.S. production conditions are as follows: diary cattle (milk), 12.2; hen (eggs), 11.0; broiler (meat), 11.9; pig (pork), 6.1; and beef cattle (meat), 2.3.

The output of protein in milk by a cow producing 13,608 kg of milk/yr is 21 g/1 Mcal of digestible energy; thus, the efficiency depends on level of output.

A large proportion of the energy required by the dairy cow (40 to 100% depending on production level) can be provided by forage; therefore, the fossil energy cost of producing the diet of the dairy cow (as well as that of other ruminants) is considerably less than that of animals such as the pig and the chicken which ingest diets consisting mainly of cereal grains and oilseed feeds.

The 7.3-fold improvement in milk production by the dairy cow that occurred during the centennial period is largely attributable to the research and extension effort exerted since 1876; indeed, 57.2% of the improvement occurring during the century was effected since 1950.

It is estimated that the following proportions of the total improvement after 1950 are ascribable to the research and extension concerned with the following subjects: genetics and breeding, 30%; increased rates and improved effectiveness of concentrate feeding, 40%; improved forage quality, 20%; and improved management and disease control, 10%.

The improved efficiency of dairying occurring between 1950 and 1976 in the United States has resulted in a reduction of the production cost of milk of approximately $0.11/kg (applying 1976 prices and costs to both years).

14

One Hundred Years of Chemicals in the Poultry Industry

Milton L. Sunde

The expression "Better Living Through Chemistry" was familiar to many of us in the 1950's as we listened to some of our favorite radio or television programs. It is entirely fitting that we reflect on this and perhaps record in a simple and direct way how this expression applies to even a small segment of our economy. Poultry raisers, as well as producers, of exhibition type birds during the last 100 years have increased their usage of chemicals in many ways. Because of their willingness to try "things" and because they have had the "eagerness" to experiment and because the individual value of at least some of their stock was not great, the progress was remarkable. The "things" they learned were quickly adopted by others and often were used by people raising other kinds of animals.

As progress was made, the poultry industry and the scientists it attracted became better known, and more and more confidence was earned. During the 1930's, data accumulated with chickens began to be utilized even by the medical profession and by clinical nutritionists.

Changing the "art" of growing and raising chickens or turkeys from a backyard enterprise to a "science" utilizing mass production and high efficiency has involved chemistry in so many ways. As will be indicated in the chapter Tables, hundreds of chemicals and mixtures of chemicals have helped the poultryman make the change from art to science.

During earlier times, no doubt, the barbers, who were also the physicians, were an important source of information for poultrymen wanting treatments for this or that disease or parasite. Many of these materials were chemicals made available by the chemists and alchemists.

The discoveries of many of the amino acids and vitamins during the first 50 years of the 20th Century were made by chemists and nutritionists at various universities. Working with the biochemists, the other two groups have developed disease controls, antibiotics, vitamins, amino acids and mixtures that have converted the poultry industry to the tremendous innovative one that we know today.

All of these marvelous discoveries have been built upon knowledge benchmarks or using either materials or ideas developed earlier by keen observers in the chemical industry.

In the original tabular material published in books, magazines, journals and bulletins, compounds reported as useful are listed by dates. The dates shown by no means are the first time these materials were actually used in the poultry industry. However, these published data do indicate early recommendations by the authors and show that these compounds were used in that manner at that time. Probably, the usage goes back at least a number of years prior to the date in these publications.

As would be expected, most of the chemicals used were involved in either disease or nutritional problems.

Many of the early writings describe, in detail, certain aspects of the art of raising poultry. Certain statements made long ago by very observant people, have since been shown to have a scientific basis. Varro (116-27 BC) stated, "The chicks ought to be driven into the sunshine and on the dung heaps to wallow, for by so doing they grow stronger and this holds good not only for chicks, but for all poultry, both in summer and whenever the weather is mild." His observation of birds made him realize that sunshine was important; 2,000 years later it was learned that this was vitamin D-3. Still later, it was found that manure is a good source of vitamin B-12.

Aldrovandi (1600) (see Table 14.1, Footnote 1) lived from 1522-1605. He wrote in his book, on page 173, "If the capons are suffering from epilepsy, smear their eyes with mother's milk; or portulaca juice; or sal ammoniac, cumin seed and honey pounded together in equal portions."

Salon Robinson (1865) (see Table 14.1, Footnote 3) states, on page 133, "This weed (tobacco), in my case, has never failed in answering all practical purposes; and this fact goes far to show that it was intended to act out higher and nobler ones than are commonly assigned to it. The fine cut is the best kind, and in using it, spread it thickly over the surface of the nests, scatter it upon the floor, and suspend large leaves about the different parts of the house." No doubt this solved some lice problems.

Very few early researchers were interested in the chicken. A few did use starlings, finches, canaries, pigeons or doves. One of the earliest to do nutrition work was a Dr. Fordyce, according to Mc-

TABLE 14.1

COMPOUNDS REPORTED AS USEFUL BEFORE 1900

Ammonium chloride, p. 173[1]
Sal ammoniac, p. 173[1]
Creosote, p. 170[2]
Flour of sulphur, p. 164[2]
Tobacco dust, p. 144[3]
Calomel, p. 186[4]
Carbolic acid, p. 32[4]
Castor oil, p. 186[4]
Crude petroleum, p. 195[4]
Mercurial ointment, p. 195[4]
Pure lard, p. 195[4]
Sulphate of iron, p. 32[4]
NaCl addition to egg yolk
    before drying[5]
Salt[6]
Calcium[7]
Thiamin[8]

Ammonium carbonate, p. 51[9]
Boric acid, p. 39[9]
Camphor, p. 44[9]
Creolin solution, p. 85[9]
(NH4)2CO3, p. 51[9]
Oil of turpentine, p. 57[9]

Potassium bromide, p. 79[9]
Potassium chloride, p. 49[9]
Sulphur ointment, p. 192[4]
Chloroform, p. 382[10]
Cod liver oil, p. 376[10]
Epsom salts, p. 374[10]
Glycerine, p. 374[10]
Iodine, tincture of, p. 378[10]
Oil of vitriol, p. 89[10]
Potassium iodide, p. 376[10]
Tobacco water p. 377[10]
Turpentine, p. 91[10]
Zinc ointment, p. 86[10]

Pyrethrum, p. 22[9]
Quinine, p. 103[9]
Sal ammoniac w/honey, p. 50[9]
Salicylate of soda, p. 103[9]
Saltpetre, p. 51[9]
Silver nitrate, p. 94[9]
Whiskey, p. 51[9]

[1] Aldrovandi, Ulisse (1522-1605)—1600—As translated and edited by L. A. Lind, 1963, Univ. of Oklahoma Press, Norman, Oklahoma.
[2] Jennings, V. S. 1864. Sheep, Swine and Poultry. John E. Potter and Co., No. 617 Sansom Street, Philadelphia.
[3] Robinson, Salon. 1865. Facts for Farmers, Vol. I. Johnson and Ward, Publishers, 113 Fulton Street, New York, N.Y.
[4] Wright, L. 1872. Illustrated Book of Poultry. Cassell, Potter and Galpin, Publishers, London, Paris and New York.
[5] LaMont, C. A. (1883). U.S. Pat. 283, 618.
[6] Felch, I. K. (1889). Poultry Culture. W. H. Harrison, Jr., Publishing Co., Chicago.
[7] Collier, P. (probably prepared by W. P. Wheeler) (1892). N.Y. Agric. Exp. Stn. Bull. 38.
[8] Eijkman, C. (1897). Arch. Path. Anal. (Virchow's) 148, 523.
[9] Pearson, Leonard. 1897. Diseases of Poultry. State Printer of Pennsylvania, Harrisburg, Penn.
[10] Fulton, Robert. 1876. The Book of Pigeons. Cassell, Potter and Galpin Publishers, London, Paris and New York City.

Collum (1957). Dr. Fordyce fed one group of canaries only canary seeds and another group was fed the seeds plus old plaster. The second group "broke down the mortar" and ate it. The first group, fed only the seeds, died. The plaster provided calcium. This experiment was conducted in about 1791.

A glance at Table 14.1 shows that most of the chemicals used early were used for treatment of various parasites and disease problems of different kinds. The oil of vitriol (sulphuric acid) was used to treat canker in pigeons. Phenol, a standard germ killer, was made before Pasteur did his work (Skinner 1974).

In the 14th Century, the condenser was developed to distill alcohol. By the 16th Century, alcohol was heated with sulfuric acid and ether was discovered. The ether and sulfuric acid have been important to the feed industry since the mid 1800's. Dumas (1835) and, later Kjeldahl (1884) developed a method of nitrogen analysis which is still used today to determine the amount of protein in feedstuffs. Ether is used for fat analysis and dilute acid and base for crude fiber.

Feed ingredients were analyzed early. The feeding values of the feedstuffs used were known reasonably well by 1873 (Wright) (see Table 14.1 Footnote 4). A comparison of the values from that time and about 100 years later has been made by Sunde (1974) in the American Poultry History book edited by Skinner (1974). The comparison revealed that the amount of protein, fat, fiber and ash could have been predicted very accurately from the Tables available. Laying rations lacked the minerals and vitamins which came later.

The information the chemists had developed was accurate. However, the chemist of that day could not believe that they could not identify all components. The value of the trace minerals and vitamins was unknown and it wasn't until much later that the classical experiments designed by Babcock and reported by Hart et al. (1911) paved the way for further identification of critical compounds—both vitamin and mineral.

The first commercial feed mill was established in 1875 by John Barwell. Salt was known to be essential, yet most early reports concerned toxic rather than needed levels.

The use of chemicals by the feed trade increased slowly. A survey of the literature by Sunde (1974) (cf Skinner 1974) revealed that as few as 5 or 6 feed ingredients were recommended in 1921. No vitamin or minerals were added. In 1936, this had stayed at the same 6 feed ingredients, but also included 4 minerals and vitamins. In 1949, there were 8 ingredients and 6 vitamins and minerals. By 1972, 5 more ingredients and 23 vitamins and minerals were added. This reflects the changes.

From 1880 to 1900 (Table 14.1), most compounds were still concerned with treating infections of various kinds of parasites.

Shortly after 1900 (Table 14.2), the research into nutritional factors began to produce results. The term "vitamine" was coined by Funk (1912) from "vital amines." The "e" was dropped a number of years later from vitamine. During the late 1800's, most of the amino acids were isolated and characterized by chemists and biochemists in

TABLE 14.2

COMPOUNDS REPORTED AS USEFUL BETWEEN 1900 AND 1930

Bone meal. 1904. P. J. Schaible. Poultry Feeds and Nutrition, 1970, AVI Publishing Co., Westport, Conn.
Formaldehyde. 1911. R. Pearl, F. M. Surface and M. K. Curtis. Poultry Diseases and Their Treatment. Maine Agric. Exp. Stn. Bull.
Vitamines. 1912. C. Funk. J. State Med. 20, 341.
Fat-soluble A. 1914. E. V. McCollum and M. Davis. J Biol. Chem. 19, 245.
Lysine. 1914. T. B. Osborne and L. B. Mendel. J. Biol. Chem. 17, 325.
Sodium hydroxide. 1914. B. F. Kaupp. Poultry Diseases. Am. J. Vet. Med. Publisher, Chicago.
Tryptophan. 1914. T. B. Osborne and L. B. Mendel. J. Biol. Chem. 17, 325.
Carotenoids. 1915. J. Palmer. J. Biol. Chem. 23, 261.
Carbolineum. 1915. R. Pearl, F. M. Surface and M. R. Curtis. Diseases of Poultry. Macmillan Co., New York.
Tobacco stems. 1916. W. B. Herms and J. R. Beach. Calif. Exp. Stn. Circ. 150.
Waterglass egg preservative. 1917. J. B. Hayes and F. E. Mussehl. Univ. Wisc. Coll. Agric. Circ. 74.
Sodium fluoride. 1917. F. C. Bishopp and H. P. Wood. USDA Bull. 801.
Naphthalene. Abbott. 1919. Econ. Ent. 12, 397.
Cod liver oil (vitamin D). 1922. E. B. Hart, J. G. Halpin and H. Steenbock. J. Biol. Chem. 52, 379.
Carbon tetrachloride. 1923. M. C. Hall and J. E. Shillinger. J. Agric. Res. 23, 163.
Oil of chenopodium. 1923. M. C. Hall and J. E. Shillinger. J. Am. Vet. Med. Assoc. 62, 623.
Vitamin A. 1923. A. D. Emmett and G. Peacock. J. Biol. Chem. 56, 679.
Vitamin D. 1925. H. Steenbock and A. Black. J. Biol. Chem. 64, 263.
Nicotine sulphate. 1925. B. A. Beach. Circ. 177, Univ. Wisc. Coll. Agric.
Argyrol. 1926. E. E. Tyzzer. Cornell Vet. 16, 221.
Sodium chloride (salt). 1926. H. H. Mitchell and G. C. Carman. J. Biol. Chem. 68, 165.
Penicillin discovered. 1929. A. Fleming. British J. Exp. Path. 10, 226.
Copper required. 1929. C. A. Elvehjem, E. B. Hart and A. R. Kemmerer. J. Biol. Chem. 84, 131.
Iron required. 1929. C. A. Elvehjem, E. B. Hart and A. R. Kemmerer. J. Biol. Chem. 84, 131.
Phenol. 1929. T. Van Hellsbergen. Ferdinand Enke, Stuttgart, Germany.
Proteins (numerical rating). 1929. C. W. Ackerson, M. J. Blish and F. E. Mussehl. Poultry Sci. 9, 112.
Iodine required. 1930. B. W. Simpson and R. Strand. New Zealand J. Agric. 40, 403.
Rotenone. 1930. W. H. Davidson. J. Econ. Ent. 23, 868.

the United States and Europe. New methods of fumigation were developed using formaldehyde and potassium permanganate. Several compounds were put into use to treat for mites, lice and worms. These include carbolinium, sodium fluoride and nicotine sulphate.

In order to indicate the use of the large number of compounds listed in Tables 14.2 through 14.7, it is necessary to group the compounds into categories. The importance of these materials in disease and parasite control has already been indicated. A number of compounds will fit into two, or even more, categories. Norris (1958) prepared a review of many of the discoveries in the poultry nutrition field.

TABLE 14.3

COMPOUNDS REPORTED AS USEFUL BETWEEN 1931 and 1940

Riboflavin isolated. 1933. R. Kuhn, P. Gyorgy and T. Wagner Jauregg. Ber. Dtsch. Chem. Ges. *66B*, S1034.
Vitamin K required. 1934. H. Dam and F. Schonheyder. Biochem. J. *28*, 1355.
7-dehydro-cholesterol (provitamin D-3). 1935. A Windaus, H. Lettre and F. Schenck. 1935 Ann. der Chem. *520*, 98.
Arginine essential. 1936. A. Arnold, O. L. Kline, C. A. Elvehjem and E. B. Hart. J. Biol. Chem. *116*, 699.
Manganese prevents perosis. 1936. H. S. Wilgus, L. C. Norris and G. F. Heuser. Science *84*, 252.
Riboflavin required. 1936. S. Lepkovsky, and T. H. Jukes. J. Nutr. *12*, 515.
Vitamin E isolated. 1936. H. M. Evans, O. H. Emerson and G. A. Emerson. J. Biol. Chem. *113*, 319.
Vitamin A synthesized. 1937. R. Kuhn, and C. J. O. R. Morris. Bericht. deutsch Chem. Ges. *70*, 853.
Pyridoxine (Vitamin B-6) isolated. 1938. S. Lepkovsky, Science *87*, 169.
Thiamin required. 1938. A. Arnold and C. A. Elvehjem. J. Nutr. *15*, 403.
Pantothenic acid essential. 1939. T. H. Jukes, J. Am. Chem. Soc. *61*, 975; and D. W. Woolley, H. A. Waisman and C. A. Elvehjem. J. Am. Chem. Soc. *61*, 977.
Pure vitamin K isolated. 1939. H. Dam, A. Geiger, J. Glavind, P. Karrer, W. Karrer, E. Rothschild and H. Salomon. Helv. Chem. Soc. *61*, 1295.
Pyridoxine essential. 1939. D. M. Hegsted, J. J. Oleson, C. A. Elvehjem and E. B. Hart. J. Biol. Chem. *130*, 423.
Sulfanilamide. 1939. P. P. Levine. Cornell Vet. *29*, 309.
Vitamin E needed. 1939. H. Dam, and J. Glavind. J. Skand, Arch. Physiol. *82*, 299.
Biotin needed. 1940. D. M. Hegsted, J. J. Oleson, R. R. Mills, C. A. Elvehjem and E. B. Hart. J. Nutr. *20*, 599.
Choline required. 1940. T. H. Jukes. J. Nutr. *20*, 445.
Glycine needed. 1940. H. J. Almquist, E. L. R. Stokstad, E. Mecchi and P. D. V. Manning. J. Biol. Chem. *134*, 213.
Phenothiazine. 1940. E. C. McCulloch, and L. G. Nicholson. Am. J. Vet. Med. *35*, 398.
Sulfapyridine. 1940. P. P. Levine. J. Parasit. *26*, 233.

TABLE 14.4

COMPOUNDS REPORTED AS USEFUL BETWEEN 1941 and 1950

Copper sulfate. 1941. E. S. Weisner. Mich. State Coll. Vet. *1*, 10.
Cystine essential. 1941. J. W. Hayward and F. H. Hafner. Poultry Sci. *20*, 139.
Potassium required. 1941. Ben Ami Ben Dor. Proc. Soc. Exp. Biol. Med. *46*, 341.
Methionine essential. 1941. A. A. Klose and H. J. Almquist. J. Biol. Chem. *138*, 467.
Sulfaquanidine. 1941. P. P. Levine. Cornell Vet. *31*, 107.
Sulfathiazole. 1941. J. P. Delaplane and H. O. Stuart. J. Am. Vet. Med. *99*, 41.
Tryptophan essential. 1941. H. J. Almquist and E. Mecchi. Proc. Soc. Exp. Biol. Med. *48*, 526.
Lysine essential. 1942. H. J. Almquist and E. Mecchi. Proc. Soc. Exp. Biol. Med. *49*, 174.
Magnesium essential. 1942. H. J. Almquist. Proc. Soc. Exp. Biol. Med. *49*, 544.
Niacin required. 1942. G. M. Briggs, R. C. Mills, C. A. Elvehjem and E. B. Hart. Proc. Soc. Exp. Biol. Med. *51*, 59.
Sulfadiazine. 1942. C. Horton-Smith and E. L. Taylor. Vet. Rec. *54*, 516.
Sulfamethazine. 1942. C. Horton-Smith and E. L. Taylor. Vet. Rec. *54*, 516.
Folic acid important. 1943. J. J. Pfiffner, S. B. Binkley, E. S. Bloom, R. A. Brown, O. D. Bird, A. D. Emmett, A. G. Hogan and B. L. O'Dell. Science *97*, 404.

TABLE 14.4 (Continued)

D.D.T. (dichlorodiphenyltrichloroethane). 1944. H. S. Telford. Soap San. Chem. 20, 113.
Glutamic acid. 1944. H. J. Almquist and C. R. Grau. J. Nutr. 28, 325.
Histidine. 1944. H. J. Almquist and C. R. Grau. J. Nutr. 28, 325.
Isoleucine. 1944. C. R. Grau and H. J. Almquist. Poultry Sci. 23, 486.
Leucine. 1944. H. J. Almquist and C. R. Grau. J. Nutr. 28, 325.
Phenylalanine. 1944. H. J. Almquist and C. R. Grau. J. Nutr. 28, 325.
Streptomycin. 1944. D. Jones, H. J. Metzger, A. Schatz and W. A. Waksman. Science 100, 103.
Threonine. 1944. H. J. Almquist and C. R. Grau. J. Nutr. 28, 325.
Valine. 1944. H. J. Almquist and C. R. Grau. J. Nutr. 28, 325.
Gentian violet. 1945. H. P. Treffers. Yale J. Biol. Med. 18, 609.
Penicillin used in poultry. 1945. Nobrega and R. C. Bueno. Arg. Inst. Biol., San Paulo 16, 15.
Sulfamerizine. 1945. J. J. Severens, E. Roberts and L. E. Card. Poultry Sci. 24, 155.
Sulfaquinoxaline. 1945. J. P. Delaplane. Am. J. Vet. Res. 6, 207.
Antibiotic gives growth response. 1946. P. R. Moore, A. Evenson, T. D. Luckey, E. McCoy, C. A. Elvehjem and E. B. Hart. J. Biol. Chem. 165, 437.
Folic acid isolated. 1946. B. L. Hutchings, E. L. R. Stokstad, N. Bohonos, N. H. Sloane and Y. Subbarow. Ann. N.Y. Acad. Sci. 48, 265.
Halogenated arsonic acids. 1946. N. J. Morehouse. J. Parasit. 32, 8.
Sulfapyrazine. 1946. Horton-Smith and R. Boyland. Brit. J. Pharm. Chem. 1, 139.
Chlordane. 1947. J. T. Creighton, L. A. Hetrick, P. J. Hunt and D. U. Duncan. Poultry Sci. 26, 674.
High energy diets. 1947. H. M. Scott, L. D. Matterson and E. P. Singsen. Poultry Sci. 26, 554.
Methoxychlor. 1947. J. T. Creighton, L. A. Hetrick, P. J. Hunt and D. U. Duncan. Poultry Sci. 26, 674.
Nitrofurazone. 1947. P. D. Harwood, D. I. Stunz and R. Wolfgang. J. Parasit. 33, No. 2, Suppl. 14.
Toxaphene. 1947. J. T. Creighton, L. A. Hetrick, P. J. Hunt and D. U. Duncan. Poultry Sci. 26, 674.
Aureomycin. 1948. P. A. Little. Ann. N.Y. Acad. Sci. 51, 246.
Hexachlorophene. 1948. Kerr. Poultry Sci. 27, 781.
Pectin addition to dried eggs. 1948. S. M. L. Tritton. U.S. Pat. 2,452,506.
Vitamin B-12 required. 1948. W. H. Ott, E. L. Rickes and T. R. Wood. J. Biol. Chem. 174, 1047.
Nitrophenide. 1949. E. Waletzky, C. O. Hughes and M. C. Brandt. Ann. N.Y. Acad. Sci. 52, 543.
Nitrothiazole. 1949. J. K. McGregor. Can. J. Comp. Med. Vet. Sci. 13, 257.
Quaternary ammonium. 1949. C. W. Barber and T. A. S. Hayes. Poultry Sci. 28, 830.
Chloromycetin. 1950. A. R. Whitehill, J. S. Oleson, and B. L. Hutchings. Proc. Soc. Exp. Biol. Med. 74, 11.
Neomycin. 1950. O. Felsenfeld, I. F. Volini, S. J. Ishihara, M. C. Backman and V. M. Young, J. Lab. Clin. Med. 35, 428.
New Unidentified factors. 1950. M. L. Sunde, W. W. Cravens, C. A. Elvehjem and J. G. Halpin. Poultry Sci. 28, 204.

TABLE 14.5

COMPOUNDS REPORTED AS USEFUL BETWEEN 1951 and 1960

Bacitracin. 1951. R. L. Davis and G. M. Briggs. Poultry Sci. 30, 767.
Dihydrostreptomycin. 1951. Chin-Min Hsiang, A. Packchanian and C. E. Weakley. Texas Rep. Biol. Med. 9, 34.

TABLE 14.5 (Continued)

Neoarsphenamine. 1951. Chin-Min Hsiang, A. Packchanian and C. E. Weakley.
 Texas Rep. Biol. Med. *9*, 34.
Piperazine. 1951. B. B. Riedel. J. Parasit. *37*, 318.
Surfactants. 1951. C. M. Ely. Science *114*, 523.
Butynorate (di-n-butyl tin dilaurate). 1952. K. B. Kerr. Poultry Sci. *31*, 328.
Ipronidazole (2-isopropyl-1 methyl-5 nitroimidazole). 1952. C. M. Kirkpatrick,
 H. E. Moses and J. T. Boldini. Am. J. Vet. Res. *13*, 102.
Nitro-benzenearsonic acid. 1952. W. C. McGuire and N. F. Morehouse. Poultry
 Sci. *31*, 603.
Chloramphenical. 1953. S. C. Wong, and C. G. James. Poultry Sci. *32*, 589.
Hydrogen peroxide, catalase and glucose oxidase combination. 1953. R. R.
 Baldwin, H. A. Campbell, R. Thiessen, Jr., and G. J. Lorant. Food Technol.
 71, 275.
Polymyxin B. 1953. S. C. Wong and C. G. James. Poultry Sci. *32*, 589.
Triethyl citrate. 1953. H. J. Kothe. U.S. Pat. 2,637,654.
Animal fats improve efficiency. 1954. M. L. Sunde. J. Am. Oil. Chem. Soc. *31*,
 49.
Butylated hydroxyanisol. 1954. L. R. Dugan, Jr., L. Marx, P. Ostby and O. H. M.
 Wilder. J. Am. Oil. Chem. Soc. *31*, No. 2, 46.
Butylated hydroxytoluene. 1954. L. R. Dugan, Jr., L. Marx, P. Ostby and
 O. H. M. Wilder. J. Am. Oil. Chem. Soc. *31*, No. 2, 46.
Furazolidone. 1954. L. C. Grumbles, K. F. Wills and W. A. Boney, Jr. J. Am.
 Vet. Med. Assoc. *124*, 217.
Malathion. 1954. L. E. Vincent, D. L. Lindgren, and H. E. Krohne. J. Econ. Ent.
 47, 942.
Phosphorus required. 1954. W. F. O'Rourke, H. R. Bird, P. H. Phillips and W. W.
 Cravens. Poultry Sci. *33*, 1117.
Aureomycin for meat spoilage. 1955. A. R. Kohler, W. H. Miller and H. P.
 Broquist. Food Technol. *9*, 151.
Nicarbazine. 1955. A. C. Cuckler, C. M. Malanga, A. J. Basso and R. C. O'Neill.
 Science *122*, 244.
Tetracycline. 1955. Olson *et al.* Poultry Sci. *34*, 1214.
Dinitrobenzamide (unistat). 1957. N. F. Morehouse and W. C. McGuire. Poultry
 Sci. *36*, 1143.
Erythromycin. 1957. A. H. Hamdy, L. C. Ferguson, V. L. Sanger and E. H. Bohi.
 Poultry Sci. *36*, 748.
Nithiazide. 1957. D. E. Fogg. Proc. Merck Poultry Nutr. Health Symp., St.
 Louis.
Nystatin. 1957. H. S. Yacowitz, S. Wind, W. P. Jambor, R. Semar and J. F.
 Pagano. Poultry Sci. *36*, 1171.
Zinc needed. 1957. B. L. O'Dell and J. E. Savage. Proc. Fed. Soc. *16*, 394.
Nihydrazone. 1958. A. M. Pilkey and B. S. Pomeroy. Turkey Day Report, Univ.
 Minnesota, St. Paul.
Novobiocin. 1958. M. L. Miner, R. A. Smart and W. W. Smith. Proc. 2nd Natl.
 Symp. Nitrofurans.
Proline. 1958. D. N. Roy and H. R. Bird. Poultry Sci. *37*, 1238.
Trithiadol. 1958. F. Coulston, E. F. Waller and S. A. Edgar. XI World's Poultry
 Congress, Mexico City, Sept. 21–28.
Bifuran. 1959. C. Horton-Smith and P. L. Long. J. Comp. Path. Therap. *69*, 192.
Glycarbylamide. 1959. C. Horton-Smith and P. L. Long. Brit. Vet. J. *115*, 55.
Sevin. 1959. P. Kraemer and D. P. Furman. J. Econ. Ent. *52*, 170.
Tylosin. 1959. R. R. Chalquist and J. Fabricant. Avian Dis. *3*, 257.
Amprolium. 1960. A. C. Cuckler, M. Garzillo, C. Malanga and E. C. McManus.
 Poultry Sci. *39*, 1241.
Dinitro-o-toluamide. 1960. E. H. Peterson. Poultry Sci. *39*, 739.
Hygromycin B. 1960. R. G. Foster, C. B. Ryan, R. D. Turk and J. H. Quisen-
 berry. Poultry Sci. *39*, 492.
Ronnel. 1960. F. W. Knapp and G. F. Krause. J. Econ. Ent. *53*, 4.
Zoalene. 1960. E. H. Peterson. Poultry Sci. *39*, 739.

TABLE 14.6

COMPOUNDS REPORTED AS USEFUL BETWEEN 1961 and 1970

Carbarsone. 1961. C. J. Welter and D. T. Clark. Poultry Sci. *40*, 144.
Dimetridazole (1,2-dimethyl-5-nitroimidazole). 1961. J. M. S. Lucas. Vet. Res. *73*, 465.
Coumaphos (coral). 1962. R. A. Hoffman. J. Econ. Ent. *49*, 347.
Aklomide. 1963. R. R. Baron, M. W. Moeller and N. F. Morehouse. Poultry Sci. *42*, 1255.
Spiramycin. 1963. J. K. A. Cook, J. M. Inglis and W. G. C. Parker. Vet. Rec. *75*, 215.
Ethopabate. 1964. M. L. Clarke. Vet. Rec. *76*, 818.
Sodium silico aluminate. 1964. R. H. Forsythe, L. G. Scharpf and W. W. Marion. Food Technol. *18*, 153.
Linoleic acid needed. 1965. H. Menge, C. C. Calvert, and C. A. Denton. J. Nutr. *85*, 115.
Moenomycin (bambermycin) (flavomycin). 1967. F. Bauer and G. Dost. Antimicrobial Agents and Chemotherapy. American Society for Microbiology, Detroit.
Buqinolate. 1966. C. F. Spencer, A. Engle, C. N. Yu, R. C. Finch, E. J. Watson, F. F. Egetino and C. A. Johnson. J. Med. Chem. *9*, 934.
Decoquinate. 1966. S. A. Edgar and C. Flanagan. Poultry Sci. *45*, 1081.
Clopidol (coyden) (3,5-dichloro-2-6-dimethyl-4-pyredinol). 1967. B. L. Stock, G. T. Stevenson and T. A. Hymas. Poultry Sci. *46*, 485.
Lincomycin. 1967. M. A. Zavala and E. Guerra. Poultry Sci. *46*, 1342.
Monensin. 1967. R. F. Shumard and M. E. Callender. Antimicrobial Agents Chemotherapy. American Society for Microbiology, Detroit.
Selenium essential. 1967. M. L. Scott and J. N. Thompson. Feedstuffs *39*, No. 51, 20.
Monosodium phosphate to scrambled eggs. 1968. R. G. L. Chin and S. Redfern. U.S. Pat. 3,383,221.
Spectinomycin. 1969. A. H. Hamdy and C. J. Blanchard. Poultry Sci. *48*, 1703.
Virginiamycin. 1969. P. L. Van Dick and G. Van Braekel. Chemotherapy *14*, No. 2, 109.
Algin, carrageenan, agar and starch combination. 1970. R. L. Hawley. U.S. Pat. 3,510,315.
New vitamin D-3 effective. 1970. G. Ponchon, H. F. Deluca and T. Suda. Arch. Biochem. Biophys. *141*, 397.
Nickel essential. 1970. F. H. Nielson and H. E. Sauberlich. Proc. Soc. Exp. Biol. Med. *134*, 845.
Salt particle size important. 1970. B. C. Dilworth, C. D. Schultz and E. J. Day. Poultry Sci. *49*, 183.
Sodium hexametaphosphate. 1970. P. K. Chang, W. D. Powrie and O. Fennema. Food Technol. *24*, 63.

TABLE 14.7

COMPOUNDS REPORTED AS USEFUL BETWEEN 1970 and PRESENT

Calcium particle size important. 1971. M. L. Scott, S. J. Hull and P. A. Mullenhoff. Poultry Sci. *50*, 1055.
Robenidine. 1972. S. Kantor, R. L. Kennett, A. L. Shor and E. Waletsky. Poultry Sci. *51*, 1823.

AMINO ACIDS AND PROTEINS

Early work with proteins and amino acids was begun by Osborne and Mendel in 1914. These men found by working with a protein in corn called zein, that it was actually quite deficient in two amino acids, lysine and tryptophan. These two amino acids had been isolated by chemists in 1889 (Drechsel) and in 1901 (Hopkins and Cole), respectively (cf Gortner 1949, page 280 for these references). It wasn't until 1929 that scientists first arrived at some biological rating for the value of protein supplements for chickens. In fact, it wasn't until 1936 when the Wisconsin group first found that arginine was, indeed, essential for the chick. This amino acid was discovered by Hedin in 1895 (Gortner 1949). People knew that certain amino acids were essential for chicks much earlier than this, but it really had not been demonstrated scientifically until that time. During the 1940's, various workers throughout the United States found most of the amino acids shown in the Tables to be essential for the chick. Much of the early work with chicks was done at California. The groups at New Jersey and at the University of Wisconsin reported that most of these same amino acids were essential for the hen. Perhaps, one should differentiate between the essential amino acid and the so-called nonessential. The essential amino acids must be added preformed to the diet. The nonessential are also "essential," but the chick itself can synthesize these. These two terms have been largely substituted with the terms "indispensable" and "dispensable."

Another event of tremendous importance to the animal industry occurred in 1932 when it was discovered by Vestal and Shrewsbury (1932) that it is possible to heat the soybean protein and to improve its nutritional value considerably. Earlier, this was not a problem because the oil was removed from soybean meal by the expeller process. This created heat and the intensity was sufficient to improve the amino acid availability and destroy certain enzymes. As more and more soybeans were produced this protein became much more important. In fact, it is now possible to take a material such as chicken feathers and, by heating them with moist heat, make them into a feedstuff that has very high protein content but the protein quality is poor.

The use of heat on feedstuffs is not new. De Reaumur (1750) wrote the following. "The several grains are not always given to poultry in their dry form; we commonly boil that with which we have a mind to fatten them; we boil it in water, until it is grown soft enough to be easily mashed between our fingers; the water makes it swell to such a degree, that the flower, by dilating, obliges the skin which contained it to burst and split; boiling any grain to that degree is what we call bursting it."

Once these various amino acids had been found to be essential and were made available to the nutritionist, diets could be formulated with no protein feedstuffs at all but only proper levels of amino acids to provide the nitrogen source. When it was possible to grow chicks and to keep hens in production on these purified diets, then the progress became much more rapid and scientists began to learn more about the effects of amino acids balance. The use of the D-enantiomorphs and keto acids of the amino acids for poultry was reviewed by Sunde (1972A).

In fact, it wasn't long until computers were being used to calculate the diets, based upon the amino acid composition of the protein feedstuff that was being used as a major part of the diet. One could then take diets composed of relatively small amounts of protein and put them together and add amino acids to make up for the rest of the diet. All large feed companies now use computers for formulations. Nutrient inputs, as well as restrictions and requirements, are also necessary. This means that nutritionists can also then blame the computer for errors.

ENERGY FEEDS—GRAINS, FATS AND FATTY ACIDS

Waste feed material that had been going into our rivers and streams became available to the poultry industry. It was found that many of these materials (corn gluten feed, wheat bran, etc.) had relatively high protein quality and could be used as part of the protein supplement or as a source of fats and energy.

The traditional thing in feeding poultry has always been to include corn, oats or wheat in the diets of the chickens. In fact, in the late 1880's and 1900's, it was considered normal to add a certain amount of the whole grains to the litter for the chickens or, in the case of the younger birds, to either grind it and have it in a cracked form or in a meal form.

The development of hybrid corn cannot be ignored in the expansion of the poultry industry. Poultry diets contain from 30-80% corn. When corn yields were improved, corn prices decreased and this enabled the industry to produce more edible products at reduced cost. High levels of corn in the diets result in improved feed efficiency because of the increased energy content.

By 1947 it was found by H. M. Scott and his co-workers at the University of Connecticut that high energy diets were very practical for broiler chicks. They came up with a formula that was much higher in energy than most of the diets that had been used previously. This higher energy diet resulted in improved growth rate and in tremendous improvements in efficiency of production. By 1952 Lillie *et al.* (1952) and others working at USDA had also produced data to indicate that the laying hen would respond in a similar way.

In 1954, it was first advocated that animal fats be added to poultry diets. This came from some of the very low quality fats which were available from the packing industry and considered unfit for human consumption. About a year or so later, it was discovered that the ratio of the protein to the energy content of the diet was also of much importance and that one had to keep the amount of the amino acids and, thus, the protein and the energy in good balance.

This brought the computer back into the picture and now it became necessary not only to calculate the amino acids, but also the amount of energy in the diet. Initially, the amount of energy was determined using a term called "net energy" or "productive energy" to make the determinations. As time went on, it was found that an energy value called metabolizable energy was easier to determine in the laboratory and was more easily reproduced with more precision between different laboratories. At the present time, most of the calculations are based upon the metabolizable energy rather than the productive energy. From the practical standpoint, it is actually productive energy that is of most interest but, because of the difficulty with which this is determined, it has been more practical to generally use the metabolizable energy figures.

In 1965, Menge and others working at USDA determined that linoleic acid was an essential fatty acid for the chicken and that this must be considered in diet formulation. It was found that using diets very low in this particular fatty acid could influence the size of the egg considerably.

VITAMINS

After Funk had coined the term "vitamines," work in this area began to move ahead rapidly. Earlier efforts by McCollum and Davis (1914) with vitamin A were the first steps in this direction. After this had been discovered, it didn't take long before Steenbock, Hart, Halpin and others had found that vitamin D was also an essential factor for the young chick. Prior to this time, it was not possible to raise chicks in winter because of the lack of vitamin D. They would go down on their legs in what was called rickets or "leg weakness disease." It is rather interesting that the early efforts in vitamin D were made in the 1920's and yet, even in the 1970's, progress is being made with vitamin D. Dr. DeLuca and his group at the University of Wisconsin have isolated and also synthesized another form of vitamin D called "25-hydroxy vitamin D." Still later, another compound called "1,25 vitamin D-3" was discovered. In addition to vitamins A and D, which were extremely important, a whole series of other vitamins have been discovered and, as each one was discovered, a new assortment of growth factors was reported and, eventually, it has been possible to isolate and synthesize and characterize a number of compounds that have vitamin activity. Vitamin K, for instance, is necessary to prevent bleeding. When birds are maintained without sufficient vitamin K, they develop hemorrhages in the tissues, the eyes and around internal organs. Vitamin E was also found to be necessary for the young chick.

One of the major breakthroughs following the vitamins A and D was riboflavin. This has been termed "vitamin B-2" and also "vitamin G," depending on whether the statement is made in the United States or in other parts of the world. The lack of riboflavin resulted in chicks that developed curled toes and extremely slow growth rates.

After riboflavin was found to be essential, and began to be used, a series of other vitamins were isolated and it was found that most of these were required in very, very minute amounts in the diet of the chicken. One of them happens to be choline, which was isolated way back in the late 1800's, but had not been thought to be useful to the chick. A whole series of compounds with vitamin activity are shown in the Tables; i.e., pantothenic acid, biotin, folic acid, and finally, in 1948, vitamin B-12 was isolated and characterized. This is of extreme interest because many of the vitamin B-12 supplements also contain an antibiotic. Now fermentation vats produce most of our vitamin B-12 and an antibiotic at the same time from cheap carbohydrates and mineral products. Vitamin deficiencies show up in chickens in a number of ways. Almost always the growth rate is greatly reduced; in some, the feathering becomes abnormal. In several deficiencies, specific symptoms help in determining which one of these many vitamins is the real problem. Hogan (1957) has reviewed the vitamin history rather well.

MINERALS

Very early it was known that chickens needed calcium in order to make normal bones, and, also, to produce eggs with strong shells. It was also known a long, long time ago that salt (NaCl) was extremely necessary for the chick. However, along with a lot of other minerals, the actual requirement was not discovered until fairly recently.

As early as 1929, Bethke *et al.* (1929) reported that it was important to have the ratio of calcium to phosphorus about at 1.5:1 for the young chick. Still later, it was discovered that some of our grains contained phosphorus in an organic form called phytin-phosphorus.

This phytin-phosphorus was not utilized by the chick nearly as well as some of the inorganic phosphates. Ironically enough, one of the earliest sources of phosphates or inorganic phosphorus was raw bone and a number of raw bone grinders were on the market for use on the farm in the late 1880's to be used for poultry.

The next major breakthrough with trace minerals was manganese. It was discovered that baby chicks would develop perosis or a very abnormal hock problem if they did not get a sufficient amount of manganese. This discovery was made in 1936 by the Cornell group composed of Wilgus, Norris and Heuser. Within a year or two it was also discovered that this same mineral in very, very low amounts was essential for the hatching of eggs. There are also a number of other minerals that were found to be required in very, very low amounts; zinc, for instance, was found to be needed to get normal feathering in baby chicks. A very minute amount of zinc (1-2 mg), even in the water, will promote normal feathering (Sunde 1972B). A good review of the early work in minerals was prepared by Branion (1937).

Fluorine is also a problem with some of our phosphate supplements and, at the present time, the raw rock phosphates are heated to remove a great deal of fluorine from this material. In addition to fluorine, requirements have been established for iron, copper, and iodine. Another mineral, selenium, was found to be essential for the chick in 1967 by Scott and Thompson. Here is a mineral which for many, many years was considered to be very toxic and to be avoided if at all possible. Now, this mineral in very, very low amounts is actually considered essential.

Still later, another mineral, nickel, was found to be essential by Nielsen and Sauberlich (1970). Even the particle size of some of these minerals was found to be important; for instance, birds respond better to a fairly large particle than they do to the finely ground (NaCl) salt. The same is true with the calcium source for the laying hen. The coarser bits of calcium tend to do better than the fine, powdered calcium carbonates that were used for a number of years in our poultry feeds. The ratio of calcium to phosphorus has already been mentioned. Vitamin D is also important in this relationship between calcium and phosphorus. Another interesting relationship between a vitamin and a mineral has been found in the vitamin E-selenium situation. It is found that these two work together and that one can relieve part of the essential need for the other if provided in proper amounts. Work with mercury poisoning, especially with methyl mercury from fish, has also indicated that the level of selenium in the diet (Ganther and Sunde 1974) is important. Even the level of vitamin E in the diet can help to relieve some of the toxicity of the methyl mercury. Arsenic has been known for a long time to be important in preventing toxicity.

ANTIBIOTICS

Sulphur was used in the 1930's as a preventative against a disease called coccidiosis. Following this, early in the 1940's, sulfa drugs were developed. These sulfa drugs had been used primarily for human purposes but were also found to be effective against a number of the diseases important in poultry production. Later, it was discovered that antibiotics would be helpful in the poultry industry for young chicks. It is rather interesting regarding this discovery, that while an initial observation had been made in the laboratories of Dr. Elvehjem in 1946, it wasn't until several years later that it was recognized by the industry. Following the 1948 discovery of vitamin B-12, a number of fermentation products were put on the market as a source of this particular vitamin. It was soon discovered by poultrymen that one particular vitamin B-12 supplement was much more effective than another commercial product. This was shown later by the research workers to be due to the antibiotic the B-12 supplement also contained. It was found that the vitamin B-12 product one company sold had some of the antibiotics, aureomycin left in it. Following these field reports and a report by Stokstad and Jukes (1950) in March, people all over the United States began to use an antibiotic in most of the diets for young growing chicks. This work has been reviewed rather extensively by White-Stevens (1975). It is extremely interesting reading for anyone who might want to review this situation. Initially, these antibiotics had been used in the poultry industry to control diseases of one type of another and it wasn't until 1950 that the industry began to use them almost exclusively as a means of relieving stress for the chickens or simply to improve their growth rate and feed efficiency. It was also discovered that these antibiotics had an effect on reducing the thickness of the intestinal wall and, therefore, the utilization of some of the nutrients was somewhat better than had previously been observed.

ARSENICALS

The arsenicals have been used in the poultry industry for more than 25 years. They are used by nutritionists to improve intestinal conditions, to help utilization of protein in low protein diets and to help in control of intestinal and cecal protozoa.

COCCIDIOSTATS, WORMERS AND OTHER CHEMICALS

The antibiotics were found to be quite effective against a number of diseases and, at the same time that this was being developed, the sulfa drugs were being used as the coccidiostats. However, as our chemistry improved, more and more specialized materials were developed for the control of coccidiosis. There are really two different types of coccidiostats; one which is an effective compound and one which could be described as a very effective compound. The effective compounds are quite good for birds which are going to be used for egg laying. In this situation, a certain amount of immunity is developed in the birds so that they could be housed on floor pens. The other groups, the very effective compounds, are best used for broiler production. Immunity is not developed but, rather, the disease is prevented entirely. One of the very effective compounds for this was amprolium. This eventually came to be used in conjunction with another material called "ethopabate." The amprolium really functioned as an antithiamin compound. The need of the coccidia for thiamin was reasonably high and, therefore, one could include this antithiamin compound in an amount sufficient to prevent the reproduction of the coccidia but not enough to interfere with the normal development of the bird. Subsequently, another class of compounds, one of which is buquinolate, was also developed. This particular compound prevented certain of the phases that the coccidia must go through in order to become an effective agent and, thus, produce the disease. A number of the compounds in the so-called effective class for coccidiostats contained arsenic in various forms and a number of our coccidiostats, especially some of the older ones, do have these compounds involved with them.

The number of compounds used to control this disease is almost endless and includes the following compounds not previously mentioned: nitrofurazone, nicarbazine, zoalene, nihydrazone, aklomide, clopidol, decoquinate, and monensin. The last named compound is especially interesting because in the spring of 1976 it was advertised as an additive for beef cattle. The name had been changed to one including a reference to the rumen.

Wormers have been used for a long time in poultry production. One of the most successful of these is piperazine. This has been used in many situations and in various combinations in the poultry industry. Another parasitic disease of poultry is called blackhead. This affects primarily turkeys, although on occasion it may affect chickens. A series of compounds has been developed for the control of this condition. One of the early effective ones was enheptin. It is now called "cyzine." In Table 14.8 a number of the commercial and chemical names for some of these compounds have been listed. Some poultrymen use the chemical name and others use the trade name or the name that is used by the industry. A new drug for the control of blackhead was introduced just a few years ago. This one is sold under the name of "Emtryl." This is a very effective compound and more recently has been found to be effective in and is being used for swine, game birds and pigeons. Another compound that is used frequently in turkey production is a compound called "tylosin," which is injected into the head or the neck. This particular compound is used to prevent certain bacterial diseases and mycoplasma conditions that occur in the early growth of the young turkey.

In addition to the compounds that have been discussed, there are a large number of disinfectants. They are listed in the Tables. Note that many of these disinfectants were used before 1880 and, as time has changed, the type used has resulted in more purified compounds. Some of them are called quaternary ammonias, organic iodines, and the chlorines. Another type of compounds, the surfactants, were used briefly but, at present, are not being used. These compounds were basically sulfated fatty alcohols; in fact, some of the detergent soaps were used for a few years in the poultry broiler feeds in order to stimulate growth. At the present time, this has been completely discontinued.

EGG AND POULTRY MEAT

For at least 75 years, the egg industry has been using sugar, salt, or glycerol to put with egg yolks when they are going to be frozen to prevent a putty-like condition from developing. More recently, BHT (butylated hydroxy toluene) or butylated hydroxy anisol has been used as an antioxidant not only for poultry meats, but also for the inedible animal fats that were discussed earlier. In addition, citric acid and citrates and other types of buffer systems have been used in our frozen eggs. They prevent scrambled eggs from turning slightly greenish after they have been cooked. Antibiotics have also been used for some time in the meat processing industry, especially in the mid 1950's. Aureomycin was the one that was used most. It was mixed in the cooling water and ice in which carcasses had been placed to preserve the carcass so that is could be transported unfrozen for longer distances without a loss of quality. However, it wasn't long until the trucking industry also noticed that they could travel for longer periods of time and the grocer observed that the

TABLE 14.8

REPRESENTATIVE CHEMICALS AND NAMES USED

Name	Chemical Name	Trade Name
2-Acetylamino-5-nitrothiazole	2-Acetylamino-5-nitrothiazole	Cyzine
Aklomide	Aklomide (2-chloro-4-nitrobenzamide)	Aklomix
Amprolium	1-(4-amino-2-n-propyl-5-pyrimidinylmethyl)-2-picolinium chloride hydrochloride)	Amprol 25%
Arsanilic acid	p-Aminobenzenearsonic acid	Pro-Gen
Butynorate	Butynorate (dibutyltin dilaurate)	Polystat
Carbadox	Methyl 3-(2-quinoxalinylmethylene) carbazate-N^1, N^4-dioxide	Mecadox
Carbarsone	Para-ureidobenzenearsonic acid	Carb-O-Sep
Chlortetracycline	7-Chloro-4-dimethylamino-1,4,42,5,52,6,11,12a-octohydro-3,6,10,12,12a-pentahydroxy-6-methyl-1,11-dioxo-2-napthacenecarboxamide	Aureomycin
Clopidol	3,5-Dichloro-2,6-Dimethyl-4-Pyridinol	Coyden
Coumaphos	[0, 0-Diethyl 0-(3-chloro-4-methyl-2-oxo-(2H)-1-benzopyran-7-yl) phosphorothioate]	Meldane
Deccox-decoquinate	Ethyl 6-(decyloxy)-7-ethoxy-4-hydroxy-3-quinolinecarboxylate	Deccox
Dimetridazole	Dimetridazole (1,2-dimethyl-5-nitroimidazole)	Emtrymix
Erythromycin	Erythromycin thiocyanate	Gallimycin
Furazolidone	3-[(5-nitrofurfurylidene) amino]-2-oxazolidinone	nf-180-Furox
Ipronidazole	2-Isopropyl-1-methyl-5-nitroimidazole	Ipropran
Monensin sodium	C$_{36}$H$_{61}$O$_{11}$Na	Coban
Nequinate	Methyl 7-(benzyloxy)-6-butyl-1, 4-dihydro-4-oxo-3-quinoline carboxylate	Statyl
Nicarbazin	4,4'-Dinitrocarbanilide 2-hydroxy-4,6,-dimethylpyrimidine	Nicarb "25%"
Nihydrazone	Acetic acid 5-nitrofurfurylidenehydrazide	Zonifur
Nitarsone	Nitarsone (4-nitrophenylarsonic acid)	Histostat-50
Nitrofurazone	Nitrofurazone	nfz
Oxytetracycline	Oxytetracycline hydrochloride	Terramycin
Phenothiazine	Phenothiazine	Phenothiazine
Ronnel	0,0-Dimethyl 0-(2,4,5-trichlorophenyl) phosphorothioate	Trolene
Roxarsone	3-Nitro-4-hydroxyphenaylarsonic acid	Dawe's Nitronic Powder
Zoalene	3,5-dinitro-o-toluamide	Zoamix

153

shelf-life was extended. In the long-run, this did not help the industry very much because many became more careless.

Water-binding carbohydrates are added to whites to prevent weeping and rubberiness of the cooked white after freezing. The lactic acid-aluminum sulfate helps stabilize the white during pasteurization. The H_2O_2 is added to reduce the temperature needed for pasteurization. This decreases processing problems. An enzyme system such as catalase and glucose oxidase increases the stability of the dried white. Triethyl citrate and sodium hexametaphosphate are used as whipping aids for egg white products. The sodium silicoaluminate improves flowability of the dried egg products.

VACCINES

Vaccines have been developed for the poultry industry. One unique innovation involves the vaccination against a parasitic disease called "coccidiosis." This is a disease of the intestinal and cecal parts of the intestine. This vaccine was developed by Edgar (1955) at Auburn University for use with poultry and a number of strains have been developed for different types of coccidia. This particular vaccine is quite effective when used as directed. More recently, a vaccine for Marek's disease has been developed. This disease produces tumors in the body. The strange thing about this is that the vaccine was developed from a virus that is not particularly infective to poultry but it is so closely related to it that the body does not really know the difference and an immunity is developed. This persists and can prevent the chicken from getting the leucosis type infection that has been prevalent in the past. This particular disease has caused tremendous losses to the poultry industry and now is under good control. A number of electrolytes have been developed and are used in the poultry industry for various situations. When birds are not drinking as they should, these electrolytes have been quite beneficial in maintaining the proper mineral levels in the plasma of those birds. In the event that the animal has been deprived of water due to storms or failure of mechanical equipment, it is possible to go in with one of these electrolytes and make it easier for the birds to get over this rather trying period when they've been without water too long. An example of this would be long distance moving of started pullets.

EFFECTS ON INDUSTRY

The increased use of chemicals and scientific knowledge has helped the industry utilize the improved strains of birds developed by geneticists and breeders. In addition, the new and improved equipment developed has helped increase the efficiency of production. Egg production per hen has increased from 83 eggs in 1909 to 222 eggs in 1972 and 232 in 1975. Broiler weights have increased from weighing 250 g at 9 weeks on a 1907-type diet to 1457 g in 1957 and to 1800 g in 7 weeks in 1975.

Turkey production has been improved equally. In 1934 (Anon. 1955) 34 weeks were required to grow a turkey tom to 20 lb. In 1955, this was accomplished in 24 weeks. Now it is possible to get that bird to 30 lb in 24 weeks. The amount of feed has been reduced so that instead of about 6 lb being required per pound gain only 3.3 lb are needed. Additional information on the improvements in production and efficiency have been reported by Feedstuffs (Anon. 1969), Sunde (1974) and Gordy (1974) (*see also* Skinner 1974).

The data in Fig. 14.1 show that only one new chemical has been adapted or recommended by the poultry industry since 1970. This is due primarily to recent regulations by the Food and Drug Administration and other governmental agencies. Chemical companies have found it to be too expensive to work on new compounds. It has been

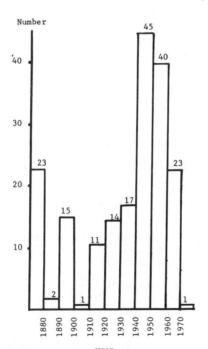

FIG. 14.1. CHEMICALS USED BY POULTRY INDUSTRY BY 10-YEAR PERIODS

estimated that more than $5,000,000 is required to develop a new drug. Unless there is a chance that this much money can be made from the venture, no drug company will take the gamble. In fact, only the broiler or laying hen industry is considered big enough to be attractive even if the potential chemical was a "sure thing" as far as acceptance by the industry and the governmental agencies. Therefore, new drugs for other species of poultry in the future will be those that can be "borrowed" or adapted after they have been approved for other species.

No one man or no one research group or industry leader was responsible for these dramatic changes. What has happened developed because the industry was willing to accept change and to adapt to it quickly. The scientist also was willing to adapt and modify his work to fit into new patterns dictated by new housing and equipment developed by industry.

The cooperation of these groups has enabled the consumer to buy poultry products at prices that would have been "good buys" in 1940. Chemicals used by the industry have helped this to happen.

ACKNOWLEDGEMENT

Research supported by the College of Agricultural and Life Sciences, University of Wisconsin, Madison.

BIBLIOGRAPHY

ANON. 1955. Important milestones in poultry feeds and nutrition. Feedstuffs May 21, 42, 100.
ANON. 1969. Turkey feeds—1969 vs. 1930. Feedstuffs 41, 36; Sept. 6, 44, 78, 87, 155.
BETHKE, R. M., KENNARD, D. C., KICK, C. H. and ZINZALIAN, G. 1929. The calcium-phosphorus relationship in the nutrition of the growing chick. Poult. Sci. 8, 257–265.
BRANION, H. D. 1937. Minerals in poultry nutrition. Sci. Agric. 18, 217.
DE REAUMUR, M. 1750. The art of hatching and bringing up domestick fowls (of all kinds at any time of the year). Printed for C. Davis, over-against Gray's Inn Gate, Holburn, A. Millar and J. Nourse, opposite Katherine-Street, in the Strand, London, England.
DUMAS, J. B. A. 1835. Traite de chimie appliques aux arts. Cited by McCollum, 1957, in A History of Nutrition, Riverside Press. Cambridge, Mass.
EDGAR, A. 1955. Coccidiosis vaccine. J. World Poult. Sci. 11, 29.
GANTHER, H. E. and SUNDE, M. L. 1974. Effect of tuna fish and selenium on the toxicity of methylmercury: A progress report. J. Food Sci. 39, 1.
GORDY, J. F. 1974. Broilers. In American Poultry History: 1823-1973. J. L. Skinner (Editor). American Printing & Publishing, Madison, Wisconsin.
GORTNER, R. A. 1949. Outlines of Biochemistry. John Wiley & Sons, New York.

HART, E. B., MCCOLLUM E. V., STEENBOCK, H. and HUMPHREY, G. C. 1911. Physiological effects on growth and reproduction of rations balanced from restricted sources. Wisc. Exp. Stn. Bull. *17.*

HOGAN, A. G. 1957. Vitamins required by the chick. Mo. Agric. Exp. Stn. Res. Bull. *634.*

KJELDAHL, J. 1884. Neue methode zur bestimmung des stickstoffs in organischen köpern. Z. Ana. Chem. *22*, 366. (German)

LILLIE, R. J., SIZEMORE, J. R., MILLIGAN, J. L. and BIRD, H. R. 1952. Thyroprotein and fat in laying diets. Poult. Sci. *31*, 1037-1042.

MCCOLLUM, E. V. 1957. A History of Nutrition. Riverside Press, Cambridge, Mass.

NIELSON, F. H. and SAUBERLICH, H. E. 1970. Evidence of a possible requirement for nickel by the chick. Proc. Soc. Exp. Biol. Med. *134*, 845-849.

NORRIS, L. C. 1958. The significant advances of the past 50 years in poultry nutrition. Poult. Sci. *37*, 256-274.

SKINNER, J. L. 1974. American Poultry History 1823-1973. American Printing & Publishing, Madison, Wisconsin.

STOKSTAD, E. L. R., and JUKES, T. H. 1950. Further observations on the "animal protein factor." Proc. Soc. Exp. Biol. Med. *73*, 523-528.

SUNDE, M. L. 1972A. Utilization of D- and DL-amino acids and analogs. Poult. Sci. *51*, 44-55.

SUNDE, M. L. 1972B. Zinc requirement for normal feathering of commercial Leghorn-type pullets. Poult. Sci. *51*, 1316-1322.

SUNDE, M. L. 1974. Nutrition. *In* American Poultry History: 1823-1973. J. L. Skinner (Editor). American Printing & Publishing, Madison, Wisconsin.

VESTAL, C. M. and SHREWSBURY. 1932. The nutritional value of soybean with preliminary observations on the quality of pork produced. Am. Soc. Anim. Prod., p. 127.

WHITE-STEVENS, R. H. 1975. Antibiotics curb diseases in livestock, boost growth. Yearbook of Agriculture, U.S. Dep. Agric., 85-98.

15

A Century of Progress in
Swine Production

Virgil W. Hays

In an early edition of Henry and Morrison (1923) the authors had this to say about the pig: "So swift is his career that he usually breaks into life with the spring flowers, plays the gormand in the summer and yields his unctious body a sacrifice to man's necessities with the dropping of the leaves in the fall. The pig is the poor man's reliance and the opulent farmer's gold mine. Of all the domestic animals he is the most prolific, and his possibilities in multiplying are the delight of the city man in his ecstatic dreams of land owning and raising his meager investment in a single mother pig to the nth power through her precocious progeny."

Have those words ever been more true than for the years 1975 and 1976? The swine producers have enjoyed one of the most profitable years ever, while the fortune seekers are gearing up just in time to experience the phenomenon that all seasoned swine growers know will follow a period of respectable profits—a period of much reduced or no profits. The cyclic prices in response to numbers is probably the only phenomenon associated with the pork industry that has not changed appreciably in the past century. The hog number or price cycle has little to do with the contributions of chemistry to the changes in swine production. Yet it may have traumatic effects on the chemistry of the individual until he has experienced a number of cycles sufficient to condition his hormonal mechanisms to the anxieties associated with the unpredictability of the numbers and prices.

SWINE BREEDING

Early efforts in swine production in the United States were devoted to improving the breed or breeds of pigs available. Emphasis was placed on type or phenotypic characteristics of the animals with little concern for the performance traits. The white hogs of Chester County, Pennsylvania, the Jersey Reds and the Poland Chinas developed from the imported animals of Europe and Asia were the forerunners of common breeds today.

In the first half of the century, the major cured pork product was salt pork, with premiums being paid by the packers for heavy, fat hogs. In those days, it was common to market entire herds of hogs weighing 500-600-lb body weight with some even weighing up to 900-1000 lb. Ferrin (1939) makes reference to records of Poland China hogs weighing as much as 1400 lb and Jersey Reds' hanging carcasses as heavy as 1350 lb. Farmers competed for prizes for the heaviest pig and were paid premiums for heavy, excessively fat pigs.

The late 1800's saw a trend to the other extreme—small, early maturing pigs, but still with excessively fat carcasses. The changes in type were probably more a result of fads and fancies of the breeders and judges of that day than market demand. Regardless of the reasons behind changes in type and size, it is evident that breeders, then as now, could make rapid changes. The United States was supplying a major portion of the British import needs during the late 1800's and this undoubtedly had some influence on the market demand. In the years 1895-1900 the United States supplied more than 275 million pounds of pork annually to Great Britain, which accounted for approximately half of their total imports. Selection for the smaller, earlier maturing type continued until 1910 or so. The score card for the Poland Chinas at that time (1895-1912) specified that 2-yr old boars should weigh no less than 500 lb, and sows of the same age should weigh 450 lb. Other breeds followed the same trend, but the extreme was probably attained only in the Poland China breed. The emphasis again shifted toward the larger type, resulting in rapid-gaining pigs who were slow to fatten. Since high yield of lard was a desirable trait, the breeders shifted again during the 1930's toward the short chuffy pig which continued to be the predominant type during the 1940's and early 1950's.

Our present-day emphasis on meat-type pigs that yield a high percentage of meat and little fat may be attributed to the marked reduction in demand for lard as a cooking fat. One can readily grasp that the swine industry has experienced several "type" cycles during the past century. Chemistry has directly or indirectly contributed to these cycles. Curing methods, hydrogenation of vegatable fats, calorie demands of man and many other factors have been involved.

The most suitable type of hog continues to be a question today, as it was in the early part of this century. Much of the experimental work in the early 1900's was devoted to comparing breeds and types.

In the 1930's, the researchers in this country (Craft 1958) recognized the need for a more organized system of swine breeding. They organized the Regional Swine Breeding Laboratory at Ames, Iowa with W. A. Craft as its director; that laboratory coordinated the swine breeding research in this country for about a third of a century.

The major research efforts in these early years were directed toward developing pure breeds; however, the geneticists were investigating the advantages and disadvantages of crossbreeding. As early as 1922, Hammond reported on the effects of crossing. Lush and co-workers (1939) at the Iowa Station reported on several years of crossing and found improvements in litter size, rate of gain and feed conversion. The early work of Hammond (1922), Roberts and Laible (1925), Winters *et al.* (1937), Lush *et al.* (1939) and others provided the basis for today's widespread application of crossbreeding.

Though numerous systems of progeny testing and performance testing have been proposed and used to some extent, the introduction of Boar Testing Stations, which has been accredited largely to the Iowa State University researchers (L. N. Hazel, R. Durham and

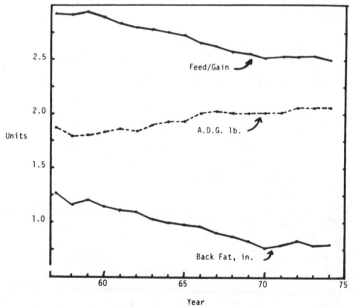

FIG. 15.1. CHANGE IN FEED/GAIN, AVERAGE DAILY GAIN AND BACK FAT THICKNESS OF BOARS TESTED FOR THE YEARS 1956 TO 1974

others) seemed to have developed into a widely-accepted method of evaluating breeding stock. Adequate facilities exist for testing only a very small fraction of the annual need for seed stock; however, these stations do provide a model for on-the-farm testing by commercial breeders.

Rather rapid changes can be made in the meatiness of pigs produced, their rate of gain and the efficiency at which they convert feed to body weight gains. This is illustrated by the data of Fig. 15.1, which summarizes the changes in performance of boars entered in several performance testing stations. From such data as provided in Fig. 15.1, it is apparent that rather rapid changes in carcass composition and performance can be accomplished, if breeders are provided the incentive to do so. In the years of 1956–1960, the tested boars gained at a rate of 1.85 lb per day and required 2.90 lb of feed per pound of gain. Since 1970, they have gained at a rate of more than 2.00 lb per day and averaged 2.52 lb of feed per unit of gain. An even greater percentage change has resulted in fatness of carcass as illustrated by the decline in back fat measurements. Thus, it is understandable that breeders of the past have made rapid adjustments in the size and fatness of pigs with anticipated changes in market demands.

SWINE NUTRITION

A century ago, good legume pastures, corn and water were the major ingredients in the diets of pigs. Skim milk was recognized for its high nutritive value, but the supply was variable and limited. By the turn of the century, little was still known about nutrient requirements except for broad generalities. Kellner (1910) concluded that there had been few useful investigations to throw light on nutrient requirements. Thus, the experience of practice was to be relied on. It was recognized that certain feedstuffs, particularly corn, were deficient in calcium. Most of our knowledge regarding nutrient needs of pigs and the means for meeting those needs are accomplishments of this century.

Some of the early experiments with pigs showed quite satisfactory performance, even by today's standards, and might raise questions in some minds as to how much progress really has been made. Tables 15.1, 15.2 and 15.3 show a comparison of swine diets for growing-finishing pigs (Tables 15.1 and 15.2) and the performance of the pigs in 1900 and 1976 (Table 15.3). Note the wide variety of feedstuffs used in 1900 to accomplish the results, and the relatively few natural

TABLE 15.1

1900 DIET FOR FINISHING PIGS

Ingredient	Dry Matter (%)
Wheat shorts	10.85
Oats	8.25
Oil meal	5.00
Corn	57.53
Wheat	4.22
Whole milk	2.16
Skim milk	2.04
Buttermilk	7.85
Potatoes	0.60
Sweet corn fodder	1.10
Green clover	0.40
Total	100.00

SOURCE: Curtis and Craig (1900).

TABLE 15.2

SAMPLE DIETS FOR 1976

| Ingredient[1] | Weight of Pig | | | |
	12–30 lb (%)	30–60 lb (%)	60–125 lb (%)	125–220 lb (%)
Ground yellow corn	49.10	72.57	78.85	84.20
Soybean meal	27.80	24.10	18.20	12.80
Dried whey	15.00	—	—	—
Sugar (cane or beet)	5.00	—	—	—
Calcium carbonate	0.80	0.80	0.85	0.80
Dicalcium phosphate	1.05	1.20	1.05	1.15
Iodized salt	0.25	0.50	0.50	0.50
Trace mineral mix[2]	0.10	0.08	0.05	0.05
Vitamin mix[3]	1.00	0.75	0.50	0.50
Antibiotics	+	+	+	+
Totals	100.00	100.00	100.00	100.00

[1] Ingredients may be varied so long as nutrient requirements as presented in Tables 15.5 and 15.8 are met.
[2] 0.10% should provide the following in ppm: copper, 10; iodine 0.3; iron, 100; manganese, 60; and zinc, 100.
[3] 1.0% should provide approximately 2000 IU of vitamin A, 400 IU of vitamin D, 4 mg of riboflavin, 10 mg of pantothenic acid, 20 mg of niacin and 10 mcg of vitamin B-12 per pound of diet.

ingredients in today's diets. Dietrich, writing in 1912, pointed out the importance of variety of feedstuffs in the swine ration. He stated that although protein and carbohydrate needs can be met by two feeds, corn and soybeans, "such a mixture is never as good as one that contains more feeds." The mixture of feedstuffs in the 1900 diet, in Table 15.1, contains a variety of feedstuffs that do much to balance the nutrients. Milk products provided vitamins, minerals and

TABLE 15.3

SWINE PERFORMANCE

Item	1900[1]	Year 1921[2]	1976[3]
	Reproductive Performance		
No. litters	45	343	612
Pigs born/sow	—	8.10	10.50
Pigs weaned/sow	5.50	6.10	8.30
	Performance From Weaning to Market Weight		
No. pigs	156	—	1477
Avg initial wt (lb)	36.80	—	35.20
Avg daily gain (lb)	0.94	—	1.69
Feed/gain[4]	4.82	—	3.12

[1] Curtis and Craig (1900).
[2] Wiley (1921).
[3] Cromwell et al. (1976).
[4] Air dry basis.

quality protein to complement the deficiencies of the cereal grains. The green plant materials provided B vitamins and the precursor of vitamin A. Pigs had free access to sunshine to provide vitamin D. One can use the same mixture today and realize quite satisfactory performance. Though the 1976 diets presented include mainly corn and soybean meal and supplements of essential vitamins and minerals, it should be acknowledged that any number of other feedstuffs could be used with excellent results.

The importance of calcium in animal rations was demonstrated well before the present century. Henry (1898) in his first edition of *Feeds and Feeding* reported on the value of bone meal as a supplement to corn. In 1908, Burnett reported on excellent research work showing that mineral supplements were important in improving growth rate and skeletal strength. He used X-rays of bones, bone size and bone breaking strength to evaluate the need for minerals. He compared skim milk, tankage and bone meal as sources of supplemental minerals. Obviously, the milk and meat products provided factors other than calcium and phosphorus, but it was equally obvious that minerals were essential to rapid gains and strong skeletons. Hart *et al.* (1909) added to these observations and further documented that the response to organic phosphorus and inorganic phosphorus differed. Thus, the first decade of nutrition research of this century was largely devoted to calcium and phosphorus as supplements to cereal diets. Wood ashes, coal, charcoal, soil and other mineral sources were tested and used in practice. We now recognize their use was beneficial for their trace mineral contributions. Hamilton (1932) concluded there was seldom, if ever, a need to add trace minerals to diets of pigs. He acknowledged that it had been demonstrated that iron, iodine, manganese, sodium, potassium, magnesium and chlorine were all essential elements, but there was no experimental evidence indicating the general necessity or desirability of supplementing ordinary rations with any others, with the possible exception of sodium and chlorine, the ordinary elements of salt.

As late as 1929, Forbes saw little need for supplementing the pigs' diets with minerals. He presented a resolution to the American Society of Animal Production that read as follows: "be it resolved that the American Society of Animal Production disapproves of the employment of mineral substances, such as common salt, bone meal, limestone or other inorganic elements, compounds or mixtures as components of commercial mixed feeds." He recognized the need for certain minerals under some conditions but suggested that the minerals should be sold by themselves on their own "merit." The resolution was referred to a committee; after a year, the committee recommended the postponement of the matter. Such deliberations are not greatly different from the difficulties of convincing researchers and the public of the essentiality of selenium in the 1970's.

Evvard (1914) proposed that a free-choice system of feeding pigs was a practical way of meeting the pig's nutrient needs. Allowing the pig free access to corn, meat meal and minerals resulted in performance superior to that of animals fed diets mixed according to the feeding standards of that time.

Significant contributions to diet formulations were provided by the Wisconsin researchers in their "Trinity" mixture, which included 50% tankage, 25% linseed meal and 25% alfalfa meal. This combination of supplements to corn resulted in marked improvement in performance of pigs, compared with that resulting from a single ingredient. Iowa research provided further improvement through development of the Big 10 supplement, which included mineral supplements and a mixture of high protein feedstuffs (Table 15.4). These two supplements, Wisconsin Trinity Mixture and Iowa Big 10 supplement, or variations of these, served as the basic supplements to grains until supplemental B-vitamins came into the picture.

Modification of Evvards' "free-choice" system of feeding pigs is used today. The composition of the supplements has changed and the number of choices are usually limited to only two; but many farmers find that grain and supplement offered free-choice is an economical and labor-efficient system of production. Some also allow free access to a mineral supplement which is even one step closer to Evvards' system. Evvard stated: "The appetite of the pig may possibly, fairly quickly, tell us more about which of the amino acids should be present in his ration and in what amounts, much

TABLE 15.4

BIG-10 SUPPLEMENT

Ingredient	%
Tankage	40.000
Cottonseed meal	20.000
Peanut oil meal	9.000
Linseed oil meal	15.000
Alfalfa leaf meal	12.800
Salt	1.000
Limestone	1.500
Iron oxide	0.098
Hardwood ashes	0.500
Potassium iodate	0.002

SOURCE: Evvard *et al.* (1922).

more efficiently than long, laborious, standard, ordinary chemical and physiological research. At any rate, the indications are that the pig can come nearer to selecting his protein diet than can technically trained men." We hope this statement is not accurate today. However, we do know that our precision of estimating the pigs' needs for amino acids and other nutrients needs extensive improvement.

The 1930's and 1940's belong to the B-vitamin and amino acid (Rose 1938) era of swine nutrition. Most of our basic knowledge regarding the essentiality of nutrients is provided by data with laboratory animals and chicks. Niacin (Chick *et al.* 1938), riboflavin (Hughes 1940A), thiamin (Hughes 1940B) and pantothenic acid (Hughes and Ittner 1942) were determined as essential nutrients; and preliminary data regarding the minimum dietary needs were established. The later developments in microbiological and chemical synthesis of these vitamins have added greatly to practical swine production. Extensive use of synthetic vitamin supplements did not materialize until the middle of this century. Natural ingredients and by-product feeds were used to provide these essential nutrients.

The identification of vitamin B-12 (Smith 1948; Rickes *et al.* 1948) and its commercial production provided the final key to evaluating protein sources on the basis of their amino acid composition, rather than on their combined contribution of amino acids, vitamins and "unidentified factors."

Much of the work establishing the essentiality of the amino acids was with laboratory animals (Rose 1938) and chicks. It was well recognized that amino acid balance of proteins was important for many years; but the research aimed at establishing the optimum levels began at the Purdue and Cornell research laboratories in the late 1940's (Beeson *et al.* 1949; Bell *et al.* 1950). The progress in this area has been somewhat limited, even until this date, by the cost of producing or isolating the amino acids. We now have, at reasonable costs, synthetic methionine and lysine to use as supplements to the natural ingredients. The availability of these has further reduced the dependency on the use of a variety of feedstuffs. Recommended levels for each of the essential amino acids are presented in Table 15.5.

Soybean meal was not a major contributor to swine production until the 1940's. Though some soybeans had been grown previously and Robison in 1924 demonstrated the advantages of heat-treating soybeans, they really did not become a significant factor in swine production until years later. Looking back on significant contributions of chemistry to swine production, one would have to agree that the rather simple "Urease Test," used in monitoring the adequacy of heat-treating soybean meal, has contributed greatly to the nutrition of farm animals. Soybeans provide the major source of supplemental protein to animal feeds today. The rapid progress made in this important crop, both from a production as well as a processing point of view, has had a major impact on modern-day swine and poultry production.

The 1950's and 1960's may be considered as the antibiotic or chemotherapeutic era of swine production. Since the introduction of antibiotics in swine production (Jukes *et al.* 1950; Luecke *et al.* 1950; Braude *et al.* 1953), numerous antibacterial agents have been

TABLE 15.5

AMINO ACID ALLOWANCES FOR GROWING SWINE

Weight Range (lb) Protein Level[1]	Prestarter -12 24	Stage of Production Starter 12-30 20	Grower 30-60 17	Developer 60-125 15	Finisher 125-220 13
Arginine	0.43[2]	0.36	0.29	0.25	0.22
Histidine	0.36	0.30	0.24	0.21	0.18
Isoleucine	0.84	0.70	0.56	0.49	0.42
Leucine	1.10	0.92	0.74	0.64	0.55
Lysine	1.20	1.00	0.80	0.70	0.60
Methionine + cystine[3]	0.79	0.66	0.53	0.46	0.40
Phenylalanine	0.60	0.50	0.40	0.35	0.30
Threonine	0.84	0.70	0.56	0.49	0.42
Tryptophan	0.19	0.16	0.13	0.11	0.10
Valine	0.74	0.62	0.50	0.43	0.37

[1] Approximate protein levels needed to meet minimum levels of all essential amino acids when corn and soybean meal are used as major ingredients. Lysine and methionine are usually the limiting amino acids in diets composed of natural feedstuffs.
[2] Percentage of total diet.
[3] Approximately 1/2 as methionine if adequate cystine is present.

TABLE 15.6

THE EFFECT OF ANTIBIOTICS AT BREEDING TIME ON FARROWING
RATE AND LITTER SIZE

No. Sows	Farrowing Rate Control (%)	Treated (%)	Live Pigs/Litter Control	Treated
377	68.5	82.9[1]	9.8	10.1[1]
59	–	–	7.1	9.7[1]
96	87.5	95.8[2]	9.0	10.3[2]
182	60.9	70.0[2]	9.8	10.0[2]
249	66.9	75.4[1]	9.9	10.2[1]
192	93.8	91.6[3]	10.9	11.3[3]
Weighted Avg.	73.0	81.7	9.8	10.3

SOURCE: Hays (1976).
[1] Chlortetracycline, 0.5 to 1.0 g/sow/day
[2] Chlortetracycline, sulfamethazine and penicillin at 0.5, 0.5 and 0.25 g/sow/day, respectively.
[3] Tylosin phosphate, 0.6 g/sow/day

tested and found valuable in improving performance of swine. An example of the beneficial effects on conception rate and litter size are presented in Table 15.6 and an illustration of their effects on growth rate and feed conversion is presented in Table 15.7.

During this antibiotic era, we have also experienced major changes in methods of housing and managing swine. Prior to 1950, most pigs were reared in pasture pens and allowed adequate space. In order to relieve valuable land and maximize use of labor, housing and equipment, many pigs today are reared in complete confinement. Our systems have rapidly changed from a pasture system of production

TABLE 15.7

RELATIONSHIP BETWEEN GROWTH RATE OF CONTROL PIGS AND PIGS FED A
COMBINATION OF PENICILLIN AND STREPTOMYCIN

Daily Gain in Weight of Controls (g)	No. of Comparisons	Improvement Over Controls By Pigs Fed Antibiotics Gain in Weight (%)	Feed Efficiency (%)
91–182	2	22.0	8.2
182–272	3	27.0	4.5
272–363	4	20.4	5.6
363–454	7	16.1	11.1
454–545	9	12.3	6.4
545–636	9	9.4	1.9
636–726	20	5.6	4.7
726 +	7	3.8	1.8
Average Improvement (%)		10.7	5.1

SOURCE: Data summarized from agricultural experiment station reports, 1960 to 1967.

to confinement systems. Also, the number of pigs produced per farm has markedly increased in these past two decades. There is good reason to believe that without such protective chemicals to control diseases, the production systems would need to be markedly changed (Hays 1976).

Accurate growth curves for pigs are difficult to find. Morse (1912) stated that "in scientific publications one can find normal growth curves for guinea pigs, white rats, dog fishes, salamanders, lobsters, caterpillars and many other animals; but, growth curves for common farm animals are so rare as to excite wonder." The data are certainly available but few people take the time to document the rates of growth. Figure 15.2 presents data of Ellis and Zeller (1939) which illustrate that it required about 200–210 days for pigs to reach 220-lb body weight. This is a substantial improvement over the 15 months or so required in the late 1800's. The developments since 1940 allow for pigs to reach 220-lb body weight by 160 days or less (Fig. 15.3).

It would be impossible to sort out the individual reasons for being able to produce market-weight pigs in 1/3 or less of the time required in 1876 or 20% fewer days than in 1940. Improvements in genetics, disease control, nutrition, housing and management have all had sig-

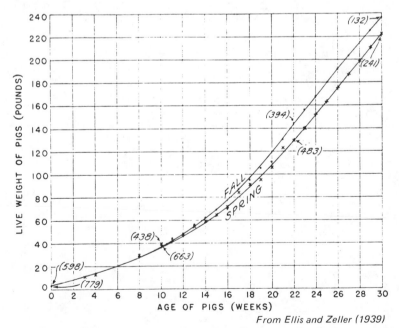

From Ellis and Zeller (1939)

FIG. 15.2. RELATIONSHIP OF LIVE WEIGHT AND AGE OF PIGS

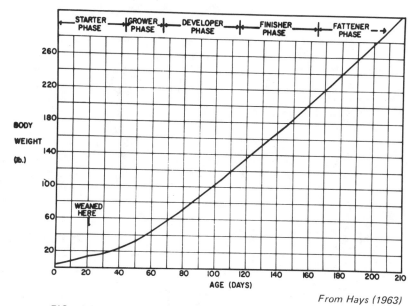

From Hays (1963)

FIG. 15.3. RELATIONSHIP OF LIVE WEIGHT AND AGE OF PIGS

nificant impacts. Certainly, the various fields of chemistry have been a vital part of this progress. Chemical identification of the nutrients, chemical synthesis of nutrients, chemical modification of ingredients, chemical monitoring of nutrients, synthesis of disease-protective chemicals, etc., have all contributed to providing the proper environment, including nutrition, for efficient swine production.

The major strides in improved performance have been accomplished during the early stages of pig growth. Hanson (1958) presented data of pigs fed diets common to the years 1910, 1930 and 1953. All pigs were fed modern diets to 56 days of age. By 140 days, pigs weighed 108, 132 and 200 lb respectively for diets 1910, 1930 and 1953. If pigs were fed a common diet to 90 days of age, the differences in performance were markedly less.

We still use a rather extensive number of ingredients in the feeding of pigs; but we do so for greatly different reasons than we did 100 years ago. The nutrient requirements have been estimated, as well as the nutrient composition of most feedstuffs. In 1876 we used a variety of those feedstuffs to provide the necessary nutrients; needless to say, milk products, meat meal, green plants and other ingredients critically important for correcting the nutrient deficiencies of cereal grains were not available year around. Today, we select the economically advantageous source of nutrients, whether it be a natural ingredient or chemically synthesized nutrient. Thus, the

TABLE 15.8

RECOMMENDED NUTRIENT ALLOWANCES FOR SWINE[1]

Stage of Production	Wt of Pig, Lb	Percent of Diet Protein[2]	Ca	P	Units/Lb of Diet Vitamin A[3]	D	Mcg/Lb of Diet Vitamin B-12	Mg/Lb of Diet Ribo-flavin	Panto-thenic Acid	Niacin	Choline	Gm/Ton Anti-biotic[4]
Boars and Gilts												
Developer	200 to - - -	15	0.80	0.60	2500	300	10	2.5	8	15	400	- - -
Sows												
Pregestation and gestation	- - - - - -	15	0.80	0.60	2500	300	10	2.5	8	15	400	- -[5]
Lactation	- - - - - -	15	0.80	0.60	2500	300	10	2.5	8	15	400	50-100[6]
Young Pigs												
Prestarter	- - - to 12	22-24	0.70	0.60	2000	400	10	4.0	10	25	500	100-250
Starter or creep	12 to 30	18-20	0.70	0.60	2000	400	10	4.0	10	25	500	100-250
Growing-Finishing Pigs												
Grower	30 to 60	16-18	0.65	0.55	1500	300	8	3.0	8	20	400	50-100
Developer	60 to 125	14-16	0.60	0.50	1000	200	5	2.0	5	15	300	20-50
Finisher	125 to 220	12-14	0.60	0.50	1000	200	5	2.0	5	15	300	20-50

[1] The nutrient allowances are suggested for maximum performance, not as minimum requirements. They are based on research work with natural feedstuffs and have been found to give satisfactory results.

[2] See Table 15.5 for amino acid allowances.

[3] About 3.0 times as many units of pro-vitamin A (carotene) are needed as compared with true A.

[4] Arsenicals, nitrofurans or other chemotherapeutics may be used instead of or in combination with antibiotics. Levels and combinations used and stage of production for which they are used must comply with current Food and Drug Administration regulations.

[5] Under high disease level conditions, antibiotics may be beneficial in diet used just prior to breeding (1.0 gm/sow/day, See Table 15.6).

[6] With low disease level conditions, antibiotics may not be needed during lactation.

modern diet may include several feedstuffs or it may contain only two or three. In all cases, vitamins, minerals and sometimes amino acids and antibacterial agents are added in amounts sufficient to meet the needs of pigs as listed in Table 15.8. In addition, chemical agents may be used to enhance the palatability and flavors, or to maintain quality, such as antioxidants.

All the potential progress has not been exploited. Data on the better individual animals today suggest that even more may be accomplished in the coming century, particularly as related to efficiency of swine production.

BIBLIOGRAPHY

BEESON, W. M., MERTZ, E. T., and SHELTON, D. C. 1949. Effect of trytophan deficiency on the pig. J. Anim. Sci. 8, 532.

BELL, J. M., WILLIAMS, H. H., LOOSLIE, J. K., and MAYNARD, L. A. 1950. The effect of methionine supplementation of a soybean oil meal-purified ration for young pigs. J. Nutr. 40, 551.

BRAUDE, R., WALLACE, H. D., and CUNHA, T. J. 1953. The value of antibiotics in the nutrition of swine: A review. Antibiot. Chemother. 3, 271.

BURNETT, E. A. 1908. The effects of food on breaking strength of bone. Nebr. Agric. Exp. Stn. Bull. 107.

CHICK, H., MACRAE, T. F., MARTIN, A. J. P., and MARTIN, C. J. 1938. Curative action of nicotinic acid on pigs suffering from the effects of a diet consisting largely of maize. Biochem. J. 32, 10.

CRAFT, W. A. 1958. Fifty years of progress in swine breeding. J. Anim. Sci. 17, 960.

CROMWELL, G. L., HAYS, V. W., OVERFIELD, J. R., and LANGLOIS, B. E. 1976. Effects of antibiotics, season and parity on reproductive performance of sows. J. Anim. Sci. 43, 250 (abstr.)

CURTIS, C. F., and CRAIG, J. A. 1900. A study of pork production from the standpoint of the farm and the market. Iowa Exp. Stn. Bull. 48.

DIETRICH, W. 1912. Swine Breeding, Feeding and Management. Sanders Publishing Co., Chicago.

ELLIS, N. R., and ZELLER, J. H. 1939. Nutritive requirements of swine. In Food and Life: Yearbook of Agriculture, U.S. Dep. Agric.

EVVARD, J. M. 1914. The free choice system of feeding swine. Proc. Am. Soc. Anim. Prod., 50.

EVVARD, J. M., CULBERTSON, C. C., and HAMMOND, W. G. 1922. Feeding minerals to pigs fattening on rape. Proc. Am. Soc. Anim. Prod., 23.

FERRIN, E. F. 1939. A factual basis for changing swine type. Proc. Am. Soc. Anim. Prod., 125.

FORBES, E. B. 1929. The inclusion of mineral substances as components of commercial mixed feeds. Proc. Am. Soc. Anim. Prod., 37.

HAMILTON, T. S. 1932. Under what conditions are mineral supplements other than salts of calcium and phosphorus necessary in feeding farm animals. Proc. Am. Soc. Anim. Prod., 317.

HAMMOND, J. 1922. Relative growth and development of various breeds and crosses of pigs. J. Agric. Sci. 12, 387.

HANSON, W. E. 1958. Fifty years of progress in swine nutrition. J. Anim. Sci. 17, 1029.

HART, E. B., MCCOLLUM, E. V., and FULLER, J. G. 1909. The role of inorganic phosphorus in the nutrition of animals. Wis. Agric. Exp. Stn. Res. Bull. 1.
HAYS, V. W. 1963. How much feed is required to produce a pound of pork. Proc. Iowa Swine Producers Conf., Iowa State Univ.
HAYS, V. W. 1976. The role of antibiotics in efficient livestock productions. Proc. Int. Symp. Nutr. Drug Interrelations, Nutr. Sci. Council, Iowa State Univ.
HENRY, W. A. 1898. Feeds and Feeding. Published by author, Madison, Wisconsin.
HENRY, W. A., and MORRISON, F. B. 1923. Feeds and Feeding, 18th Edition. Morrison Co., Madison, Wisconsin.
HUGHES, E. H. 1940A. The minimum requirement of riboflavin for the growing pig. J. Nutr. 20, 233.
HUGHES, E. H. 1940B. The minimum requirement of thiamine for the growing pig. J. Nutr. 20, 239.
HUGHES, E. H., and ITTNER, N. R. 1942. The minimum requirement of pantothenic acid for the growing pig. J. Anim. Sci. 1, 116.
JUKES, T. H. et al. 1950. Growth-promoting effect of aureomycin on pigs. Arch. Biochem. 26, 324.
KELLNER, O. 1910. The Scientific Feeding of Animals. (Translated by W. Goodwin.) Macmillan Co., New York.
LUECKE, R. W., MCMILLEN, W. N. and THORPE, F. JR. 1950. The effect of vitamin B_{12}, animal protein factor and streptomycin on the growth of young pigs. Arch. Biochem. 26, 326.
LUSH, J. L., SHEARER, P. S., and CULBERTSON, C. C. 1939. Crossbreeding hogs for pork production. Iowa Agric. Exp. Stn. Bull. 380.
MORSE, E. W. 1912. Some suggestions concerning the planning and reporting of feeding trials. Proc. Am. Soc. Anim. Nutr., 4.
RICKES, E. L. et al. 1948. Crystalline Vitamin B_{12}. Science 107, 396.
ROBERTS, E., and LAIBLE, R. J. 1925. Heterosis in pigs. J. Hered. 16, 383.
ROBISON, W. L. 1924. The influence of the method of oil extraction on the feeding value of soybean oil meal. Proc. Am. Soc. Anim. Prod., 60.
ROSE, W. C. 1938. The nutritive significance of the amino acids. Physiol. Rev. 18, 109.
SMITH, E. L. 1948. Purification of anti-pernicious anemia factors from liver. Nature 161, 638.
WILEY, J. R. 1921. Hog production record demonstrations. Proc. Am. Soc. Anim. Prod., 55.
WINTERS, L. M., JORDAN, P. S., KISER, O. M., and COMSTOCK, R. E. 1937. The Minnesota studies of crossbreeding swine. Proc. Am. Soc. Anim. Prod., 196.

16

One Hundred Years of Progress in Sheep and Goat Production

Carl S. Menzies

Sheep and goats have provided mankind with nutritious food and quality clothing for centuries. Because of their useful products, small size, adaptability, and noncompetitiveness with man, they were probably among the earliest animals domesticated. Both species were introduced to North America by early explorers and colonists.

GOATS

Goats are under-appreciated and markedly under-exploited as an animal food resource. In the United States, there are three distinct types: dairy goats, raised under intensive management primarily for milk, "Spanish" or Mexican goats, produced under extensive range conditions for meat; and Angora goats, also managed rather extensively, primarily for fiber. The total goat industry is very minor compared to other livestock industries, and population statistics are not available except for Angoras.

Most of the progress in productivity made over the past century has been through the adaptation of technology developed for other species. Very little research and only limited product development efforts have been conducted directly for the goat industry. However, goat production has benefited greatly from the development of anthelmintics, insecticides, antibiotics, equipment and general knowledge concerning animal nutrition, reproduction, genetics and disease control.

Goats for milk production were brought to the United States by early colonists, but it was not until 1893 that the first purebred dairy goats (Toggenburg) were introduced (Briggs 1969). There is no good way to determine the extent of the U.S. dairy goat industry, but interest appears to be increasing. In 1975, the American Dairy Goat Association, Spindale, North Carolina, registered 26,644 head, which represents a considerable increase over the 3611 recorded in 1955. This association has around 7500 members, and 1000 of these flocks are on an official milk production testing program (Wilson 1976).

In addition to being efficient milk producers, dairy goats are fertile, often giving birth to twins. They have a short generation interval because of a 5-month gestation and early sexual maturity. Excess males and cull goats are used for meat.

There are five main recognized purebred breeds of U.S. dairy goats—Toggenburg, Saanen, Nubian, French-Alpine and American LaMancha (Colby et al. 1966).

Angora and Spanish goats are reared in the Southwestern States, primarily in Texas. On brushy ranges, they complement cattle and sheep by eating large amounts of browse, thereby improving pastures for these species, as well as providing a means for making more efficient utilization of the total available range forage. Research has indicated that cattle, sheep and goats are all more productive when grazed in combination on brushy ranges than when grazed at the same grazing pressure as single species (Merrill et al. 1966). In descending order of preference, cattle eat grass, browse and forbs; sheep consume forbs, grass and browse; and goats prefer browse, forbs and grass.

Goats are an effective and profitable biological brush control tool. A combination of chaining or dozing brush down, followed by heavy grazing with goats, will materially reduce and control many types of brush plants.

Spanish—often called Mexican goats—appear to be increasing in numbers in the Southwest. These multicolored goats are descendents of goats introduced by early Spaniards. They are hardy and fertile, often producing twins and kidding more than once a year under favorable conditions. Kid crops of 130–140% per year have been obtained in research flocks managed in an extensive manner. Kids are often slaughtered when 3–5 months of age and weighing from 25 to 50 lb. However, many goats are older and heavier when marketed; but most, except aged cull goats, are slaughtered when less than a year of age. There are no well-developed markets and no official grading standards.

Angora goats were first imported to South Carolina in the United States from Turkey in 1849 (Briggs 1969). Today, Texas produces approximately 97% of the U.S. production of mohair, which approximates 30% of the world's supply. Angoras are sheared twice yearly and produce an average of around 7 lb per year; however, selected flocks will produce twice this amount. Castrated males (muttons) are commonly kept for several years for hair production; and because of their larger size and hardiness, are especially effective for brush control and are less subject to predation.

The improvement in productivity of the U.S. Angora flock is given in Table 16.1 and is derived from USDA Statistical Reporting Service data. Hair production per goat has increased from 3.6 lb in 1920 to 7.1 lb in 1974, which is equal to an increase of 98.6%. Several factors have been responsible for this marked improvement, including supplemental feeding and improved insecticides and anthelmintics; but the major reason is probably because producers have selected for increased hair production. Mohair production traits, such as pounds and length, are fairly easily measured and are highly heritable.

The intensive selection and resulting increase in fiber production may have adversely affected reproduction, as poor reproduction is one of the major problems affecting Angoras. Most flocks are kidded under range conditions, and death losses of kids born are often high, due to adverse weather and predators. Also abortion rate is high in Angora does, which can be due to insufficient feed or diseases, but some of the abortion problem is now believed to be due to an adrenal hormone insufficiency (Shelton and Groff 1974). Other problems are high susceptibility to freeze losses after shearing and the erratic mohair market. In recent years, mohair has been in high demand and has brought record prices.

SHEEP

Columbus, on his second voyage in 1493, is credited for bringing the first domesticated sheep to America (Briggs 1969). Subsequent

TABLE 16.1

U.S. MOHAIR PRODUCTION BY SELECTED PERIODS

	Avg Goats Shorn Per Year (1,000 hd)	Avg Mohair Per Year (1,000 lb)	Avg Mohair Per Goat (lb)	% Increase Per Goat (by periods)	% Increase Per Goat 1920-1974
1974	1,175	8,400	7.1	+ 4.4	
1970-74	1,812	12,278	6.8	+ 3.0	
1960-69	4,048	26,718	6.6	+15.8	
1950-59	2,078	16,484	5.7	+16.3	
1940-49	3,917	19,263	4.9	+11.4	+98.6
1934-39	3,826	16,694	4.4	+ 7.3	
1930	4,241	17,583	4.1	+13.9	
1920	2,367	8,556	3.6		

SOURCE: USDA Statistical Reporting Service. Data reported for Texas only, beginning in 1963.

Spanish explorers brought sheep to the New World, and during the 16th, 17th and 18th Centuries, sheep spread up through Pan America and Mexico into what is now New Mexico and California. The Navajo sheep, still raised by Indians in the Southwest, are descendants of these coarse-wooled, shallow-bodied, variable "Churro" sheep. Also, the small so-called "native sheep," found in Southeastern United States, trace back to the Churro, introduced by the Spanish.

Sheep importations to the Eastern colonies date back to 1609, when sheep were brought to Jamestown, Virginia. The first Spanish Merinos were not introduced until 1801, and the English Medium Wool Breeds (Shropshire, Southdowns, Hampshires, etc.) were not introduced until after 1850. E. N. Wentworth, in his book, *America's Sheep Trails*, published in 1948, describes the interesting history of sheep in America.

According to USDA Statistical Reporting Service data, 100 years ago, in 1876, there were approximately 37 million sheep in the United States. For 67 years, numbers fluctuated between this figure and a peak of over 56 million, which was attained in 1942. Numbers declined sharply from 1942 until 1950, when they stabilized at around 30 million head for an 11-year period until 1961, when the current decline started. On January 1, 1976, the U.S. flock was estimated at only 13.3 million, the lowest number on record since census records were started in 1867. The number of U.S. farms with sheep have also declined by 44% since 1965, from 241,590 to 135,330 in 1975.

What has caused this decline? There is no one reason, but following is a listing of probable major factors: (1) Good cattle prices in the 1960's and early 1970's, as cattle feeding increased rapidly, caused a shift away from sheep to cattle. During the past 2 years, the price situation has changed markedly and now favors sheep. (2) Predators—a real problem—have put quite a few people out of the sheep and goat business. Predators cause economic losses to the industry in two ways: direct deaths; and indirect losses, such as increased costs of attempting to prevent predation and when producers sell out, the inability to raise sheep in the area. Losses to predators have been increasing. A USDA Economic Research Service Survey of sheep losses in the 15 Western States, during 1974, indicated that more than 8% of the lambs born and about 2.5% of the mature stock sheep were lost to coyotes. The lamb losses to coyotes amounted to 735,000 head, equal to about 1/3 of all losses (Magleby 1975). A real blow came following the Cain Report released in 1972, when by Presidential order, toxicants were prohibited for predator control. The use of the very specific M-44 coyote-getter, which uses cyanide, is again approved for use. (3) Labor requirements (herding, drenching, shearing, etc.) are becoming an increasingly important problem, especially for larger producers. (4) The sheep business is becoming a minority industry, which causes problems in marketing products, availability of supplies, and research and development de-emphasis by governmental and commercial concerns. (5) Other problems, such as toxic plants, have had local effects. Low demand for wool has not been a major problem because of the USDA support price.

The past 100 years can be characterized as a period of intensification of production, with a corresponding shift of product emphasis from wool to lamb meat. Donald S. Bell (1970) described some of the changes made. This trend will undoubtedly continue into the future. To intensify, does not necessarily mean rearing sheep in confinement. Range sheep production can also be intensified. It simply means getting more production from a flock by changing production practices. It requires some input (labor, equipment, material, feed, management practice, etc.), and the resulting increase in production must exceed the cost of the input for it to be adopted. Following are steps in the general intensification in sheep production in the United States beginning with the early 1800's:

1800-1900: In the early 1800's, the Merino was the most popular breed. They spread across the country from the Eastern coast, and during the period from 1870 to 1900, there was a rapid expansion in the Western range area. Wool was the primary salable product.

1850-1900: English-medium wool breeds (Hampshire, Shropshire, Southdown, etc.) were introduced and became popular in the East, as demand for mutton increased.

Early 1900's: Lamb feeding industry developed, first around flour mills at St. Paul and Chicago to use by-products, then in the Midwest, where farmers marketed feed through lambs by purchasing feeder lambs in the fall off of Western ranges.

1900-1950: Replacement of the small, wrinkled and unprolific Merino, with Rambouillets, and the development and introduction of the Medium-Wool crossbreds (Columbia, Targhee, Corriedale, etc.)

1950's: Crossbreeding in Western ranges using Hampshire and Suffolk rams and development of spring lamb production programs (lambing in fall) to produce milk-fat lambs in the spring in Arizona, California, Kansas, Kentucky, Tennessee and Texas.

1960-1970: Increased emphasis on lamb production: lambs produced year-round, multiple births desired, development of high concentrate rations for feeding lambs, younger slaughter ages but heavier weights.

Most of the U.S. sheep are located in Western Range States, and these ten states, all Western except Ohio, have 70% of the U.S. sheep. These States, with their respective percentage of the U.S. flock, are Texas, 19.5%; Wyoming, 9.5%; California, 7.9%; Colorado, 6.9%; South Dakota, 5.1%; Montana, 4.8%; New Mexico, 4.4%; Utah, 4.4%; Idaho, 4.0%; and Ohio, 3.7%.

Approximately 85% of the income from the nation's flock in 1975 was derived from lambs, compared to only 45% in 1940. One of the accomplishments of the sheep industry has been the production of a more even supply of slaughter lambs throughout the year.

There is no doubt that considerable progress and changes have occurred in U.S. sheep production over the past 100 years, even though data to support progress is difficult to obtain because no control flock is available. Using USDA Statistical Reporting Service data, wool production per sheep shorn has increased by approximately 20% since 1909 (see Table 16.2). Production per ewe has remained relatively constant since the 1950's. This possibly is due to several developments: shearing more lambs, shearing more frequently and a shift to breeds that produce less wool. Meat productivity (pounds of lamb and mutton produced, divided by number of ewes 1 year and older) per ewe in the breeding flock is given in Table 16.3 and has increased over 66% since 1930—from 23.83 to 39.65 lb in 1975. This rather remarkable increase is due primarily to two factors: (1) an improvement in percentage of lamb crop raised and (2) carrying lambs to heavier slaughter weights. Improving the number of lambs reared per ewe per year provides the greatest potential for improving production efficiency. Number of lambs raised can be increased by reducing lamb mortality and increasing number of lambs born during the year. Lamb mortality amounted to an estimated 2 million head

TABLE 16.2

U.S. WOOL PRODUCTION BY SELECTED PERIODS

	Avg Sheep Shorn Per Year (1,000 hd)	Avg Wool Per Year (1,000,000 lb)	Avg Wool Per Sheep (lb)	% Increase Per Sheep 1909–1975
1975	14,347	119	8.31	
1970–74	18,099	155	8.54	
1960–69	25,253	214	8.49	
1950–59	28,251	237	8.40	
1940–49	39,377	315	8.00	+19.9
1930–39	45,304	361	7.98	
1920–29	34,964	264	7.56	
1910–19	37,786	265	7.01	
1909	44,758	310	6.93	

SOURCE: USDA Statistical Reporting Service.

TABLE 16.3

MEAT PRODUCED PER EWE ONE YEAR AND OLDER,
BY PERIODS

(lb)		
1975	39.65[1]	
1970-74	40.95	
1960-69	37.76	
1950-59	36.14	+66.4%
1940-49	29.63	
1930-39	27.01	
1930	23.83	

[1] Calculated on total commercial production and does not include meat produced for home consumption.

in the 15 Western States in 1974 or nearly 25% of those born (Magleby 1975). USDA Statistical Reporting Service data show that the lamb crop reared per 100 ewes 1 year and older has improved some 15% since 1925 and now averages around 96% for the United States (see Table 16.4). Some commercial flocks have produced in excess of 200% lamb crops, and some States, where there are limited predators and sheep receive excellent care, produce high lambing percentages; examples for 1975 (USDA Statistical Reporting Service) are: Kentucky, 125%; Virginia, 120%; Minnesota, 116%; North Dakota, 116%. Compare these to Arizona, 79%; Wyoming, 80%; Texas, 88%; Utah, 90%; and New Mexico, 91%.

Information presented in Table 16.5 shows that carcass weights of lambs have increased. Since 1925, the average weight has increased over 30%, from 39 lb (80 ± lb live weight) to 51 lb (about 105 ± lb live weight). This increases the pounds of meat produced per breeding ewe. The major cost in producing a lamb is not in costs associated

TABLE 16.4

U.S. LAMB CROP AS PERCENTAGE OF EWES BY PERIODS

	Avg Ewes One Yr. & Older (1000 hd)	Avg. Lamb Crop (1000 hd)	Lamb Crop As % Of Ewes	
1975	10,107	9,858	98	
1970-74	12,731	12,189	96	
1960-69	19,341	18,519	96	
1950-59	21,123	19,636	93	+15.3%
1940-49	30,846	26,685	87	
1930-39	35,723	29,829	83	
1925-29	28,768	24,815	86	
1925	25,999	22,195	85	

SOURCE: USDA Statistical Reporting Service.

TABLE 16.5

AVERAGE U.S. CARCASS WEIGHT OF SHEEP AND LAMBS,
BY PERIODS

(lb)		
1975	51.0	
1970-74	51.6	
1960-69	48.8	
1950-59	46.3	+30.8%
1940-49	42.2	
1930-39	39.2	
1925-29	38.9	
1925	39.0	

SOURCE: USDA Statistical Reporting Service.

in feeding the lamb but in fixed costs associated with keeping the ewe for 12 months. Therefore, it is desirable to produce as many pounds of lamb per ewe per 12 months as possible. This can be accomplished (a) by producing more lambs; by increasing twinning; by lambing ewes more frequently than once every 12 months; by reducing lamb deaths; by lambing ewes first at 1 year of age, rather than at 2 years; and (b) by selling lambs that are heavier at slaughter.

Another major accomplishment has been the realization that lambs can be weaned at early ages and the development of schemes and rations for successfully doing this from about 30 days of age. Not many lambs are early weaned at less than 60-75 days of age. The traditional scheme was to wean lambs when 5-7 months of age, to slaughter those fat, and go to the feedlot or grain fields with the feeders. Ewe milk production drops off after 6-8 weeks of lactation, and it is more efficient to feed concentrate feeds directly to the lamb, rather than using it to attempt to maintain high milk production. Ewes can then be placed on maintenance rations. Under range conditions, early weaning is not as common, but lambs are being weaned at earlier ages; especially in times of droughts, it has been practiced widely. Lambs weaned at early ages when still healthy and in good condition make very efficient gains on high concentrate rations, requiring only 3-4 lb of feed per pound of gain. Advantages usually given for early weaning include lower ewe feed cost; salvage range lambs during droughts; parasite control; to prevent needle and spear grass problems (range lambs); to rebreed ewes; and for overall improved lamb management.

Maybe you've bottle fed a lamb. It is an exciting experience for young people for a few times but soon gets to be a chore, and the successes are few. In the late 1960's, satisfactory formulas for ewe milk replacers were developed. They especially need to be higher in fat than replacers for calves. In addition to good milk replacers, USDA researchers developed systems for feeding lambs using cold milk with formalin added. Based on sales of lamb milk replacer powder, it is estimated that the number of U.S. lambs reared artificially was about 41,700 in 1971; 70,000 in 1972 and 100,000 in 1973. It requires about 20 lb of milk replacer to rear a lamb to 28-30 days of age, when it can be weaned to dry feed. Starvation is a major cause of lamb deaths. Most of these lambs could be saved. In addition, if the industry adopts the use of the more prolific breeds, the milk-replacer system of feeding lambs will be necessary to save all lambs produced. Ken Frederiksen (1976) of the Agricultural Research Service estimates that 891,000 of the lambs lost in 1975 could have been saved with milk replacers.

During the past century, a lot of knowledge has been gained on the inheritance of important sheep traits, and selection and breeding programs for using this knowledge have been developed. Clair E. Terrill (1958) reviewed 50 years of progress in sheep breeding for the 50th anniversary of the Journal of Animal Science.

Important sheep traits can be grouped according to their relative heritability, with those having heritability estimates from 40 to 50% classified as highly heritable, 25 to 40% moderate and less than 25% as low. With this classification, reproductive traits are low, growth traits are moderate, carcass traits are moderate, wool traits are high, face cover is high and skin folds are moderate to high. Generally, those traits that are not highly heritable have high heterosis, and crossbreeding programs have been used extensively by the industry to improve percentage of lamb crop.

Unquestionably, selection has been used by producers to bring about much change in the nation's flock. This can be visually seen by looking at pictures of sheep taken at different periods of time. For example: Rambouillets today are obviously more open faced, freer from wrinkles and larger in size. Performance has also undoubtedly improved. Good evidence is the results of the Rambouillet Ram Performance Test, conducted for the past 28 years by the Texas Agricultural Experiment Station at Sonora (1976). A.D.G. has improved 64%, clean wool production by 42%, staple length by 40%. Today's rams have less wool on their face and are freer from wrinkles than those tested 28 years ago. Part of the improvement is undoubtedly due to improved environment, but much genetic change has evidently occurred. Commercial fine-wool lambs coming off the ranges are much heavier at weaning than they were 25 years ago. Part of this increase is undoubtedly due to genetic improvement for growth.

There are over 25 U.S. breeds of sheep but only a few of these play an important role in sheep production. The Rambouillet is by far the most numerous breed, and the nation's commercial ewe flock consists largely of fine wool ewes of predominately Rambouillet breeding. This breed produces fine wool, has strong herding instincts, is hardy, and will breed out of season. Sheep of the medium-wool crossbred type are probably the second most numerous. These sheep were developed in the early 1900's, primarily for the Northwest, to provide a sheep that was more prolific and larger than the Rambouillet. They include the Columbia and Targhee, developed in the United States, and the Corriedale, introduced from Australia and New Zealand. Suffolks and Hampshires are important as sire breeds for siring market lambs. They were introduced from Great Britain and are large, growthy breeds, with good carcass characteristics. They are not important as commercial sheep except as sires. One of the limitations of the nation's sheep that has become more evident in recent years because of the emphasis on increased production has been in prolificacy. Selection practiced by most producers has really been against this trait, rather than for it. There is interest in introducing into the United States new breeds that are prolific. U.S. breeds now that are considered important primarily because of this trait include the Dorset (introduced from England but changed considerably in recent years) which will breed out of season, is fairly prolific and is listed as potentially becoming more important for producing crossbred commercial ewes. The Finn sheep, native of Finland, introduced first in 1968 from Ireland and through Canada in 1970, is the most prolific U.S. breed today. The breed averages up to 3.5 lambs per ewe in Finland. It probably has little value as a purebred, except as a means for producing crossbred ewes that are more prolific. A general rule is that lambing percentage can be increased by the approximate percentage of Finn breeding in the ewe flock. Two lambs per ewe are desired by most producers, not 3 or 4; therefore, only a 1/4-1/2 Finn breeding is usually desired. Results of a commercial producer, John Wichern, who is a cooperator with the Pipestone, Minnesota, Vocational Technical sheep project, support the claim for high lamb production potential from Finn Cross ewes. From crossbred Finn X Rambouillet ewes, he marketed lamb crops of 162%, 214% and 261%, when ewes were 12, 24 and 36 months of age, respectively (Holaway 1976).

The Barbado breed is receiving some research interest in the South. It was introduced to the United States through zoos and then to Southwestern U.S. ranges as an exotic game animal. It is a woolless sheep and is prolific; it will breed out of season and is hardy. Barbado X Finewool ewes, when mated to Suffolk or Hampshire rams, produce an acceptable lamb.

Crossbreeding has been used extensively by sheep producers, especially since the 1950's. The main advantages have been to utilize the strong characteristics of both breeds. (Example: Finewool ewe with high wool value, breed out of season, good mothers, availability, mated to a Suffolk or Hampshire ram with good growth rate and carcass desirability.) Also, the crossbreeding improves percentage of lamb crop through heterosis. Early crossbreeding used only single crosses; later crossbred ewes and rams were used. Now multiple breed crossing systems are being proposed. The increased lambing percentage is the main advantage of using the crossbred ewe. USDA, Agricultural Research Service (1963) workers reported 2%, 14% and 27% more lambs from 2-way cross, crossbred ewes X another ram breed, and 3-way crossbred ewe X another ram breed, respectively.

Much progress has been made in feeding sheep during the past 100 years. Among major accomplishments are: (1) Establishment of nutrient requirements for sheep fed for different production purposes. (2) Determination of nutrient content of feedstuffs. (3) Development of feed supplements for feeding range sheep. (4) Establishment of the lamb feeding industry and learning that high concentrate rations could be fed. (5) Development of the use of nonprotein nitrogen in protein supplements. (6) Increased knowledge gained about nutritionally-related diseases. (7) Development of improved pastures and pasture and range grazing systems. (8) Mechanization of feed harvesting, storage, processing, transportation and feeding.

The lamb feeding industry developed in the last 75 years. It has changed considerably, since it started around Minneapolis, St. Paul and Chicago to utilize flour milling by-products. It then developed in the Corn Belt, Kansas, Nebraska and Colorado, where wheat pasture, beet tops and stubble fields, as well as corn and grain sorghum, were marketed through lambs. Drylot rations were high in roughage, and lamb gains were slow and inefficient compared to performance of lambs in today's feedlots. Pelleted rations and high concentrate rations, vaccine for enterotoxemia, antibiotics and growth promotants have been developments that have greatly benefited lamb feeding. Rations and lamb performance presented in Table 16.6 are typical for 1935 (Cox and Wagner 1935) and 1975 (Calhoun 1976) and give the marked changes in rations and lamb performance made during the 40-year period.

Today, more lambs are being fed near the areas where they are produced because feeders are heavier and transportation more expensive. In farm-flock areas, lambs are creep fed, weaned early, and fed out on high concentrate rations. These lambs perform exceptionally well, often reaching 100-lb weights in 125–150 days of age

TABLE 16.6

TYPICAL LAMB RATIONS AND PERFORMANCE FOR 1935 VS 1975

	1935[1]	1975[2]
No. days on feed	120	47
Initial wt (lb)	60	72
Final wt (lb)	93	102
A.D.G. (lb)	0.27	0.64
Feed/cwt gain (lb)	1254	476

Hand-Fed Ration (lb):	Self-Fed Mix (%):
2.1 ground sorghum fodder	70.6 ground sorghum grain
1.0 ground milo	10.0 cottonseed hulls
0.2 cottonseed meal	12.4 cottonseed meal
+ limestone	4.0 molasses
	3.0 minerals + vitamin and antibiotic premix

[1] Kansas Agricultural Experiment Station (Cox and Wagner 1936).
[2] Texas Agricultural Experiment Station (Calhoun 1976).

and with feed conversions of 3–6 lb of feed per pound of gain. More range lambs destined for market may be weaned earlier and fed these type rations in the future. Lambs are valuable, and several bad things can happen to them on the range—coyotes, needle grass, stomach worms, becoming stunted during droughts, are examples. The relative prices of concentrates, lambs and pasture will determine future feeding systems for finishing lambs for slaughter.

The main goal for producing sheep is to produce a quality meat that will be highly acceptable to the consumer. Carcass quality grades and yield grades have been developed. Yield grades have not been implemented by the meat processing and marketing industries to a great degree, but lambs today have more acceptable amounts of fat than they had years ago. Trimmer and longer bodied carcasses are now desired.

Ralgro (Zearanol) and low levels of antibiotics are the only growth promotants available to sheep producers. Diethylstilbestrol, which proved to be an excellent growth promotant for feeder lambs (especially wethers), is no longer used. One of the real problems facing the sheep industry is getting new drugs developed and cleared by the Food and Drug Administration for use because of the high expense in doing so, compared to the potential of the market.

Considerable advancement has been made in sheep health. New anthelmintics, antibiotics, drugs, and insecticides have meant millions to the industry. Vaccines have been developed for contagious ecthyma, enterotoxemia, bluetongue, epididymytis, vibrionic abortion, and anthrax.

Useful information has been obtained on other infectious diseases, such as foot rot, pneumonia, listeriosis, chlamydial diseases, and mastitis.

In the South, one of the most significant developments aiding sheep and goats, during the past century, has been the initiation of the screwworm control program. This program, which started in 1962, has been worth millions to both sheep producers and consumers. Without this program, sheep and goat numbers would have declined drastically in the South.

Rearing sheep in confinement has received some interest. Northern farm-flock sheep have for years been maintained in drylots and barns during winter months, but sheep have been successfully maintained in total confinement for several years. Total confinement has some advantages, such as total control of nutrition, prevention of foot rot or internal parasites. The closer observation and management provides an opportunity for increased production. The question is, will the benefits exceed the costs? In areas where environment is adverse and adequate feed is available, confinement may be considered. Partial confinement during critical periods of production, such as at lambing, is a good management practice for many producers. However, the main place where sheep excel is on the range or in the pastures, where they harvest feeds directly.

Without improvements in wool harvesting, the sheep industry would have ceased to exist long ago. This has progressed from hand shears to human-powered clippers to gas- or electrical-powered clippers, to self-contained electrical clippers and recently to newly developed air-powered clippers. Chemical shearing, using a drug developed from cancer research, which disrupts cell growth, has been investigated rather intensively (Shelton 1971). It has not been approved for use by the Food and Drug Administration, and there are some management limitations to its use. Further advancements in shearing are urgently needed.

During the past 100 years, many more advancements have been made that have benefited sheep and goats. Mechanization, especially in the area of transportation, has affected the industry greatly. Gooseneck trailers, large triple-decked trucks for hauling live animals, and refrigerated trucks for transporting carcasses, are indispensable.

The single greatest advancement made by the sheep industry has been to recognize that we have not even come close to reaching production limits of these animals that can be adapted to many varying circumstances. The sheep industry will continue to change and become more productive. Sheep and goats are efficient in converting feedstuffs (not in direct competition with man) to useful products.

These small ruminants could become more important, as the food and fuel shortages become more critical.

BIBLIOGRAPHY

BELL, D. S. 1970. Trends in the sheep industry of the United States. Ohio Agric. Res. Dev. Cent. Res. Circ. *179*.

BRIGGS, H. M. 1969. Modern Breeds of Livestock. Macmillan Co., Collier Macmillan Canada, Toronto, Ontario.

CAIN, S. A. *et al.* 1972. Predator control—1971. Report to the Council on Environmental Quality and The Department of the Interior. Inst. Environ. Quality, Univ. Mich.

CALHOUN, M. C. 1976. Unpublished data. Tex. Agric. Exp. Stn., San Angelo.

COLBY, B. E. *et al.* 1966. Dairy goats—breeding, feeding and management. Coop. Ext. Serv. Univ. Mass.

COX, R. F., and WAGNER, F. A. 1935. Sorghum roughages in lamb fattening rations. 2nd Annu. Lamb Feeders' Day Rept., Kans. Agric. Exp. Stn.

FREDERIKSEN, K. R. 1976. Personal Communication. U.S. Sheep Exp. Stn., Agric. Res. Serv. Dubois, Idaho.

HOLAWAY, D. 1976. Personal communication. Pipestone, Minnesota.

MAGLEBY, R. S. 1975. Sheep losses due to predators and other causes in the Western United States, 1974. USDA Econ. Res. Serv. Rept. *616*.

MERRILL, L. B., REARDON, P. and LEINWEBER, C. L. 1966. Cattle, sheep, goats . . . mix'em up for higher gains. Tex. Agric. Progr. *12*, No. 4, 13-14.

SHELTON, M. 1971. Prospects for chemical shearing. Wool Symposium, Sheep Industry Development Program, Lubbock, Texas, Nov.

SHELTON, M., and GROFF, J. L. 1974. Reproductive efficiency in Angora goats. Tex. Agric. Exp. Stn. Bull. *1136*.

SIDP. 1968. Genetics Symp. Proc. Sheep Industry Development Program, 200 Clayton Street, Denver, Colorado, April.

SIDP. 1968. Physiology of reproduction in sheep. Proc. Symp., Sheep Industry Development Program, July.

SIDP. 1968. Sheep diseases and health. Proc. Symp., Sheep Industry Development Program, Sept.

SIDP. 1968. Sheep nutrition and feeding. Proc. Symp., Sheep Industry Development Program, Oct.

SIDP. 1968. Production and business management. Proc. Symp., Sheep Industry Development Program, Dec.

SIDP. 1970. The Sheepman's Production Handbook, 2nd Edition. Sheep Industry Development Program.

SIDP. 1971. Wool Symp. Proc., Sheep Industry Development Program, Nov.

SIDP. 1972. Intensive sheep management. Symp. Proc. Sheep Industry Development Program, June.

SIDP. 1973. Profitable range sheep production. Symp. Proc., Sheep Industry Development Program, Oct.

SIDP. 1975. Sheep breeding and feeding for profit. Symp. Proc., Sheep Industry Development Program, Aug.

TERRILL, C. E. 1958. Fifty years of progress in sheep breeding. J. Anim. Sci. *17*, 944-959.

TEXAS AGRICULTURAL EXPERIMENT STATION. 1976. Improvement of sheep through selection of performance-tested and progeny-tested breeding animals, 1975-76. Tex. Agric. Exp. Stn. Sonora.

USDA. 1963. Hybrid lambs. USDA Agric. Res. Serv. Spec. Rept. *22-32*.

USDA. 1958. Livestock and Meat Statistics, 1957. USDA Statist. Rept. Serv. Bull. *230*.

USDA. 1963. Livestock and Meat Statistics, 1962. USDA Statist. Rept. Serv.
   Bull. *333.*
USDA. 1973. Livestock and Meat Statistics. USDA Statist. Rept. Serv. Bull. *522.*
WENTWORTH, E. N. 1947. American Sheep Trails. Iowa State College Press.
   Ames.
WILSON, D. 1976. Personal Communication. Box 865, Spindale, North Carolina.

Section 4

Challenges of Food Processing and Flavor: Past, Present, and Future

17

Introduction

George E. Inglett

The challenge of change is the introductory theme of this session on "Challenges of Food Processing and Flavor: Past, Present, and Future" of the symposium on "One Hundred Years of Agricultural and Food Chemistry." The time is appropriate for reviewing the past 100 years of agricultural and food chemistry and for probing present problems and planning for the future. The food processing industry is confronted with greater challenges today probably than any other time in history. Some problem areas are with labor organizations, regulatory restraints, research costs, greater merchandising expenses, more quality control, and greater consumer considerations.

In a 1-day session, only the high points relating to some chemically related areas can be covered.

The Agricultural and Food Chemistry Division of the American Chemical Society is fortunate to have represented in this section of the book the following: A. S. Clausi, General Foods Corp. (Chap. 18); J. J. Albrecht, Nestlé Enterprises, Inc. (Chap. 19); R. L. Opila, Globe Engineering Co. (Chap. 20); R. L. Hall, McCormick Co. (Chap. 21); W. E. Hartman, Worthington Foods (Chap. 22); and a host of other personages prominent in the field of food processing.

18

Food Ingredient Challenges

A. S. Clausi

It is surely a pleasure for a long-time member of the American Chemical Society, who has spent his entire career in the food industry, to have the opportunity to participate in this Centennial Symposium. And I am especially pleased to be able to talk about food ingredient challenges because, although relatively few people seem to realize it, all foods *are* chemicals.

Back at the beginning of time, nature began putting chemical substances together in ways that were attractive and nutritious for man and beast. Only in recent decades has man intervened significantly with nature to make food more abundant, more useful, more convenient and available, more affordable. In fact, there is still so much that can be done through the application of advancing technology to food ingredients that I believe we can truthfully say—with apologies to Shakespeare for paraphrasing—that what's past is merely prologue.

It has only been within the last third of this past century, actually, that America's food industry has begun to meet vital challenges in the area of food ingredients. Until the World War II era, most of the packaged groceries which helped to mark our country's transition from an agricultural to an industrial society were by no means complex foods compounded of multiple ingredients. A quick look back over the last century will underscore what I mean.

100 Years Ago

When the American Chemical Society was founded, most farm products were consumed in their natural state, very close to where they were grown. Prevention of spoilage was the paramount concern. Although Gail Borden had successfully canned evaporated milk and the Walter Baker Company had been packing cocoa in tins for more than a century, there were relatively few food products sold in packages. Most foods sold in stores were still measured out or ladled from bulk containers.

75 Years Ago

Sending raw foods from farms to distant markets was still a major problem because of spoilage—since chemical reactions made foods stale, or rancid, or just plain bad. Salt, spices and vinegar were still used to hide off-flavors and odors. But as urban populations grew, some enterprising manufacturers began to develop ways to get food products to city dwellers.

Flour millers and sugar refiners started packing measured weights of processed flour and sugar in paper bags, rather than delivering these staples to stores in bulk. Canning as a commercial preservation method was coming into its own. John Dorrance experimented, and brought out Campbell's canned condensed soups. Henry J. Heinz was offering his growing line of pickles, catsup, relishes and baked beans in cans and jars. Van Camp's Pork and Beans in Tomato Sauce could be found on store shelves.

In addition, makers of the new ready-to-eat cereals were marketing their wares in paperboard cartons. Battle Creek, Michigan, was on its way to becoming the cereal capital of the world, thanks to Kellogg's new Corn Flakes and Post's Grape-Nuts. Jell-O, billed as "America's Most Famous Dessert," was being compounded and packaged in upstate New York. A Minnesota grocer was blending maple and sugar cane syrups, which he was selling in a little log cabin-shaped tin to honor his boyhood hero, Lincoln. Uneeda Biscuits and other crackers, now packed in cartons, were making the cracker barrel obsolete.

Only 50 Years Ago

Canning was still the principal commercial method of food preservation. Consumers could purchase everything from cut green beans to tins containing a whole chicken in aspic. As refrigeration became more widespread in American homes, however, some food processors developed new products or new forms of packaging. For example, J. H. Kraft changed from selling his homemade cheeses off the back of a wagon to marketing pasteurized process cheese in a tin-foil package. And some new preservation techniques—the first commercial application of the quick-freezing process by Clarence Birdseye plus modern methods of dehydration—were ushering in the age of specially designed and formulated foods.

25 Years Ago

New preservation techniques that had been developed enabled the food industry to broaden its perspective on food ingredients dramatically. Instant, premixed and aerosol forms of products, as well as better flavors and new textures were introduced. Processors began to consider raw farm produce not merely as major product ingredients but more as building blocks around which to develop the food products consumers were indicating they wanted and needed.

With more women joining the work force, there was a growing craving for convenience, for foods that were quick and easy to prepare. Not only did processors provide consumers with "maid service": cleaning, cutting, shelling, washing and so on; but they also began blending and mixing so that all the preparation that was needed in the home kitchen was heat and serve, boil-in-bag, or mix and chill.

Even more recently, in the last 15 years or so, processors have complied with homemakers' desires for additional quality and values—for "chef service" in addition to "maid service." This has meant new combinations of familiar grains, vegetables and fruits; a great deal of precooking in processing plants; the addition of seasoning packets or sauce cubes to stir in; new forms of meat products; and a host of innovations that take the chores out of meal preparation yet provide guaranteed satisfaction in use.

In its desire to fulfill consumer needs, the food industry's rapid advances in food technology during the last few decades have stepped up its intervention with nature. Aided by sophisticated tools, such as the mass spectrometer and gas chromatograph, the ability of flavor chemists to analyze and synthesize flavors has expanded markedly. New knowledge on the texture or mouth feel of various foods has been developed, as processors learned that texture is as vital to product acceptance as are flavor and color. Processing techniques are no longer simple mixing operations. They involve ultrasonics, reverse osmosis, microwaves, irradiation and extremely intricate mechanical assembly systems that make certain specially-designed foods are economically feasible.

Today

There is no question that the food industry is well into an era of specially-formulated or fabricated foods, foods which consist not only of naturally derived ingredients but also of synthetic components which occur in nature and are identical to those in nature. Most of the new products now appearing on supermarket shelves come under this heading. And ingredient selection is probably the first—and surely one of the most important steps—in the industry's product development activities.

Obviously, when a new product concept is developed, a wide range of ingredients must be explored to determine how well they measure up to desired functionality, cost guidelines, and availability—both now and in the future. That last point has taken on immense significance in recent years, not only because we have been experiencing shortages of some of the building blocks of new food products, but also because it has become increasingly difficult to guess when one ingredient or another may become questionable in the eyes of government regulatory agencies or consumers—or both.

So in the selection of product ingredients, just checking availability is no longer enough. In addition to locating alternate suppliers for first-choice ingredients, it's also necessary to identify replacement materials so there are options to meet possible future needs. This is particularly true of colors, flavors, antioxidants and emulsifiers.

Then, there is the need for assurance of regulatory acceptance at the use level planned. Obtaining clearance for the use of any *new* ingredients being considered for a product can be a long and costly procedure. Demonstrations of ingredients and product safety, i.e., the dollars spent for testing, safety and toxicology as well as for quality control and quality assurance, are taking a bigger bite of the food industry's research and development dollars every year. As a matter of fact, when results of the 1975 Delphi survey on trends and changes in the food business during the next decade were presented at the Food and Drug Law Institute's Fourteenth Food Update meeting last year, it was obvious that most processors felt the expense and difficulty of demonstrating safety of new ingredients or additives were major impediments to success. And the high priority placed—of necessity—on establishing ingredient safety has resulted in pushing product and process development, as well as work in the basic sciences, farther down on the industry's list when it comes to funding.

The question of ingredient safety also has tended to foster the promulgation of stricter food laws and regulations and their interpretations. In addition, it has led to what I suppose could be called consumer concerns. This is especially true about the broad public attitude toward additives—that collective noun used for such very useful ingredients—that has taken on a somewhat undesirable connotation in the last few decades.

It is unfortunate that too few consumers and not many consumer activists, either, know very much about additives. They read the ingredient line on a package and are mystified—perhaps even apprehensive—about some of the long, chemical-sounding names they find there. And for some people, if it's a chemical, it's naturally suspect, since few realize that all foods are compounded of chemicals. Even their own bodies are nothing but a group of assorted chemicals.

One reason consumers show concern and apprehension about additives is that they don't consider them just another ingredient. They fail to understand that food additives, preservatives and other useful chemicals bring many benefits by permitting the food industry to maintain its nationwide distribution channels and provide consumers with tremendous value and variety, regardless of the season. Not to mention the bacterial safety and nutritional values that are protected.

Most critics of additives tend to gloss over an unavoidable fact of life: since raw food products are no longer produced close to where food is consumed in our highly urbanized society, we need additives to extend the keeping qualities of food—to vitalize logistics, you might say. Additives also do much more.

Despite the food industry's continuous efforts in the last few decades to provide consumers with information about the role additives play, it is the exception rather than the rule when consumers understand their usefulness. Few people are aware that additives are what improve flavor, texture, appearance and add color as well as nutrients; that additives are what keep some foods creamier and smoother, others crisper and fresher. Nor do they realize that additives help speed the homemaker's preparation of daily meals, and protect the products she buys, from the time they leave the factory until she fixes them in her kitchen. Of even more significance, very few people know that the components of additives, with minor exceptions, are exact matches of those found in nature.

If I were to name one of the most important challenges our industry faces in the next few decades, it would be to try to compound new food products with as few ingredients and additives as possible. For unless we can do a much better job of educating consumers to the need, the usefulness and safety of combining both natural ingredients and so-called additives to give them the food products they desire, our industry will have to give deep consideration to the length of the ingredient line on its packages of new products, and how consumers may react to it.

It is really coming down to this: society must assess whether man's intervention with nature—as exemplified by the American food industry—is good, or whether it is not. Those of us in the industry are sure we must continue to "manipulate" nature, if you will. But the votes that really matter from millions of consumers are still to be counted.

Scientists both within and outside the food industry realize that it's often necessary to use quite a few ingredients to give the desired end-product, just as it would be true if a consumer followed a recipe at home. Two of my own company's frozen vegetable items provide a good example of what I mean.

The ingredient line on the label of Birds Eye French Style Green Beans, which provides the "maid service" I mentioned earlier—in other words, provides beans that are ready to put into the pot and cook—reads simply: "Harvested at just the right level of maturity to bring out full flavor—French cut for extra tenderness." And that's all. No words to cause a consumer concern. But there are three times as many words in the ingredient copy for the Birds Eye Bavarian Style Beans and Spaetzle package, because this product offers "chef service" as well. The product allows the homemaker to serve an international style vegetable that has been combined imaginatively and seasoned with a distinctive sauce, just as a gourmet chef would (and, as a matter of fact, *did*) design them. The four lines of ingredient copy say:

"Ingredients: Green beans, noodles, sauce made from: margarine, hydrogenated coconut and palm kernel oils, salt, sugar, imitation bacon bits". . . . these, by the way, are explained in parentheses. . . . "(soy flour, hydrogenated coconut oil, salt, dried yeast, flavorings, caramel and artificial coloring), hydrolized vegetable protein (enhances flavor), parsley, dried green onions, spice."

Although the aim of every company in the food industry is to use as few ingredients and food additives as practical, when a firm is developing new fabricated foods, the number of ingredients often mounts. And when a company's product line is wide and diverse, its ingredient shopping list is necessarily lengthy, too. General Foods' more than 400 domestic grocery products, for instance, call for the purchase of 1700 different ingredients, natural and man-made, which include the following categories:

(1) There are sugars and syrups, as well as artificial sweeteners.

(2) We purchase some 60 different dairy products and derivatives, including whey and milk solids, lactose, casein, sodium caseinate, dried buttermilk and dehydrated cheeses. But not one ounce of fluid milk.

(3) We also purchase various fats and oils, emulsifiers, antioxidants.

(4) More than 250 flavorings and 100 colorings are on our shop-

ping list, together with a number of chemical compounds; gelling and thickening agents; starches; grains and flours; vitamins and minerals; and more than 170 seasonings and spices.

(5) Our products require more than 140 different chocolate ingredients and some 40 kinds and cuts of coconut.

(6) We use dozens upon dozens of eggs and egg products; tons upon tons of vegetables and fruits, meat and fish and their by-products; coffee, tea, yeast, nuts.

A 1700-item shopping list for one company may seem incredible. But if our industry is to continue providing consumers with the quality, convenient, value-added products that marketing research tells us they want, we are going to have to continue using a wide range of ingredients—and find better ways than we have used up to now to reassure consumers of their safety and purity.

Another major challenge the American food industry faces in the years ahead concerns the nutritional modification of foods. This, too, importantly involves food ingredients.

Although members of the food industry would be the first to admit that nature is wonderful, our industry knows that Mother Nature is far from perfect when it comes to supplying optimum nutrition (both quantity and quality) in the foods people want to eat. There are still various nutrient deficiencies in some areas of the United States today, as well as in specific segments of the population. Malnutrition is by no means wiped out, even in this best-fed and most prosperous of all nations. The U.S. food industry sees ample opportunity to intervene meaningfully to modify our food supply nutritionally—in several ways.

The first way—and this is the type of nutritional modification that almost everybody is already familiar with—involves the fortification of conventional foods. Best examples are the enrichment of bread and the addition of Vitamin D to milk, which had their beginnings more than a generation ago, as you know. Since that time, ready-to-eat cereals have been fortified with vitamins and minerals, for example. Other cereal products such as baked foods and macaroni have had their protein levels increased.

This type of nutritional modification is practiced rather widely, but in some cases not too wisely, I fear. For a few manufacturers seem determined to fortify their products with the entire Recommended Daily Allowance of certain vitamins or minerals. What is needed is a careful analysis of food consumption trends in this country to make certain that the food industry builds into the right foods the right levels of nutrients that are needed and that are compatible.

Food analogs provide another way to modify foods nutritionally. The food industry is only in the early stages of manufacturing counterparts of nature, making them taste just as good as the natural foods but without many of nature's deficiencies. An important point to remember in doing this, however, is to assure that the new foods are at least the equal of nature's products in the principal nutrients expected of that food, and that they are more uniformly so than nature provides.

One of the best examples of food analogs are the sometimes-maligned synthetic breakfast drinks which can replace traditional orange juice. It's undeniable that they are successful in filling an important consumer need. Also, they have the advantages of offering uniformly good taste and often are functionally better than the natural product. They also offer the major nutritional values expected of breakfast juice drinks: Vitamins C and A.

A third way to modify foods nutritionally involves the opportunity to make nutritional adjustments to reduce the occurrence of chronic diet-related diseases. For example, the food industry is controlling the high levels of saturated fats in meat and dairy products—producing lean meat and nonfat milk. It is also learning how to eliminate or reduce some nutritional problems that man has created for himself: problems such as high sugar or fat intake, generally excessive calories, high salt intake, and so on. The combined talents of the medical profession, the nutrition scientists and the food technologist are being applied to develop foods which eliminate these man-made problems, yet do not force consumers to give up the fun of eating the foods they like.

These three major ways of modifying foods nutritionally are all useful. But they all call for the application of sound judgment and restraint lest some over-zealous marketers of fortified products start a horsepower race. For we know that too much of some nutrients can be as harmful as not enough.

It's obvious, I believe, that the food industry faces quite a challenge as it bends its efforts to the development of nutritionally-modified foods. And that challenge significantly involves food ingredients; for we face some important technological problems:

(1) We need to obtain quantities of protein for our fortification programs. So far, this is the most costly nutrient and the most difficult to find with both proper functionality and nutritional quality.

(2) A wealth of new knowledge is needed on micronutrients—trace minerals and other substances. In the case of only a few do we know how much is needed, and what is a toxic amount.

(3) Adding a broad range of nutrients is not without serious problems of taste, product performance, processing, and so on. These must be solved.

(4) Also, there's the problem of bioavailability, of assuring that the nutrients we put into modified foods are biologically active when they are consumed. This is seldom a problem with vitamins, but often one with minerals.

These problems will eventually be solved, and I feel certain that the U.S. food industry will be successful in its nutritional modification of the foods people eat—even though the added nutrients can be considered those additives which some people dislike and worry about.

I, for one, believe that the American food industry has done a remarkable job in the last century, especially during the last 30 years. But there are what I like to call "anti-development or anti-technology forces" at work today which may well have an impact on the progress we can make during the last quarter of this century. . . . and beyond.

These forces are evident in unrealistic attitudes of some Americans toward our country's food supply, attitudes which consistently thwart science and technology in their attempts to find useful solutions to both changing consumer needs and world hunger problems. A few examples will illustrate what I mean.

(1) Some U.S. citizens are "back to nature" advocates, who believe that foods right off the farm are the *only* ones fit to eat. Their feelings are understandable, but they conveniently seem to forget that in today's society very few of us live close enough to the land to do that. Rather, we must depend on an efficient food processing and distribution system to supply our needs.

(2) Then, as I mentioned earlier, there is far too much misunderstanding and consequent apprehension in America about food additives; they help make our food supply the most advanced in the world. Just in recent months we have seen what a furor can be stirred up on food colorants.

Apropos of my point about additives, I feel that it would be in the best interests of the American public if the Delaney Clause were given another look. It seems to me that it is time to review this legislation, which Congress enacted 18 years ago as a bona fide consumer health protection measure, in the light of our experience of living with it for almost two decades. Of even more significance, it should be reviewed in terms of the technological progress which has been made since its enactment. When you consider how much new knowledge on various food additives and other substances has been brought to light as well as the growing sophistication of analytical chemistry these days, there appears to be a major need to reconsider the Delaney Clause, which is largely responsible for the growing confusion and anxiety about additives among the American public.

(3) Another anti-development force—one felt particularly by agriculture—involves those opposed to herbicides and pesticides. There are some people who do not seem to care how much of the world's food supply is choked by weeds and eaten by rodents and insects, even though they must realize that these chemicals, wisely and carefully used, can help bring more food from farm to table.

Food processors are particularly careful about herbicides and pesticides, of course; and we examine every agricultural commodity we purchase for residues which may be toxic or may accumulate and become toxic with repeated exposure. My own company also insists that suppliers of commodities adhere rigorously to prescribed agricultural practices, especially in the production of certain vegetables in which there could be high levels of nitrate residues if fertilization has been excessive.

(4) One more force that also tends to negate the food industry's efforts, in my view, is the continuing controversy over natural versus processed or synthetic foods—a controversy led by some consumers who are four-square *for* Mother Nature and against food technology.

Fortunately, more and more people are coming to recognize that Mother Nature isn't always perfect. Mother Nature cannot produce low-cholesterol eggs, for instance; nor low-cost, meat-like products from texturized protein; nor more practical high-potency sources of many of our essential vitamins that are useful in a variety of foods. But synthetic chemistry and food technology can—and did.

These unrealistic attitudes, I believe, present the most significant challenges the food industry faces today. As I said earlier, man's intervention with nature is being assessed right now by our society. The major question is whether that assessment will be made in a sound, responsible, positive way, or whether, because of the histrionics of anti-development forces, our continuing progress will be slowed down, or even halted.

As I look ahead to the next 100 years of food ingredient challenges, I cannot help but wonder how long the clamor of these anti-technology forces will prevail. For it must be obvious to every thoughtful American that if people are to be adequately fed, our research and development efforts must be deeply involved in providing truly useful new food products. And these food products—like *all* foods, natural or man-made—will be compounded of chemicals.

In the years ahead, the U.S. food industry must respond positively to consumers' changing life-styles. It must refashion some of its processes, certainly much of its packaging; it must lower its energy requirements, as that commodity will become scarcer and increasingly expensive. And the food industry must develop new, even more efficient processing and distribution methods which will help people in the affluent nations eat more sensibly at the same time as we help those in the starving world survive.

The very real challenges we face—not only those concerning ingredients but much more—can only be dealt with through technology, in my opinion. And we will need national policies that foster, not inhibit, the successful application of our research capabilities toward practical solutions to worldwide food problems. We need, in the words of Dr. J. Herbert Holloman of Massachusetts Institute of Technology, "something better than a 'passive' Government attitude toward protecting, encouraging, developing, and improving technology and, through this, our competitive position in the world."

We need realistic regulations that will not demotivate the food industry, or so effectively hamper it that there is no allowance for scientific judgment. Great gains have been made in the last century—particularly the recent past—and we are not about to throw them away. Rather, we look forward to the continued application of constantly advancing technology to add greater values to raw food products and other ingredients and help to bring more and better foods to millions of dinner tables.

19

Challenges in Food Product and Process Development

James J. Albrecht

In considering today's challenges in food product and process development, one must first look at the basis of the challenges before trying to design a response. Without a doubt, the two most important phenomena affecting product and process development in the United States today are inflation and a dramatic increase in regulatory pressures imposed upon our industry.

As in other parts of the industrial community, the food industry has experienced dramatic inflation in the areas of produce costs, packaging materials, labor rates, water and energy, etc. The significance of these inflationary changes, however, is not truly appreciated until one compares those dramatic cost increases with the profit structure of the food industry. First, food processing as an industry nets less than 4% profit. In the case of canned food manufacture, this figure is less than 2%! Clearly, there is no profit cushion available to even partially absorb significant cost increases in our industry.

Second, there is relatively little price permissiveness on the part of consumers when it comes to food. There is no obvious acceptable substitute for gasoline to run automobiles, or for cigarettes for cigarette smokers, but there are many substitutes to draw from when beef or sugar prices grow significantly higher than normal. This was clearly evidenced during the last few years when consumers made significant changes in food purchases to avoid the high prices of beef and sugar-containing products. In the first instance, consumers shifted away from beef to less expensive chicken and pork products, and also to canned meats and meat extenders. In the case of dramatic increases in sugar prices, we saw significant decreases in consumption of canned fruits and confectionary products which had to be priced 20-50% higher than normal because of the sugar used in these products.

Another serious economic concern of the food industry, and of the United States as a whole, is the capital drain which has been forecast for the next ten years. The chief executive officer of the H. J. Heinz Company recently told a convention audience that the problem of capital shortage is critical for the entire U.S. industrial system. Between 1965 and 1974, this country used $1.6 trillion in capital. Over the next decade we will need almost three times that amount in new capital funds. We will fall short of our needs by about $650 billion! The capital drain in the canned foods industry is particularly serious since the only sources of capital dollars—profits and depreciation—are much scarcer in our industry because of very old plants and the previously mentioned low profit margins.

In order to meet the first challenge to our food industry we must respond effectively to these economic pressures and, at the same time, maintain (and hopefully improve) our profitability so that we can attract more investors and expand our overall base and efficiency of operations in this very competitive business.

The second major basis of challenge in the food industry today is regulatory pressures. For reasons which aren't all that obvious, the food industry seems to have come in for an undue share of regulatory pressure from both federal and local governments. It is not the purpose of this paper to fully discuss the many regulations which have been imposed on our industry, and the financial impact those regulations have had, particularly in the area of capital drain. To put it briefly, every time industry has to invest "X" millions of dollars for refinements of waste disposal systems, noise abatement, inordinate testing of new packaging systems or food additives, etc., it loses dollars which are needed for truly innovative technological approaches to problems that affect the industry's overall efficiency and productivity.

The food industry is now faced with over 2000 federal regulations! Consider the area of environmental control as an illustration of the capital requirements imposed on industry by such regulations. In 1972, the United States spent over $18.7 billion in dealing with environmental control issues; this represented 1.6% of our gross national product that year. While it is unreasonable to assume that all this money could have been spent to better advantage in initiative-oriented project work, it certainly seems logical that a significant portion of the dollars would have been better spent in trying to prevent pollution problems in the first place, rather than treating the net result of present manufacturing operations. In other words, they should have been spent trying to treat cause rather than effect.

To make this point in more basic economic terms, the following five statements illustrate the impact of regulatory activity on the food industry's ability to innovate and thereby improve efficiencies and productivity:

1. Lower prices are the result of lower costs and/or healthy competition.
2. Lower costs stem from greater supply and/or higher productivity.
3. Productivity and efficiency result from capital expenditures which are founded on research and development activities.
4. Research and development activities are funded from on-going operating revenues of the company.
5. Capital dollars are formed from net profits and depreciation.

Any unnecessary or nonproductive activities which adversely affect operating revenues and profits or use significant amounts of capital will decrease or inhibit productivity. This, in turn, will decrease or inhibit supply which will increase costs, and, in the end, cause higher prices, reduce competition and create inflation.

Government regulatory activity has already adversely affected productivity, competition and costs. When scarce capital dollars are used for nonproductive regulatory-related activities, it forces food company executives to close the gap between capital needs and sources of capital by lowering overall capital requirements. They will reduce their investments in plant and equipment; they will cut back inventory spendings; they will cut back their financial asset holdings. The result in all cases will be reduced business activity, more unemployment, slower growth and productivity, and continuing inflation and stagnation.

In the case of the canning industry, investments in new plants and equipment grew rapidly between 1958 and 1969; but after that period, and particularly in the 1970's, such investment has been virtually stagnant. Unless this picture is changed, the effect will eventually be felt all the way down to the consumer level. The inability to replace inefficient plants and machinery will reflect itself in higher prices on the grocery shelf. A shortage of money for research and development will slow down the rate of innovation that has eased the chores of consumers and done so much for the American diet. And, sadly, the smaller companies in our industry may be forced out of business . . . something nobody wants to see.

I'd like to review breifly the results of a survey which was conducted about a year and a half ago, relative to the technical expenditures of major food processing companies. Perhaps I should preface these remarks by quoting Dr. Bruce Hannay, Vice President of Research and Patents at Bell Laboratories and President of the Industrial Research Institute. In talking before the U.S. House of Representatives Science and Astronautics Committee, Dr. Hannay said:

I find some disturbing signs, such as inadequate R&D in mature industries and recent reductions in long-range industrial research. We see increasing concentration of effort in cost reduction, and product differentiation designed to increase market share, but requiring no significant innovation. The U.S. productivity growth rate is lagging. A larger fraction of R&D is directed toward our meeting new government regulations and is not necessarily productive otherwise. Along with these changes, there has been a corresponding reduction in R&D directed toward the entirely new processes, products and services that represent major steps forward.

In 1974, a survey of 38 major food companies was carried out in which R&D executives were asked for data regarding the percentage of their budgets which was spent on traditional R&D activities during the five-year period from 1970 through 1974. The companies surveyed represent 60% of the food manufacturing industry and account for annual gross revenues of more than $50 billion. The survey results indicated that of the 28 companies which responded, 10 indicated their new product development programs increased in budget from 1970 to 1974; 13 indicated decreases; and 1 indicated its spending stayed the same. Increases averaged 5.7%, with a range from 2.0 to 9.6%. Decreases were greater, averaging 13.5%, with the range from 2.6 to 40%.

New product development appears to be the most vulnerable R&D activity. As Mr. Binkerd (1975A) pointed out in his survey report, this is no surprise. The research segment of the cost involved in introducing a new product successfully is normally only a small fraction of the entire cost. In a pinch, a corporation can gain extra dollars by merely chopping off new product introductions. In addition, retailers are often reluctant to risk shelf space for uncertain new products. The Binkerd survey showed that basic research didn't fare much better than new product development when it comes to activity over the past five years. Five companies reported an increase in research spending, averaging 2.2% and ranging from 1.0 to 5.0%. Ten companies indicated that they had made cutbacks, with reductions averaging 4.1% and ranging from 1.0 to 10.0%. Seven companies reported no change.

If one looks at R&D activities that can be directly related to pressures resulting from regulatory activity, the survey indicates that nonproductive technical activity was increased by 18% of the respondents from 1970-1974; none showed a decrease; 2 remained the same; and 4 were not able to segregate the cost of this function.

Increases averaged 9.2% and ranged from 1.0 to 30.0%. Two companies reported that as much as 40% of their 1974 R&D budgets were charged to nonproductive technical activities!

Clearly Dr. Hannay's comments were corroborated by the Binkerd study. There is no indication that the trends illustrated by this study will change in the immediate future. There probably will be a continuing decline in the areas of true research and new product development. Process development probably will hold its own. Quality control, technical service and nonproductive technical activity will accelerate.

All of the above suggests that if the private sector of our economy is to remain strong, there must be changes in national direction. The U.S. government has been growing faster than the economy that supports it! This is sapping industry's strength. For all of its strong regulatory activities, our government has shown little strength regarding energy policy, incentives for research and development, consumer education, etc. Without government support in these areas, regulatory activities present an "all suction, no exhaust" dilemma.

In spite of all the negativism expressed above, the fact remains that food companies must continue to invest in both product and process development work. In the case of new products, this activity represents one of the key gaps that must be filled if a company is to deliver the profit performance expected by its stockholders. Other gaps can be filled by acquisitions, expansion of existing markets, etc.—but rarely can such activities totally substitute for the introduction of successful new products.

If new products are inevitable, one might ask what major appeals or consumer interests should be considered in developing such products. In my opinion, economy will be a major consideration in developing new products for the next five years. Why does such an assumption seem logical? The December, 1975 Opinion Research Corporation public opinion index rated 30 national issues which concern the public most. Included were the following: economic recession; drug addiction; crime and violence; the high cost of food; discrimination against women; problems of the elderly; discrimination against minorities; air and water pollution, etc. Out of this impressive group of issues, the high cost of food was the most frequently listed response (70%) when people answered the question, "Here's a list of problems facing the country today. Which of these are you personally most concerned about?"

Another factor which has gained considerable importance over the past five years is that of nutrition. People are definitely more nutrition-conscious than they were in the 1960's. As a result, we see new products that stress vitamin fortification, low or no cholesterol, high dietary fiber, use of less refined sugars, etc. We see a corollary activity relative to so-called natural foods being purchased in favor of those foods which contain many chemical additives. Nutrition and naturalness will continue to be factors worthy of consideration in product development work.

Finally, the packaging area is one which will require much more technical attention in the years ahead. The costs and availability of today's conventional packaging materials will give impetus to important new research in this area. Retortable aluminum pouches and trays, aseptic processing of particulates filled into plastics, chemically sterilized plastic containers, etc.—these are the packaging areas of the future.

One of the most difficult and technically intriguing areas of future development activity is that of food chemicals. Certainly, the regulatory pressures placed on pest control chemicals have only aggravated the fact that 1 bushel out of every 4 grown in the United States today goes to feed pests of one form or another! As we continue to restrict the use of pesticides, fungicides, etc., this figure can only get worse.

The banning of certain food colors, artificial sweeteners, flavor enhancers, animal growth stimulants, etc., has also effectively reduced the productivity of our industry. At the same time, these regulatory activities or pressures have discouraged many chemical companies from R&D activities aimed at finding new and/or better materials to replace those which have been banned. The risks and costs associated with such development programs simply preclude management allowing such work to go on.

Until consumers, government and industry can agree on a prudent risk/benefit ratio approach to assessing whether or not functional additives can be used in the production of our food supply, one can look forward to depressing statistics in the years ahead. We simply cannot allow ourselves (nor government and consumer groups) to expect perfect safety in foods. We don't have it in any other facet of our lives; why should we expect or demand it in our food supply?

In the area of grocery products, even with the high risks and high costs associated with new product introduction, one still sees quite a flurry of activity. Low-cholesterol products; soy protein-based breakfast meats; so-called new-fangled potato chips; high nutrition; very convenient breakfast items—including those aimed at the single serving market; natural cereals; yogurts;—all of these are product types that are in today's grocery stores and apparently doing reasonably well. The New Products Newsletter of January and February of a major New York advertising agency also mentioned the following products as being introduced within the last 4–6 months: glass-packed gravies; a hot "natural" cereal; canned fruit cobblers; shrimp and fish burgers; prefried quick-broiled bacon; corn chips packed stackable in canisters; new seasonings for cottage cheese; Italian dressing and pizza; a new presweetened fruit-flavored breakfast cereal; sugarless bubble gum; frozen sandwiches filled with meats; oatnut bars; 90% egg, low-cholesterol breakfast product; homogenized bacon (without nitrite); breakfast or snack bar-on-a-stick, made of eggs, orange juice and milk.

Consumer reactions to most new products have been, at best, unimpressive during the last several years. Perhaps it's because there have been some changes in the eating habits of Americans. These changes are best indicated by the homemakers using fewer convenience foods and doing more "scratch" cooking. One sees a return to basics, such as home baking, eating more stews and casserole products made from scratch in the home, etc.

The trend toward naturalness continues to be exhibited. Much of the interest in nutrition expressed by adults apparently has a direct relationship to their concern for the nutrition of their children. This, by the way, leads to a particularly important question regarding excessive uses of refined sugar in processed foods.

The high costs and risks associated with new product development work are very unpalatable for food company executives. On the other hand, there are some good incentives for working more aggressively in the area of process development since this usually involves somewhat lower risks and offers more measurable payout potential vis-a-vis the inflationary items mentioned earlier. Listed below are areas of technical development opportunity where our technical skills can be brought to bear to good advantage in the canning industry:

New protein sources—fish farming; bacterial, vegetable and yeast protein; mechanically deboned meat; etc.

Mechanical harvesting of produce and high speed processing—including at-the-site cleaning of produce, with less trash being brought into the plant.

Controlled environment agriculture.

Enzymatic conversion of cellulose in solid waste materials and/or extraction of other carbohydrate materials from these wastes.

New strains of (canning) crops—produce that is more disease-resistant and/or capable of giving higher yields. The interesting cross-breed of rye and wheat (Triticale) is a good example.

Continued research on *better seeds, more effective pesticides* (including biological systems), *improved planting and cultivation methods.* Despite the arsenal of agricultural chemicals developed to destroy them, insects, with their remarkable ability to adapt, are growing increasingly resistant to chemical pest control. More effective control methods are needed—and quickly—if the United States is to meet the growing demands on its food production capabilities.

More efficient packaging and shipping methods and materials (e.g., shrink-wrap, retortable pouches and trays).

Development of *better manufacturing techniques* for washing and peeling and/or blanching of fruits and vegetables to reduce loss of solids and minimize waste disposal problems.

More efficient utilization of crops through development of new foods from materials which are presently considered wastes, (e.g., sugar syrups from cut fruit flume water, use of green tomatoes for food, etc.). As a dramatic example, for each 1 ton of corn that enters the front end of a cannery, about 2/3 of the ton in waste materials (husks, cobs) goes out unused at the back.

Better economies regarding *water and energy usage.*

While all the above items are of immediate and great interest to the canning industry, they also illustrate the kinds of process development opportunities that exist in the food industry in general. Within each specific segment of the industry, there are pressing problems that require technical innovation for their solution. Most of these can be categorized as being related to regulatory pressures, economic constraints or the closely-related need to improve our manufacturing efficiencies and productivity in the food industry.

As time goes on and the need for increased efficiencies in food production and distribution becomes greater and greater, I hope that industry will be able to dig out from some of the regulatory pressures mentioned above, and to devote more of its creative technical abilities and capital dollars to improving overall efficiencies rather than maintaining status quo (at best) with resulting rising food costs and diminishing food varieties.

BIBLIOGRAPHY

BINKERD, E. F. 1975A. The luxury of new product development. Food Technol. *29*, No. 9, 26–27.

BINKERD, E. F. 1975B. Changing roles of industry research groups. Food Technol. *29*, No. 9, 51–54.

GOOKIN, R. B. 1976. Industry speaks. Nat. Canners Assoc. Washington, D.C. 20–21.

NATIONAL CANNERS ASSOCIATION. 1974. Impact of environmental controls on the fruit and vegetable processing industry. *D2686R*, Nat. Canners Assoc., Western Res. Lab., Berkeley, California.

OPINION RESEARCH CORP. 1975. Public opinion index. *33*, No. 23, 5.

20

Energy and Food Processing

Robert L. Opila

"WASHINGTON, D.C.—It is late September 2000. Another sub-freezing morning, the U.S. Secretary of Food Policy notes as he boards an Ottawa-bound flight with his subordinates, the Secretary of Agriculture and the Secretary of State.

"Again, the monsoon rains have failed in Asia. The U.S. corn harvest looks good by year 2000 standards. But why, the Secretary asks himself, can't somebody develop a 90-day corn that would yield the way 120-day corn did back in the 1970's? If it weren't for frosts nearly every May and September, he muses, the high-yielding long-season corn would still mature. And wouldn't it be nice to pile on the chemical fertilizer?

"In Ottawa, the entourage is met by its Canadian counterparts. After perfunctory greetings; the delegates get down to the grim, yearly task before them: rationing the food produced by their nations to the rest of the world."

Was this written by Ray Bradbury? George Orwell? No, by Joseph M. Winski, who is a staff reporter for the *Wall Street Journal*. In this article appearing on March 25, 1976, he points out some of the inter-relationships between energy consumption and food processing in the food system.

Approximately 16.1% of total U.S. energy input is utilized in the food system, as shown in Table 20.1. Food processing utilizes 28–33% of the total energy requirements for the food system consisting of farm production, food processing, transportation, marketing and distribution, and households. It should be pointed out that household refrigeration and preparation of food for final consumption in the home utilize almost as much energy as is utilized in food production. Total food-related energy requirements are estimated to be in the order of six quadrillion Btu's per year.

The Federal government's Standard Industrial Classification system identifies 44 distinct industries within the food and kindred products

TABLE 20.1

ENERGY CONSUMPTION FOR THE FOOD SYSTEM

Component	Percentage of Total U.S. Energy Input
Production	2.9
Manufacturing	4.8
Distribution	
Wholesale	0.5
Retail	0.8
Consumption	
Out-of-home	2.8
In-home	4.3
	16.1

SOURCE: Hayden (1976).

grouping. Because of this extreme diversity, energy conservation in food plant design begins with the identification of the primary users of energy in the food industry.

Table 20.2 describes the characteristics of 15 segments of the total food industry, and presents data for 1973 showing energy utilization and production for that industry. In the last column of Table 20.2, estimates of energy requirements are shown per pound for various food products.

The 15 industries contained in this study constitute slightly less than 60% of the dollar volume of shipments, and employ approximately 65% of the personnel working in the food industry.

The basic industries include meat packing, the production of sausage and other processed meats, fluid milk, canned and frozen fruits and vegetables, animal feeds, wet corn milling, the baking industry, cane and beet sugar production, malt beverages, vegetable and animal fats and oils, and manufactured ice. As can be seen by this study, the energy consumption in 1973 for these industries is shown to be 838 trillion Btu's.

The largest users of energy in the food industry are meat packers, animal feed operations and fluid milk processors which utilize considerable amounts of energy simply because they process very large volumes of output. Energy-intensive industries would be wet corn milling, and beet sugar and cane sugar refining industries.

The indicated wide variations in unit energy requirements are due in some cases to plant size. But other factors were the over-riding reasons for variation, including plant design, type of process and age of boilers. New boilers have a higher operating efficiency than old boilers.

The range of energy consumption shown in Table 20.2 can be ex-

TABLE 20.2

CHARACTERISTICS OF 15 SEGMENTS OF THE TOTAL FOOD INDUSTRY

SIC No.	Industry Description	No. of Plants	1973 Energy Consumption (Trillion Btu's)	1973 Production (Million Lb)	Estimated Energy Requirements (1000 Btu/Lb)
2011	Meat packing	2,500[1]	99	51,364	0.8–1.5
2013	Sausage and processed meats	1,350[1]	26	10,500	1.5–12
2026	Fluid milk	2,400[1]	79	69,400	0.3–3.6
2033	Canned fruits and vegetables	1,050[1]	53	28,014	1.6–2.2
2037	Frozen fruits and vegetables	640[1]	62	12,265	1.0–8.5
2042	Prepared animal feeds	2,400[1]	86	165,064	0.3–7.1
2046	Wet corn milling	17	84	15,800[2]	3.0–12.0
2051	Breads, cakes and related products	3,300[1]	69	19,410	2.4–3.1
2061	Raw sugar production	42[1]	NIL	2,700	– –
2062	Cane sugar refining	24	44	15,682	2.8
2063	Beet sugar processing	52	77	52,404[3]	0.9–1.9
2082	Malt beverages	134	75	149[4]	1.5–2.7
2092	Soybean oil mills	94[1]	56	19,376[5]	1.1–1.5
2094	Fats and oils	550[1]	24	12,520	0.7
2097	Manufactured ice	1,150[1]	5	9,900	0.5
		15,703	838		

SOURCE: Federal Energy Office (1974).
[1] Approximate
[2] Ground corn.
[3] Beets, sliced.
[4] Millions of 31-gal. barrels.
[5] Beans, crushed.

213

plained by a review of the various processes used within each individual industry. As an example, Fig. 20.1 depicts the points of usage of thermal energy in electrical power in the meat packing industry, and includes the basic steps of slaughtering, hide removal, eviscerating, the necessary trimming, cutting and boning along with edible and inedible rendering. The carcasses are chilled and held under refrigerated conditions until consumed, either as fresh products or until they have been processed further.

The symbols used in Fig. 20.1 show whether the described function requires electrical energy, heating, cooling or does not require significant quantities of energy.

Processing steps for the corn wet-milling industry are shown in Fig. 20.2. The corn wet-milling process can be broken down into

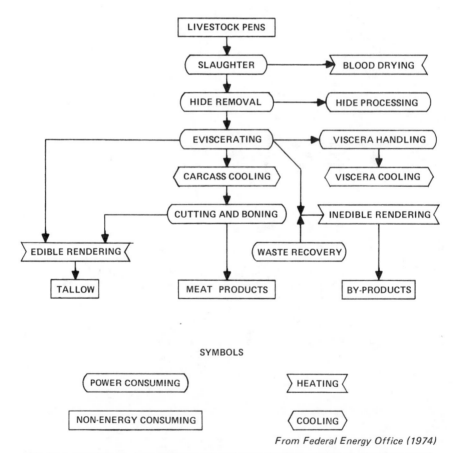

SYMBOLS

From Federal Energy Office (1974)

FIG. 20.1. MATERIAL AND ENERGY FLOW FOR BEEF PACKING PLANT (SIC 2011)

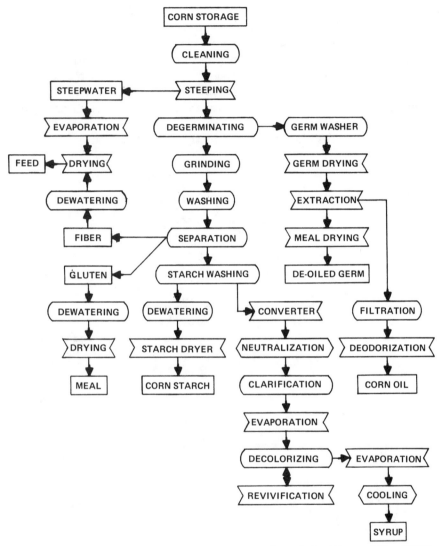

From Federal Energy Office (1974)

FIG. 20.2. MATERIAL AND ENERGY FLOW FOR CORN WET-MILLING PLANT (SIC 2046)

five separate processes for the production of cornstarch; corn syrup; the production of feed, meal and germ; and the production of steepwater. The individual processing steps include the steeping of corn followed by degermination, further grinding to separate the corn kernel into its constituents of starch, the protein fraction called gluten, corn fiber, and the steepwater produced in the steeping process. The starch can be dried or converted by acid or enzyme into corn syrup. The gluten and fiber are dewatered and dried for feeds. The steepwater is evaporated and sold as a nutrient in fermentations, or can be added to the fiber stream to increase its protein content. The germ is dewatered and dried prior to solvent extraction for the recovery of corn oil. As shown in Table 20.2, this is an extremely energy-intensive industry.

Fig. 20.3 treats the brewing industry in similar fashion, including

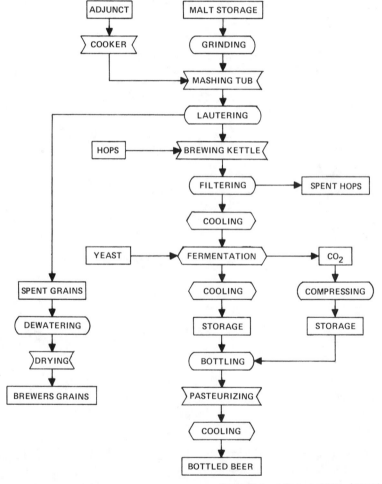

From Federal Energy Office (1974)

FIG. 20.3. MATERIAL AND ENERGY FLOW FOR TYPICAL BREWERY (SIC 2082)

the steps of mashing, brewing, fermenting, aging and filtering prior to bottling.

Fig. 20.4 shows the typical operations for the soybean oil process wherein soybeans are cracked, dehulled and conditioned, solvent extracted for the recovery of oil, and ultimate drying of the meal.

The brewing industry also dries spent grains for use as animal feed.

It should be pointed out that energy is reported only in Btu. Electrical energy is converted to Btu on the basis of an overall

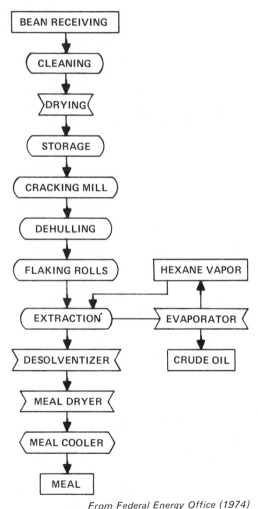

From Federal Energy Office (1974)

FIG. 20.4. MATERIAL AND ENERGY FLOW
FOR A SOYBEAN OIL MILL (SIC 2092)

conversion efficiency of 32% from the fossil fuel used to generate electricity to the utilization of electrical energy by the end user.

All of the various industries included within the 15 selected industries are analyzed from the standpoint of very basic steps of receiving, cleaning and/or storage of raw material, preparation of raw material prior to processing, the processing step itself, preserving of processed food by various means, followed by packaging and product storage.

The various aspects of receiving, cleaning and/or storage of the raw materials varies between industries and includes washing and grading, material handling, and storage in receptacles either at ambient conditions or through refrigeration or freezing to preserve the raw material for the next steps in preparation. Depending upon the industry, preparation varies according to the industry: stunning, bleeding, hide removal, eviscerating and trimming in the meat packing industry; trimming and slicing in the fruit and vegetable industry; proportioning process streams in feed-processing and sugar production; hydrolyzing the carbohydrate in the production of corn syrup, corn sugar or the wort in brewing.

Processing would include chilling and cooling carcasses in the meat industry; cooking, emulsifying and extruding of the processed meats industry; cutting and slicing of fruit and vegetables; the separation or dewatering of food products in the fruit industry; crystallization in the corn, and beet and cane sugar industries; fermentation in brewing.

After processing, preservation of the resulting food product can be done by refrigeration, freezing, aging or drying, pasteurization, evaporation, drying or baking, and cooking and cooling. The bulk of these products require packaging which for the most part uses nominal amounts of energy for the actual packaging process.

The storage of the food products—particularly fresh meats and frozen foods—requires extensive utilization of power because of refrigeration requirements.

Sanitation in the food industries utilizes considerable amounts of energy, with particular emphasis in the meat, dairy and brewing industries.

The review of the 15 basic industries from the standpoint of receiving and storage of the raw material, preparation, processing, preservation, packaging, storage and sanitation, allows the identification of certain unit operations within food processing that are extremely energy-intensive.

The energy consumption patterns can be shown to involve the basic steps of:

1. Dehydration by drying.
2. Concentration by evaporation.

3. Dewatering by centrifugation or filtration.
4. Thermal treatment by pasteurization or sterlization.
5. Clean up.

Energy requirements for dehydration can be reduced by:
1. Additional dewatering of product.
2. Higher drying medium inlet temperatures.
3. Countercurrent, as opposed to cocurrent, flow of product and drying medium.

Energy requirements for concentration can be reduced by:
1. Use of multiple effect evaporators.
2. Vapor recompression.
3. Reuse of hot condensate.
4. Use of improved evaporator types for further concentration if dehydration by drying is to follow.

Overall energy reductions in dewatering can be achieved by:
1. Replacing screens or shakers with filters or centrifuges.
2. Mechanical compression of cakes produced by filters or centrifuges.
3. Operating filters and centrifuges with extended cycles at reduced production rates.
4. In sufficiently large installations, alternating the function of motors as generators when batch centrifuges are accelerated and decelerated.

A useful concept for evaluating dewatering operations is the reporting of water on a dry basis. That is, a material such as wood pulp analyzing 10% solids, contains 9 lb of water per pound of solid; a starch cake analyzing 40% solids contains 1½ lb of water per pound of product; and a sucrose centrifuge cake analyzing 8% water contains 0.086 lb of water per pound of solids. Depending on moisture content, a 1–2% reduction in moisture content can substantially reduce the drying load.

Overall energy requirements in pasteurization or sterilization operations can be reduced by use of regenerative systems such as those included in some high temperature-short time (HTST) systems. An overall process heat balance may show other ways of reducing energy requirements.

Cleaning techniques can be modified using lower temperature techniques and/or through use of clean-in-place (CIP) systems, and, of course, using the absolute minimum of water.

Water usage, in general, must be checked closely. Improper selection of pump seals can cause dilution of process streams.

One industry requiring particular emphasis is the raw cane sugar industry. In this particular industry the only external energy utilized within the plant is for lighting, since the bagasse resulting from the cane-crushing operation is burned directly in the sugar factory boilers for the production of steam. In the sugar factory, steam is used for operation of the evaporators and steam turbines are used as prime movers to drive pumps and centrifugals. In the well-run sugar factory, excess bagasse is available which can be used for the production of by-products such as acoustical tile and insulating board.

In addition to the process studies outlined above, other means of conserving energy involve the return of condensate to the boilers where this is practical. Up to 10% of the heat required for steam generation can be recovered in this way. Where extensive refrigeration is required, consideration should be given to use of ammonia as opposed to freons for energy reduction since ammonia is thermodynamically more efficient than the freons. In any event, means for recovering useful heat from the condensers in a refrigeration system should be thoroughly investigated. Certainly, building heating could be reduced in this way.

Where storage areas are maintained at reduced temperature, as in coolers or freezers, the use of additional insulation to reduce heat loss should be investigated. Increased thicknesses of insulation can easily be justified by today's energy cost. Whereas 6 in. of insulation was standard for cooler and freezer design several years ago, 8 and as much as 10 in. can be justified today on the basis of reduced operating costs resulting from energy conservation.

Considerable energy is wasted in environmentally controlled warehouses through poor control of infiltrating air into the conditioned areas, thereby resulting in additional cooling load. Controlled access to these areas through use of air locks reduces infiltration to, at most, the volume of the air lock and the number of cycles per day. Several estimates show infiltration can be reduced to 1/10–1/6 of that normally encountered when fork lift/truck traffic moves freely in and out of environmentally controlled storage areas.

Thus far, we have been discussing, essentially, new plant design only. There are many areas in existing plants where energy can be conserved. Some of these are:

1. Check water usage closely.
2. Reduce water temperature where feasible.
3. Check operation of combustion processes and equipment to ensure good combustion without excess air usage.
4. Repair and/or replace deteriorated insulation.
5. Repair or replace leaking steam traps.
6. Minimize use of outside make-up air.

7. Turn off all process equipment when use is not required.
8. Check air seals around truck loading dock doors and keep dock truck doors closed when not in use.
9. Recharge storage batteries during off-peak demand periods.
10. Optimize truck routing and deliveries.
11. Evaluate the overall energy used in packaging.

These are just a few of the items that can be reviewed in existing plants to reduce overall energy needs. A planned energy conservation program, much like a preventive maintenance program, can have substantial impact on future energy requirements.

Where the food processor once was concerned only with costs of raw material, direct labor, packaging cost and overall product yield, now he is concerned about the cost of energy, since energy costs have quadrupled in the last two years. Many of the process improvements outlined above can be justified in new plant construction on the basis of today's energy costs alone.

The most probable new source of generating power will be nuclear reactors which currently utilize expensive uranium as a fuel. Current projections of the completion of design of the breeder reactor indicate that this reactor will not be on line until approximately 1990, and that electrical energy produced by the fusion of hydrogen probably will not be available until the year 2020. In the meantime, energy costs will continue to increase because of the very large investment required in electrical generation and distribution facilities, as well as the ever-increasing cost of fossil and nuclear fuels.

Current fossil fuel costs are in the order of $1.00-1.40 per million Btu, in comparison to a cost of $0.22 per million Btu approximately 5 years ago. And enriched uranium now costs $24.00-28.00 per lb compared to $8.00 per pound just 2 years ago.

Energy costs by the year 1985 are projected to be $0.10-0.12 per kilowatt-hour, in comparison to current costs of $0.008-0.03 per kilowatt-hour.

Energy conservation in the food processing industry is important to all of us on the basis of simple economics, in addition to assisting the United States of America to become self-sufficient from the standpoint of energy production.

BIBLIOGRAPHY

HAYDEN, C. L. 1976. How to Achieve Energy Conservation in Operating Plants. Inst. Food Technologists Meeting-In-Miniature, New York Section.
FEDERAL ENERGY OFFICE. 1974. Industrial Energy Study of Selected Food Industries. III-3, IX-4, XII-3, XVI-6.

21

Food Additives and
Their Regulation

Richard L. Hall

Food additives are not new; we have used at least smoke and salt—
and probably many others—since the Neolithic revolution. Indeed,
there are very few *new* additives, and very little that is new in their
use. But a week rarely passes without something new in their regulation. These contrasting statements are closely related. This relationship, its causes and consequences form an appropriate subject for a
centennial—and bicentennial—review.

There is no need or time to discuss here the rationale and technology of additive development, use and safety evaluation which are
covered in detail in many other forums. Instead, I wish to compare
additive use and regulation a century or two ago, and now. Both the
changes and the similarities are instructive. In the reasons for the
changes—and for the problems we now have—we can see not only the
legal and technical issues themselves, but much deeper and more
general conflicts which our professional problems merely reflect.

Two hundred years ago there was no food processing industry.
Nicholas Appert was only 24 years old, and his work in food preservation was far in the future. With the exception of a few local bakers,
food was totally processed at the point of service at home or inn. A
century ago, the industrial revolution had significantly affected agriculture, but industrial food processing had barely begun. Many of its
products fully deserved the suspicion with which they were regarded.

Additives here played a role significant both in public attitudes and
later regulation. In the home, where most food was still processed, additives, including many we use today, were regularly employed. This is a
comparison worth noting. We recognize today 40 different technical
effects (Table 21.1) for which additives are used (published in the
Federal Register *39*, No. 185, 34173, 1974). Some additives (dough

TABLE 21.1

FOOD ADDITIVE TECHNICAL EFFECTS

Classifications Used in 1976	Additives Used Prior to 1876 to Achieve These Effects		References
Anticaking agents, free-flow agents	Cornstarch		
Antioxidants	Spices		
Colors, coloring adjuncts	Green:	Unlined brass or copper utensils for making pickles	Child (1813)
		Spinach juice or grass	Dooley (1975)
	Yellow:	Marigold flowers	Dooley (1975)
		Heart of yellow lily	Child (1813)
	Red:	Cochineal, cream of tartar and alum	Child (1813)
	Blue:	Water in a plate rubbed with indigo stone (copper sulfate)	Child (1813)
Curing, pickling agents	Saltpeter (for meats)		Hertzberg *et al.* (1973)
	Salt		Hertzberg *et al.* (1973)
	Vine leaves		Child (1813)
	Roach alum		Child (1813)
Dough conditioners, strengtheners	—		
Drying agents	—		
Emulsifiers; emulsifier salts	Egg Yolk		
Enzymes	Rennet (from calves' stomachs)		Dooley (1975)
Firming agents	Alum		Hertzberg *et al.* (1973)
Flavor enhancers	Soy sauce		Anon. (Circa Mid-19th Century)
	Vinegar, lemon juice, juice of other sour fruits, used in small quantities to develop other flavors		
Flavoring agents, adjuvants	Spices, herbs, smoke, roots (sassafras, sarsaparilla)		Hertzberg *et al.* (1973)
	Rose water		Simmons (1796)
	Violet leaves and flowers		Anon. (Circa Mid-19th Century)
	Rose geranium leaves		Anon. (Circa Mid-19th Century
	Primrose petals in vinegar		Anon. (1819)
Flour-treating agents	—		
Formulation aids	—		
Fumigants	Sulfur dioxide (by burning sulfur)		
Humectants, moisture-retention agents, antidusting agents	—		

Continued

TABLE 21.1 (Continued)

Classifications Used in 1976	Additives Used Prior to 1876 to Achieve These Effects	References
Leavening agents	Pearl ash (from wood ashes)	Simmons (1796)
	Vinegar, saleratus (sodium or potassium bicarbonate)	Dooley (1975)
	Egg whites	Mann (1861)
Lubricants, release agents	Bacon fat, pork fat, salt pork	
Nonnutritive sweeteners	—	
Nutrient supplements	Cranberries and lemon juice, used as a medicine	Rundell (1808)
	Cod liver oil	Rundell (1808)
pH Control agents	Vinegar, baking soda	
Preservatives, antimicrobial agents	Sodium or potassium silicate (for eggs)	Dooley (1975)
	Sulfur dioxide (from burning sulfur)	Hertzberg et al. (1973)
Processing aids	—	
Propellants, aerating agents, gases	—	
Sequestrants	Lemon juice	
Solvents, vehicles	Alcohol, glycerine	
Stabilizers, thickeners	Pectin (from apple juice)	Hertzberg et al. (1973)
	Sassafras leaves	Dooley (1975)
Surface-active agents	—	
Surface-finishing agents	Egg, milk, fat, maple syrup, molasses	
Synergists	—	
Texturizers	—	
Masticatory substances for chewing gum	Pine resin	
Antigushing agent	—	
Component in the manufacture of other food additives	Cream of tartar, alum	Child (1813)
Washing-peeling aid; vegetable cleaning agent	Lye (for hulling corn)	Hertzberg et al. (1973)
Fermentation aid; malting aid	Mother of vinegar	Dooley (1975)
Freezing agent	—	
Nutritive sweeteners	Sugar, honey, maple syrup	
	Pumpkin or watermelon juice boiled to a syrup	Dooley (1975)
Oxidizing and reducing agents	—	
Boiler water additives	—	
Ion-exchange resins	—	

conditioners, propellants and ion exchange resins) are obviously associated with recent technology. But of these 40 technical effects, 24 are clearly recognizable in even a cursory reading of recipes of our first hundred years.

These include—albeit in cruder from—enzymes and colors, release agents and fumigants, firming agents and solvents, and 18 other technical effects. We can see that there has been growth, more understanding, and technological refinement, but no revolution in the purposes for which additives are used.

The number of additives then in use is more difficult to judge. Cookbooks were much less common; local practices and ethnic folkways seldom recorded. The glimpses which survive suggest that the number was extensive, the use frequent and the price high. Two hundred years ago, the price of salt occasionally was four times the price of beef (Anon. 1964)! As we have become more concerned with additives, our *awareness* of their number has grown, probably much more rapidly than the actual increase.

There has been some change in the identity of additives. Some have been dropped from use or replaced—for reasons of effectiveness, or cost, or aesthetics, or safety. We have dropped sassafras; more effective sequestrants have replaced lemon juice and vinegar for that technical effect; refrigeration and better distribution facilities have replaced water glass for preserving eggs. We no longer use spinach juice for green color, and marigold blossoms for yellow—but it is not yet clear that this is a permanent gain. I note with a tinge of commercial regret that the old cookbooks stressed the need for more spice to mask the off-taste from the added natural color.

These examples point up a still greater change—from the crude, variable, untested, natural mixture of chemicals of a century ago, to the isolated or synthesized, specified, evaluated single chemical entity of today. Much of the sophisticated technology devoted to additives has been devised and employed by members of this ACS division in the isolation and identification of useful natural constituents.

By far the greatest changes, however, have been in the area of processors' intent, and in consumer knowledge and confidence. Here we approach the crux of this discussion.

A century ago *caveat emptor* was still the rule of the marketplace. An alarming number of "additives" which were themselves often alarming, were used with appalling frequency and unconcern to adulterate food ingredients and finished foods. They were used in the classical, legal senses of "concealing inferiority" and "making the food appear to be of greater value." Adulteration, of course, long antedated food processing, and efforts to control it date from Plantagenet and even Roman times. But the growth of food processing, remote from the consumer, combined (to use Shaw's felicitous phrase) a maximum of temptation with a maximum of opportunity. The litany of cheap and dangerous offenses is familiar to us all: The pioneering work of Frederick Accum in England in the early 19th Century, the long line of investigators, muckrakers, and scientists (including Dr. Harvey Wiley in this country), together with our present regulatory agencies, more informed consumers, and more responsible manufacturers, have gradually removed the most offensive, most dangerous, and most blatant forms of adulteration, although it still exists in more subtle form. Today adulteration is less flagrant—not necessarily less fragrant.

Indeed, we probably will never avoid it entirely because adulteration is related inversely to confidence and information. If, in a bad year for cherries, Aunt Tess puts a little red color, a bit of cornstarch and some almond extract in a cherry pie, that is merely because she is a good cook, and she is esteemed for her resourcefulness and skill. If the Aunt Tess Pie Company does the same things, and doesn't tell you, the company is criminally guilty of adulteration. The difference, then, is first one of confidence. In Aunt Tess's dining room, that confidence was personal. In today's more impersonal marketplace, brand and company reputation are still valuable, but confidence now depends also on regulation and information.

As the center of processing of food moved slowly and inexorably from home kitchen to commissary and factory, and as technology introduced unfamiliar complexity to what had once been household practices, we became less able both to feel and to have a sense of control over our food, and the need to regulate became ever greater. We moved from virtually no regulation in the middle of the last century to the first comprehensive food law in the United Kingdom in 1875, which aimed principally at adulteration, and prohibited the sale of food which was "not of the kind, substance, or nature demanded." In this country, our first Food and Drug Act, similarly motivated, was passed in 1906 and dealt in part with misbranding, but also prohibited any "added poisonous or deleterious substances." Both of these acts left the responsibility with the government to determine if the action was harmful or the ingredient poisonous. The U.S. law classified everything as either black or white. An added substance must either be harmless or poisonous. Today, of course, as we all know, we have a complex set of laws and regulations which require a prior showing of adequate safety as a precondition to the use of a substance in food, and we are about to add to this with new legislation which will establish quality assurance practices for every food processing operation. The complexity of today's regulations is undesirable. Time prevents detailed analysis, but one can at least contrast the bulk of the dictionary definition of "additive" with that of the FDA. According to Webster's New World Dictionary, 2nd College Edition, 1968, an additive is:

A substance added to another in small quantities to produce a desired effect, as a preservative added to food, an anti-knock added to gasoline, etc.

The definition of "food additive" found in the Food, Drug, and Cosmetic Act is:

(s) The term 'food additive' means any substance the intended use of which results or may reasonably be expected to result, directly or indirectly, in its becoming a component or otherwise affecting the characteristics of any food (including any substance intended for use in producing, manufacturing, packing, processing, preparing, treating, packaging, transporting, or holding food; and including any source of radiation intended for any such use), if such substance is not generally recognized, among experts qualified by scientific training and experience to evaluate its safety, as having been adequately shown through scientific procedures (or, in the case of a substance used in food prior to January 1, 1958, through either scientific procedures or experience based on common use in food) to be safe under the conditions of its intended use; except that such term does not include—

(1) a pesticide chemical in or on a raw agricultural commodity; or
(2) a pesticide chemical to the extent that it is intended for use or is used in the production, storage, or transportation of any raw agricultural commodity; or
(3) a color additive; or
(4) any substance used in accordance with the sanction or approval granted prior to the enactment of this paragraph pursuant to this Act, the Poultry Products Inspection Act (21 U.S.C. 451 and the following) or the Meat Inspection Act of March 4, 1907 (34 Stat. 1260), as amended and extended (21 U.S.C. 71 and the following); or
(5) a new animal drug.

It makes one regret that the only Webster with which our Congress is familiar is Daniel, rather than Noah.

We have reviewed the technical effects for which we have used additives. Their use is both a result and a cause of the continuing need for variety, nutrition, safety, economy, convenience and sensory quality—precisely the criteria applied by the Panel on Food Safety of the White House Conference on Food, Nutrition, and Health (Anon. 1969). Additives are, in fact, as the Panel on Chemicals and Health has pointed out (National Science Foundation 1973), adjuncts to processing. They extend the range and flexibility of the very few available basic processes. In this country, we would not starve without our present large number of food additives, but our food would have less variety; it would be more expensive; it would be more variable and lower in quality; it would suffer in nutritional value, unless we compensated by spending even more and—if we persisted in favoring such a foolish notion—it would be less safe.

I should like now to shift to the manner in which our regulation of additives, and the concerns which give rise to that regulation, reflect far more general dilemmas of our time. We do not consider additives as a subject apart. We approach them with the same set of fears or lack of them, the same priorities and perspectives as we bring to every other area in which we are involved. The first of these dilemmas springs from the uneven advance of knowledge.

We have already mentioned that additives are essentially processing adjuncts which extend the range and flexibility of food processes. The use of additives is generally less expensive than a major process revision, not only in terms of capital cost and operating cost, but very often in terms of the indirect cost to the quality of the product or consumer acceptance. Additives, therefore, tend to be regarded with jealous concern by the food processing industry, and substantial effort goes into investigating their potential for useful technological effects. As the use of additives has shifted from home to factory, and scientific advance has increased our awareness of at least implied, if not actual, hazard our apprehensions and our safety requirements have increased sharply.

The Food Additives Amendment of 1958 fixes on industry the responsibility for developing information adequate to establish the safety-in-use of additives. If an additive is deemed to require a petition to establish its safe use, rather than being generally recognized as safe, then it is the sole responsibility of the Food and Drug Administration to pass on the adequacy of that information. The industry quite logically does not wish to press forward with an additive which involves so high a probability of risk that a regulation is unlikely to issue. Beyond public health considerations, that would be a wholly uneconomical expenditure of effort. But the industry, naturally, is primarily preoccupied with capturing the potential benefit of the additive; while the regulatory agency, on the other hand, is almost entirely concerned with adequate assurance of the absence of risk. A government official will never be criticized for the unrealized benefit, but only for demonstrated harm. Tension, therefore, develops because the risk tends to be appraised by one party from one point of view while the benefit is evaluated by another party from another vantage point.

In addition to this difference, the technology of use inevitably moves ahead of the technology of safety evaluation. Demonstration of a useful technical effect is a generally simple, precise, and objective determination. Even where it involves organoleptic effects, as with flavors, aromas and textures, at least panel data are obtainable on short notice and at relatively low cost in time or money. Industry, by competitive necessity, tries to react quickly to cost or use advantages. By contrast, safety data are rarely obtainable in any depth in short-term studies, and, in contrast to data on technical effects, cannot ethically be obtained on humans without extensive preliminary animal work. Furthermore, the precision and usefulness of the results of toxicological studies leave a great deal to be desired. In the interpretation of taste panel results or measurements of texture or viscosity, there is no equivalent to the 100-fold safety factor routinely applied to the results of toxicological studies. Thus, the technology of safety evaluation lags far behind. Industry, therefore, reacts quickly to cost or use advantages, but the larger society reacts much more slowly to remote, uncertain and delayed threats.

Worse yet, the technology of *detection*—analytical chemistry—has also moved far ahead of the technology of safety evaluation.

As a society, we have responded to all of this by instituting a complex group of laws and regulations intended to redress this imbalance and to hold these tensions in some kind of constructive relationship. It has certainly slowed down the imbalance, primarily by slowing down the rate of introduction of food additives (Table 21.2).

In summary, then, we can measure with precision that which we cannot evaluate with precision. We can sense threats we cannot appraise. If this suggests the atmosphere of Salem in 1692, the parallel is appropriate. What we require, obviously, is more effort at quantifying remote or implied risk. There is little hope for improvement in

TABLE 21.2

WHOLLY NEW REGULATED ADDITIVES INTRODUCED EACH YEAR

Year	"Natural" Flavors (121.1163)	"Artificial" Flavors (121.1164)	Non-Flavor Additives (121. - - - -)
1960	—	—	11
1961	—	—	75
1962	—	—	17
1963	—	—	9
1964	—	574	22
1965	51	2	87
1966	—	1	7
1967	7	40	47
1968	—	1	10
1969	—	—	5
1970	—	—	3
1971	—	—	4
1972	—	—	2
1973	—	14	2
1974	—	—	5[1]
1975	—	—	—

[1] 121.1258—passed 7/26/74, stayed Dec. 5, 1975.

regulatory practices—much less for rational applications or modification of the Delaney Clause—unless this can be done. This comment leads us to the next dilemma.

This is the dilemma of conflicting demands. Although there are a number of these, the one that concerns us here most is the demand for economical and high-quality food versus the demand for absolute safety. Of course, it is obvious to us all from the mere fact that we came to New York City, that we do not demand absolute safety in everything. We are quite inconsistent. Sir Edward Pochin (1975) has estimated the risk of death "as one in a million for smoking 1½ cigarettes, drinking 1/2 bottle of wine, traveling 50 miles by car or 250 miles by air, rock climbing for 1½ minutes, canoeing for 6 minutes, engaging in typical factory work for 1-2 weeks, or simply being a man aged 60 for 20 minutes." In the case of food ingredients, we only demand absolute safety with those that are added carcinogens, for which we theoretically set a safety factor of infinity. With naturally-occurring ingredients, we are much more tolerant. The safety factor for salt, at least for those who are salt hypertensives, is probably less than one. Perhaps one of the reasons we continue to perpetuate this set of conflicting demands is the relative availability and economy of food in the United States. We can afford the illusion that we can pursue absolute safety, because we have never really made an effort to calculate the direct and indirect costs of its pursuit. This is true not only of consumers and regulators, but of those in industry as well. An all-too-common reaction of food processors is that proof of safety is the job of the supplier. With a few commendable exceptions, processors have seldom played a major role in supporting the testing necessary to establish safety-in-use. Usually, they have taken the rather cavalier view that if one ingredient didn't do, they would simply shift to another, and often this has been possible. Their interest quickens, however, when alternatives begin to run out, as with artificial sweeteners, nitrite, brominated vegetable oil and red color. In fact, the value of an additive cannot be measured either by the dollar value of the sales of the additive or by sales of the foods in which it appears. The true value of an additive is the total incremental costs of shifting to the next best alternative—whatever that is. This includes not only the costs of reformulation, of differing consumer acceptance, of market shifts and shifts in product mixes, but the indirect cost attendant upon regulatory activity and industry response. It is obvious that this calculation of cost and benefit is so tedious and complex it has never been done at all. Thus, we are in the position of not only being unable to quantify the remote risk, but unable to estimate the cost of avoiding it. Nevertheless, we tend to try to avoid risk without this knowledge, and with increasing diversion of scarce technical resources. The relevance, productivity and value of this effort is difficult to measure, but it often must be marginal or negative. We need, therefore, much more effort on cost/benefit relationships. In any strict, hard quantitative sense, there can be no risk/benefit calculation with regard to food ingredients; at any rate, not as it is more rigorously applied in medicine. Among other reasons, this is because the risk, as we have seen, must be remote or implied, or both, and the benefits are often largely subjective and difficult to measure. What we can strive for instead are balanced decisions. Cost/benefit analysis can provide legislators, regulators and consumers with some basis for rational choice.

We come now to the next major dilemma. We have so many fears, so many threats, so much to learn, and we have recently become much more aware of the limits on our own resources. We must start from the sound position that we can only answer fears with facts, and we will never know enough to answer all fears. We need, then, to spend some careful thought on priorities; how much time and effort to devote to the investigation of ever more remote or improbable hypotheses—where, in short, to cut if off. This gets us back again to the quantification of risk. As an example of our need, I have heard a private, but reasonably well-informed estimate, that at one point last year the GRAS review had cost approximately $14 million. At that point, the review was probably not more than 20% complete.

Our next problem is more irony than dilemma; it is inadequacy of communication in an age of overcommunication; isolation in an age of sensory overload. The vast increase of our knowledge in nutrition and in food safety has really made this information, or at least a comfortable attitude toward it, inaccessible to most people. The information is remote, complex and formidable. In part, this is due to the failure of good scientists to communicate well, and, much rarer but much more unfortunate, instances of scientists who, for whatever reasons, communicate irresponsibly. Vice President Rockefeller pointed this out recently in saying, "Public confidence is shaken if a small minority of scientists, without adequate basis for their claims, spread unfounded fear and retard or prevent progress." But in a free society, "there must be a better method for bringing into focus for the people the facts and the informed mature, objective judgments of the scientific community." He suggests that establishing the Office of Science and Technology Policy in the White House is one way to do this (Anon. 1976). Some professional societies have begun an effort to cope with this. Among the more notable of these are the Public Information Program of the Institute of Food Technologists and the Council of Agricultural Science and Technology (CAST). The task is enormous, and the effort is still small.

A problem that must concern us is the problem of inverted priorities. I have mentioned elsewhere that the objective ranking of the six major sources of food hazards is solidly established, but many visible and vocal groups in our society, including consumer groups, Congress, the press and regulatory agencies, tend to treat these generally in inverse and perverse order of importance (Hall 1973).

As an extreme example, the maximum annual risk to our total population, to the extent it can be estimated, from DDT use and exposure (National Science Foundation 1973) was equivalent to limiting annual automobile travel by the entire population to 37,000 miles (HEW, National Center for Health Statistics). A nation that cannot come to grips with drunken driving and handgun control has no business worrying about DDT residues and Red No. 2.

It is obvious that as a result of all these—the uneven advance of knowledge, the problem of conflicting and irreconcilable demands, the limits of our resources, inadequacy of our communications, and inversion of our priorities—we have too often reacted simply by coping with crisis rather than persisting with plan. The tragic aspect, however, is not what we do to ourselves, but what we do to others. In fact, we seem often to give in to an irresistible impulse to export our priorities, including those which do not apply even to us. Examples of this are the rush of other countries to follow our peremptory example and ban DDT, even when they needed it for malaria control, or the tendency of other countries to adopt food and drug codes similar to ours, when they cannot possibly afford the enforcement machinery to make them effective, or the legislative proposal made several times here to permit agricultural imports only from those countries whose pesticide legislation is identical to ours.

How then will we resolve these dilemmas? A general answer is that we probably won't because we can afford not to. We obviously have coped to a considerable extent with the uneven advance of knowledge as it applies to food additives by slowing down virtually to immobility the introduction of new ones. This has been accomplished, not so much by conscious policy as by the indirect effects of cost and regulatory inertia. In the conflict between economy and absolute safety, we will, in a fashion typical of our history, probably resolve this by promising that which is unattainable and accepting that which is available. We are spending much more on food additive safety; and while that expenditure is sufficient to slow down innovation, it is not enough to affect in a visible way the cost of food. We will probably deal with the Delaney Clause not by modifying it, but by simply ignoring it in those instances in which its rigid application would affect a popular food, just as we have done with inconvenient and outmoded traffic laws. In a general sort of pragmatic way, we will probably recognize the limits of our resources, not so much through rational choice, as through our inability to sustain for very long a high level of concern and enthusiasm for any one fear. Familiarity breeds contempt. Certainly it has with the automobile and with alcohol. Are we now going through the prohibition era with food additives? The communications problem may in part be soluble, perhaps by efforts such as those already mentioned. Finally, we have begun to handle this problem by the transfer of dilemmas. Rather than decrease risk, we may simply reduce choice. That doesn't decrease the risk, but it may bring it more within our ability to comprehend.

BIBLIOGRAPHY

ANON. 1819. Family Recipe Book. (Publisher unknown.)
ANON. Circa Mid-19th Century. Foods and Household Management. (Publisher unknown.)
ANON. 1964. American Heritage Cookbook. American Heritage Publishing Co., New York.
ANON. 1969. Final Report. White House Conference on Food, Nutrition, and Health.
ANON. 1976. Rockefeller says nice things about science. Chem. Eng. News 54, No. 9, 6.
CHILD. 1813. American Frugal Housewife. (Publisher unknown.)
DOOLEY, D. (Editorial Director). 1975. Better Homes & Gardens Heritage Cookbook. Meredith Corp., Des Moines, Iowa.
HALL, R. L. 1973. A modern three r's—risk, reason and relevancy. Can. Inst. Food Sci. Technol. J. 6, No. 1, A17.
HERTZBERG, RUTH H., VAUGHAN, BEATRICE V., and GREENE, JANET G. 1973. Putting Food By. Stephen Greene Press, Brattleboro, Vermont.
MANN, MARY TYLER. 1861. Christianity in the Kitchen. (Publisher unknown.)
NATIONAL SCIENCE FOUNDATION. 1973. Chemicals and Health. Rep. Panel on Chemicals and Health, President's Science Advisory Committee. Nat. Sci. Found. Sci. Technol. Policy Off., Washington, D.C.
POCHIN, E. E. 1975. The acceptance of risk. Br. Med. Bull. 31, No. 3, 188.
RUNDELL. 1808. A New System of Domestic Cooking. (Publisher unknown.)
SIMMONS, A. 1796. American Cookery. (Publisher unknown.)

22

Plant Protein Foods-Their Development, Their Future

Warren E. Hartman

This subject appears to be so broad and overwhelmingly comprehensive that only the foolish would attempt to cope with it. It is not certain that these questions can be answered, but should there be accurate answers, it would take many volumes to disclose them. It should be pointed out, therefore, that the subject is being approached as based upon the experiences of Worthington Foods, now a division of Miles Laboratories.

The earliest reference as to plant protein foods appears to be in the writings of Moses in the Book of Genesis, which indicates that every green herb and plant and every plant-bearing seed shall be a food for you. However, the Lord evidently endowed man with sufficient knowledge and intelligence to make proper selections, and has probably left it up to the chemist to decide how we extract the toxic and nonnutritional, how we supplement or complement the nutritional values, how we process and modify, how we formulate and fabricate, how we fractionate the original agricultural product, how we extract its protein, how we rearrange its molecules, and how we recombine products and by-products to structure new foods.

Much of current food technology had its beginnings in foreign cultures. Missionaries, travelers and explorers have extracted technology from and injected it into the indigenous populations. Much of this basic "know-how" or technology imported from various cultures was born of necessity. This is apparent in the present food habits and food types observed in the Far East. China, India, and Japan are good examples. Aqueous and alkaline extractions of soybean have yielded soymilk-type beverages, soy cheeses, such as tofu and korio tofu, which have been consumed fresh as well as cultured, and also have been formed and dried to be rehydrated when needed.

The bland rice diet prevalent in these countries stimulated production of such things as soy sauce, hydrolyzed vegetable proteins, early impure forms of monosodium glutamate and other items which enhanced the flavor of the bland diet. These products are not new but go back many centuries in the culture, food habits, and "know-how" of these countries.

The washing of wheat flour dough or the extraction of wheat gluten also may be credited to early "know-how" in China. These small gluten portions were cooked and used centuries ago as meat replacements. Seaweed portions also were dried and flavored to resemble small tidbits of meat. There is some indication that the extraction methods used on soy and mung bean were also applied to various other plant protein sources, such as leaves and roots of certain plants. This is probably the forerunner of our present work in leaf protein concentrate, a basic concept which has been the subject of much research through the years and now challenges a fruitful completion.

Worthington Foods has been in the vegetarian protein business for approximately 35 years. The business was initially concerned with supplying such foods to consumers who had medical, philosophical or religious reasons for wanting them. The market consisted specifically of Adventist institutions, hospitals, schools and the communities surrounding them. It was also composed of a portion of the Catholic trade, the Jewish trade, and the Mormon community. It also included, on one extreme, sales to health food stores, and on the other extreme, sales to specific medical applications in controlled diets. These foods could be produced to be hypoallergenic. They could be produced with low calcium, low sodium, or with specific amino acid profiles. They could be supplemented or nonsupplemented in manners which made them adaptable to certain specific medical needs.

Early Worthington products clearly bore the mark and influence of John Harvey Kellogg and the Seventh-Day Adventists. In 1866, the Adventists established the forerunner of the Battle Creek Sanitarium and the Battle Creek Food Company in Battle Creek, Michigan. Dr. Kellogg continuously experimented with materials and processes to provide a wholesome and palatable nonflesh diet. This led to the invention or production of breakfast cereals (flakes and granola), peanut butter, cereal coffee substitutes, and even decaffeinated coffee at this early date. These items were served in the then popular Battle Creek Sanitarium or health spa. John's brother, W. K., and a patient, Post, promoted the cereals into two multi-million dollar businesses.

It is not generally realized that these products, now a regular part of the diet of both meat eaters and vegetarians alike, emanated from the vegetarian habits of the early Adventists and the pragmatic approach to research and end-product development of Dr. John Harvey Kellogg. Early meat analog products were called nutmeats. Usually nuts were finely ground into a paste or nut butter and water was then added to produce an emulsion. To this emulsion were added flours, starches, cereals or other ingredients. This white milk-like oil and water emulsion was then retorted or cooked to set up into a solid mass. This product could be diced or sliced for salads or used as a main entree. The texture of the resulting product depended mainly upon the degree of emulsification and the varying degree of particle sizes incorporated within the product.

Extracted wheat protein or wheat gluten was also used as basic raw material for the early vegetarian protein products. The gluten or protein extraction process usually starts with a high protein, hard wheat flour, which is mixed with water to form a dough in a conventional dough mixer. This dough is repeatedly washed with water to remove the starch. The carboyhdrate water is decanted and the process repeated several times until the gluten reaches the desired protein content; typically, around 72% protein on a dry basis. This dough-like elastic mass of gluten is cut into slices or cubes and cooked in boiling water where it expands and the protein simultaneously denatures, giving an expanded cellular chewy texture to the product. This product may be canned in a bouillion-type broth or may be ground through a chopper to simulate minced meat or hamburger. The wheat gluten also provided an elastic matrix for incorporation of nuts, grains, flours, legumes and other vegetable material into a mass. When mixed vigorously at high speed with considerable shear for an extended length of time, it would yield a thin, string-like, fibrous textured dough mass in which other materials could be incorporated. This product was also canned and cooked to produce a solid textured loaf-type product which served as a vegetarian entree, or was diced or ground for use in home recipes.

Similar products are being manufactured today and wheat protein or wheat gluten is still an important basic raw material for its contribution to texture. The Worthington Division of Miles Laboratories presently has a number of products based almost exclusively on wheat gluten which are peculiar, yet significantly textured protein products. Miles Laboratories also is becoming basic in the production of vital wheat gluten for utilization by industries throughout the world. Today, it would be well to look again at some of the early products and techniques, as the end-products were dependent solely upon the raw materials available and their macro and micro structure.

It appears evident today that an in-depth study of the micro and macro structures of agricultural products would serve to enhance their usage. It would also appear that many of these products could be further utilized by controlling and varying the particle sizes which are used to form different textures and consistencies. It also appears that the internal micro structure might be utilized, rather than destroying it, and then attempt to reconstruct it. It appears that extraction and fractionating, or separating techniques, would be of value in saving energy and reducing processing times presently used. Flours, isolates and concentrate must again be restructured or texturized in order to form an acceptable finished end product. It seems that raw materials in their more nearly native form could be utilized to a greater percentage in end products. One only has to visualize the structure of a simple item such as steel-cut oat groats which, when hydrated, have a chewy elastic shear quality. Groats have sufficient internal bound moisture and are bland and easily flavored. In this form and in selected particle sizes, groats presently provide one of the textural components of a finished engineered food.

Worthington had accumulated considerable formulating and flavoring "know-how" by the time the soy products such as flakes, toasted grits, isolated protein and concentrate became available. One of the first Worthington products based on soy protein was a powdered milk-like product called Soyamel, designed primarily as a hypoallergenic formula for infants and children who were allergic to milk. This spray-dried product is reconstituted with seven parts of water for use.

Worthington Soyamel was the first soy milk-type beverage to be based on or formulated primarily from soy protein isolate. There were numerous other soy milk products on the market but these were largely composed of soy flour suspended with a hydrophilic colloid of some type, or they were complete water extractions of soy similar to those used in China. More recently, the suppliers of infant formulas have shifted their formulas to bases of soy protein isolates. It is believed that these soy milk preparations, initially used for children and infant feeding, are only the forerunners of many substitute milk preparations designed to meet specific needs. It is also felt that plant protein milks can and will be produced in the future, and this will have significant economic advantage over animal-produced milk.

As flavor problems are overcome it is expected that plant protein products will be acceptable in the marketplace and in great demand in the future.

In the early work with dairy and meat analogs, one of the greatest frustrations of Worthington Foods was the lack of a commercially available, edible, soy protein isolate. Worthington spent considerable time attempting to stimulate basic companies to commercial production of edible protein isolate. Central Soya and Ralston Purina, especially, should be credited with contributions in this area. Others who contributed markedly to the advance of fabricated foods in the meat analog area were Archer Daniels-Midland and General Mills. Although Worthington had done some protein isolate work in connection with curd produced for the soy protein cheeses, Worthington was at that time financially unable to expand into commercial-scale production of such basic material, as economic feasibility was based upon extremely large production volumes.

Until recently, the majority of the U.S. population was totally unaware of the so-called vegetarian protein food products and the food industry gave them little more than passing attention. Two basic processing techniques which gave the greatest thrust to meat analog fabrication and structure were thermoplastic extrusion and wet spinning processing. The thermoplastic extrusion as first practiced at Worthington yielded a *solid* translucent rod-shaped extrusion of industrial soy protein isolate. This rod-like extrusion was approximately 1/8-in. in diameter and was cut at the die face into small discs approximately 1/16-in. thick. These were used to simulate the ground particle sizes of a minced-type meat in spaghetti sauces and various casseroles.

Later, patents were issued to Archer Daniels-Midland Company for extrusion processes and products which were "puffed extrusions." Here, the moistened proteinaceous materials in the form of a plastic mass at temperatures above 93.3°C were extruded through an orifice or die into a medium of lower pressure. They resulted in a porous protein-containing product of plexilaminar structure which had an open cell structure similar to a solid foam or sponge. It has been described in patent literature that the majority of cells have dimensions of greater average length than average width. The length of the cells is substantially aligned and substantially greater in length than in width. Extrusion products may vary greatly in their textural quality and structural integrity.

The pioneering efforts and expertise of Worthington developed along the lines of spun fiber textured basic raw materials and end products derived therefrom. Worthington was practicing the "Boyer" spinning technique long before the advent of edible soy protein isolates. The Drackett Company in Cincinnati, Ohio was producing an industrial grade soy protein isolate. Early experimental work was done with this protein in combination with other proteins which had been precipitated "in-house." Also, casein was used in early experiments to develop process technology. It formed an excellent spinning dope with no crude fiber to plug spinnerettes and the resultant tow was white in color and extremely bland in flavor.

The availability of spun protein monofilaments made it possible to fabricate much more sophisticated and complicated end products. The fibers could be dispersed, layered, crosshatched, or arranged in parallel orientation. Different patterns yielded different textural qualities in the finished product. Although there are other methods of fiber formation, wet spinning still is the most commonly-practiced means of producing monofilaments from a protein isolate. The process of wet spinning has been explained and illustrated in many instances in literature. It depends upon an aqueous alkaline solution being precipitated at its isoelectric point. Various chemicals and processing techniques are utilized to vary the desired texture in the finished monofilament.

There is a point which should be clarified regarding spun fiber textured meat analogs. Many consider such products to contain practically 100% spun soy protein fibers together with small amounts of color and flavor. However, these products may contain only 20–40% fibers. The operating principle is simply to add only that amount which is needed to contribute to optimum texture in the end product. There are many types and combinations of ingredients which are used in the so-called spun fiber products. Labels of these products clearly show the wide variations in the ingredients used. The raw materials are selected for their contribution to texture, flavor and nutritional value.

The end product may be comprised of combinations of the textural structures of many natural commodities, plus engineered or fabricated structural elements which are specifically designed, selected and formulated to achieve the end product texture desired. There may be involved the viscosity of liquids, the agglomerates of cell particles, the inclusion of fatty substances, the elasticity of gels, the tensile strength and extensibility of fibers, and the cellular voids of foams and sponges. The objective is to fabricate or compose these natural and man-made materials into a finished product simulating as closely as possible the desired natural counterpart.

Much research is still needed to determine the contribution of functionality and nutritional value of various components to the complex finished food product. The combination of materials may be synergistic or they may nullify the characteristics desired. Also, when an attempt is made to incorporate flavor, it is found that the electrolyte system markedly affects the textural qualities of the product. Consideration of this change or modification must be taken into account in the formulation or fabrication of an end product.

Presently, textured vegetable protein analogs are appearing in increasing numbers on the grocery shelves. These analogs are composed of natural substances which, by engineering and fabricating techniques, have a different texture, taste and appearance in an attempt to simulate a more traditional food. These products are of particular interest to the chemist because his knowledge of chemistry and processing techniques is what has made these foods possible. Presently, there is a vast array of texturized or structured raw materials primarily based upon soy. These are used per se or as extenders for meat. Finished product lines are now rapidly expanding. Breakfast-type items from General Foods, General Mills and Miles Laboratories are currently in the marketplace.

Presently, an enormous amount of effort is being expended in research and development in the fabricated food technology field. Government, university and industry laboratories are proceeding at a very rapid rate to multipy and expand present technology from the molecular level approach to final product formulations. This is clearly evident from the enormous number of patents that are being issued in the field almost daily. Anyone interested in fabricated foods or textured plant proteins or related products certainly should keep abreast of the current patent literature. This patent literature is the largest and most comprehensive collection of technological development in the world. Its review is certainly indicated at this time in the evolution of plant protein foods. An excellent review of this patent literature is given by M. Gutcho (1973) in a book entitled, *Textured Foods and Allied Products.* Also, the annual reviews by Hallie B. North which are titled, *Commerical Food Patents* provide a means of keeping abreast of the patent literature in this field; and the most current releases can be obtained by a check of the patent lists from *The Official Gazette.*

Plant protein foods! Where are they going from here? A recent article by J. M. Winski in the Wall Street Journal, March 25, 1976, was entitled, "By 2000—Prevention of Starvation May Be Chief Global Concern." It seems logical, therefore, to predict that the future for plant protein foods is certainly assured and presents many challenges. It cannot be ignored that the world population explosion accompanied by a skyrocketing need for food is, indeed, a reality. Many fail to realize that we are in a period of change from surpluses to shortages. It is estimated that by the year 2000, the present world population of 3.9 billion will have grown to nearly 7 billion and before the year 2050, it will double again to 14 billion. The future will see greater expansion and usage of America's three basic crops—corn, soybeans and wheat. As populations increase, there will be increasing competition for food between man and his domestic animals. We will see that currently-used animal feeds can and will be diverted to human use. To the extent that this can be accomplished, these protein resources will be extended from 5 to 10 times!

Plant protein foods will become increasingly important, for it should be remembered that photosynthesis is still the most effective and cheapest method of producing food. The importance of using solar energy as far as possible and as efficiently as possible is self-evident. Technically, it also seems obvious that legumes represent the most important source of future protein; their utilization must dominate any discussion of man's use of plant proteins. Such crops as soybean and alfalfa live symbiotically with nitrogen-fixing bacteria and efficiently produce protein with less depletion of nitrogen from the soil. Our basic agricultural seed products will remain chief potential plant protein sources. However, even forage crop plants may contribute a percentage of their protein to human usage. The conversion of plant material to animal products is inefficient, and it should be pointed out that the conversion of leaf protein into seeds is, theoretically, also a somewhat wasteful conversion. Leaf protein should not be discarded or forgotten as a potential protein source, for the primary and greatest source of protein is in the leaves of green plants. If this can be efficiently obtained and utilized, it will contribute much to better eating for the man of the future.

Increasing attention will be given to nutritional, nonnutritional and toxicological characteristics of fabricated foods. New methods of measurement of protein quality will be devised which will more accurately predict its value to the human diet. Growth rate in a selected strain of weanling rats is not sufficient to predict the effect upon longevity, vitality or functionality upon the separate organs of man. Amino acid supplementation, and vitamin and mineral fortification will become more prevalent, but will become more rigidly controlled with more specific guidelines. It will become apparent that a purely synthetic diet is not feasible. As food becomes more fabricated, long-term effect of a lack of bulk on the structures of the gut will be studied to a greater degree and the role of crude fiber in metabolism will become better known. Crude fiber or plant fibers may, of necessity, be incorporated in fabricated foods in the market-place.

Many new methods of texturizing will be developed and the flavor chemist will make great advances in duplicating meat flavors which are distinctive (beef, chicken, fish, etc.). Perhaps there will be a shift from the current approach of analysis and duplication of flavors to a better understanding of flavor perception and compatibility with plant proteins; also, compatibility with texture, shear, mouth feel, disappearance, etc., of the food in question.

Microbiological fermentation techniques will become more prevalent and more adaptable as processes for producing materials for fabricated foods. An excellent illustration of this is the recently issued patent of Solomons and Scammel, assigned to Rank Hovis, McDougall entitled, "Production of Edible Protein Substances." In this instance, a carbohydrate substrate is used to proliferate a nontoxic strain of the genus *fusarium*. This produces fungal protein in the form of mycelium. The impact of this approach is great in that it can use carbohydrate waste or any plant carbohydrate and convert it into proteinaceous material. Not only is it converted into a proteinaceous material, but the "spinning is already accomplished" in that it appears in the form of mycelium or fine thread-like portions similar to a spun protein monofilament.

In 1973, the U.S. Department of Agriculture published a marketing research report, *947*, entitled, "Synthetic and Substitutes for Agricultural Products—Projections for 1980." This report is required reading for anyone interested in the potential impact of fabricated food. An examination of some of the market penetration figures indicates that processed meat analogs will attain 16% or more of the market by 1980. This report will serve to emphasize the impact that engineered foods will have on our diet in the future. Perhaps the most startling marketing report was known as "The Stanford Institute Report," by Gentry and Connolly, which projected a 1.5–2.0 billion dollar volume for meat analogs by 1980. A Cornell University report by David L. Call indicated that meat extenders and meat analogs would probably reach 10% of all domestic meat consumption by 1985. Most all reports are in essential agreement and indicate that all of us will become more familiar with meat analogs in the near future.

Plant protein foods will be fabricated in many new and novel forms, and many new and different forms of raw materials will be available. One of the reasons that plant protein foods will become increasingly successful in the marketplace is because it is possible to formulate them to any nutritional pattern desired. Polyunsaturate to saturate fat ratios may be varied. Presently, many are in a ratio of 2:1. Calories may be varied, amino acid profiles may be varied, and vitamin and mineral fortification may be tailored to suit the particular need of the individual or the product in question. Fabricated foods present an opportunity to have increasing variety in the diet from a single food source. Actually, the entire food industry will probably give deliberate attention to broadening the base of ingredients which go into fabricated foods so as to create a variety of ingredients in single foods.

This use of a variety of food ingredients and nutrients is the reasonable approach to applied nutrition and always has been. It doesn't say there are unknown nutrients, but it does say that if there are any, then consuming a variety of foods minimizes the chance of not obtaining them. One of the reasons that plant protein foods will be successful is the flexibility in selection and combinations of raw materials to yield desirable nutritional values. I believe that the food industry will and should reorient its development and research work towards the methods that have been used with success in the chemical and petrochemical industries. The scientific and technical approaches, which led to an understanding of the nature of synthetic polymers and hence to the success of plastics and synthetic fiber, will be applied to a study of biopolymers and their structures. The plastics industry still is attempting to make changes of the monomers to make more sophisticated products with quite different properties. In this same way, the monomers with which we deal are inherently more variable and the difficulties are greater than those in the plastic field. But the success with such studies with food raw materials will occur and will have an enormous effect on the food of the future.

We will have an increased understanding of the micro and macro structure of the raw materials with which we deal, and will have a better understanding of destruction and reconstruction of these materials. We will have a better understanding of the separation and reconstitution of biologically variable raw material, and we will determine how to build structures and properties from the controlled assembly of subunits. There will be available materials of different properties rather than having all raw material sources yield a protein isolate with similar physical and chemical characteristics. It will be seen that it is desirable to have subunits with differing physical and chemical characteristics to make it possible to produce acceptable and saleable products whose properties are predictable and can be adjusted at will.

It can be said with certainty that plant protein foods are here to stay! They will survive! It remains with you, the chemist, and your scientific associates to accept the challenge of this new technology. Upon you, the scientists, rests the responsibility of how imposing the future results will be and how rapidly they will come. The variations in raw materials, functionality, new processes and finished product acceptability depends upon your ability to cope with the challenge posed. It is extremely important that we accept this challenge as another potentially large step forward in our continuing efforts to better feed the people of the world at lower net cost. We should not view these developments as a threat to any existing system. The tremendous food complex in this country was built upon the ability of products to withstand the competitive forces in the marketplace. We did not get where we are by ruling out or making illegal new forms of technology. I trust that the marketplace will continue to be the final judge of how well we, as scientists, relate to the challenge of this new technology. The field of plant protein technology has barely been scratched. It is my contention that plant protein foods and the products of microbiological fermentation techniques will provide the basis for our foods of the future.

Plant protein foods will be formulated in many new and different shapes, textures and flavors which will add interest and pleasure to eating. All facts tend to support the belief that plant protein foods will become an increasingly important segment of the U.S. and world food supply. The challenges still remain in the vast array of untried concepts, in the exploding world population, in the economics of basic agricultural products, in processes and products, in medical applications, and in the terrific technical possibilities and potentials for plant protein foods.

BIBLIOGRAPHY

ALTSCHUL, A. M. 1973. The revered legume. Nutr. Today 8, No. 2, 22.
ATKINSON, W. T. 1970. Meat-like protein food product. U.S. Pat. 3,448,1970, Jan. 6.
BOYER, R. A. 1954. High protein in food product and process for its preparation. U.S. Pat. 2,682,466, June 29.
GENTRY, R. E. and CONNOLLY, E. M. 1969. Fabricated foods. Stanford Res. Inst. Rep. 347.
GIDDEY, C. 1960. Artificial fibres. U.S. Pat. 2,947,644, August 2.
GIDDEY, C. 1960. Protein compositions and process of producing the same. U.S. Pat. 2,952,542, Sept. 13.
GUTCHO, M. 1973. Textured Foods and Allied Products. Noyes Data Corp., Park Ridge, N.J. and London, England.
HARTMAN, W. E. 1966. Method of making rehydrated gluten products. U.S. Pat. 3,290,152, Dec. 6.
HARTMAN, W. E. 1967. Vegetable base high protein food product. U.S. Pat. 3,320,070, May 16.
HARTMAN, W. E. 1971. Textured soy proteins. Proc. 3rd Int. Congr. Food Sci. Technol. Inst. Food Technologists, Chicago.
HORAN, F. E. 1974. Meat analogs. In New Protein Foods, Vol. 1A. Aaron Altschul (Editor). Academic Press, New York.
HORAN, F. E., WALKER, D. B. and BURKET, R. E. 1971. Engineered foods—the place for oilseed proteins. Food Technol. 25, 813.
INGLETT, G. E. 1975. Fabricated Foods. AVI Publishing Co., Westport, Conn.

LOCKMILLER, N. R. 1973. Increased utilization of protein in foods. Cereal Sci. Today *18*, No. 3, 77.
NORTH, H. B. 1969-75. Commercial Food Patents. *In* U.S. Annual Reports. Oliver S. North, 802 S. Ode St., Arlington, Virginia.
ROBINSON, R. F. What is the future of textured protein products? Food Technol. *26*, No. 5, 59.
SOLOMONS, G. L. and SCAMMELL, G. W. 1976. Production of edible protein substances. Assigned to Rank Hovis, McDougall, London, England. U.S. Pat. 3,937,654, Febr. 10.
THULIN, W. W. and KURAMOTO, S. 1971. Bontrae—a new meat-like ingredient for convenience foods. Food Technol. *25*, 813.
USDA. 1971. Textured vegetable protein products. Child Feeding Programs, USDA Food Nutr. Serv. Notice *219*.
USDA. 1973. Synthetics and substitutes for agricultural production—projections for 1980. USDA Market. Res. Rep. *947*.
ZIABICKI, A. 1967. Physical fundamentals of the fiber spinning process. Manmade Fibers, Vol. 1, H. F. Mark, S. M. Atlas and E. Cernia (Editors). Interscience Publishers, New York.

23

Food Preservation—
Retrospect and Prospect

Samuel A. Goldblith

I thought that I would begin my talk today on a statistical and philosophical note.

As to statistics, Table 23.1 presents some comparable data in the United States 1776 vs 1976 in terms of wealth, people and other important demographic factors.

Let us now look at another aspect. This relates to our beginnings in food processing, and, in particular, in thermal processing—man's first directed attempt to create a new method for food conservation. It was about 1796 that Nicholas Appert began his initial experiments on canning in France, a country whose friendship and alliance with the United States had much to do with the success of our becoming a sovereign state. At about the time of our Declaration of Independence, the Industrial Revolution began in Great Britain. Its initiation was due to the work of a galaxy of scientists—Cavendish, Faraday, Davy, Dalton, Watt, Newton and many others. All of these did their

TABLE 23.1

COMPARATIVE U.S. DEMOGRAPHIC FACTORS OF
1776 AND 1976

	1776	1976
Population	2.5 million	212 million
Gross National Product	$1.6 billion	$1.5 trillion
Newspapers	30	11,252
Cities (10,000 or over)	4	2,300
Agricultural workers	95 %	4%
Cost (average) of night's lodging	5¢	$18.00
Cost of gallon of milk	9¢	$1.50
Wholesale prices (1967=100)	27	171.2
Life expectancy	38 years	67 years
Wages of factory worker (average)	2.5¢	$4.45

SOURCE: Morgan Guaranty Trust Co., New York.

work outside of the universities of Oxford and Cambridge—two citadels of learning manned mainly by celibates of the holy orders who were busily engaged in peering intently into the grandeur of ancient Greece and the glory of Rome, without attempting to plant the basis for the future. It was only after Darwin, in about 1853, that these two distinguished universities became interested in science and have since contributed so much to man's welfare.

So the first century of the Industrial Revolution was due to the genius and work of the British. In terms of thermal processing, the British perfected the steel plate and canister. William Underwood, who came from England in 1819, established America's first pioneering food cannery in 1821. This has been devoted to thermal processing and has been in continuous existence since that time (Fig. 23.1).

A son of France invented the art of thermal processing, and a son of France, Louis Pasteur, discovered the scientific basis of thermal processing in 1861 to 1865. In April 1861, the General Court of Massachusetts approved the Charter of the Massachusetts Institute of Technology for the purpose of ". . . and aiding generally, by suitable means, the advancement, development and practical application of science in connection with arts, agriculture, manufactures, and commerce . . ." This Act and creation of the Massachusetts Institute of Technology with its strong liaison with industry was responsible, in part, for America leading the second century of the Industrial Revolution—a century that has witnessed remarkable advances in transportation, communications, health and food technology. At MIT, Samuel Cate Prescott and William Lyman Underwood worked together, using Pasteur's findings, to convert canning from an art based on experience to a technology based on science.

We are now about to enter the third century of America's sovereign

Courtesy of Wm. Underwood Co.

FIG. 23.1. DIORAMA OF FIRST UNDERWOOD CANNERY ON RUSSIAN WHARF, BOSTON, MASSACHUSETTS

existence and almost, coincidentally, the third century of the Industrial Revolution. Who will lead it and where will it go?

First, let us examine the past 200 years with respect to processing developments and their relation to societal conditions of that particular time.

ADVANCES IN FOOD PROCESSING
TECHNOLOGIES, 1776-1876

America, as can be seen readily in Table 23.1, was an agrarian nation. As the country grew and expanded, urbanization became a way of life. Expansion to the Mississippi River and beyond, to the West coast, ensued and a large food processing industry began to develop.

Canning expanded in volume due to the Civil War and the concomitant response to more production needs of tinned foods with the development of the calcium chloride bath in 1860-1861 by Isaac Solomon (Goldblith 1971, 1972). This enabled greater throughput in factories by being able to process at higher temperatures. The limiting factor then became the rate at which tin cans could be manufactured by hand craftsmanship.

Ice from frozen blocks taken from rivers frozen over in the winter provided the major sources of refrigeration and ice cream.

Perhaps the greatest single development in the era 1776-1876 was the discovery by Pasteur of microorganisms as the causative agents of food spoilage (Goldblith 1968). For a review of Pasteur's work on food preservation, see Goldblith (1968). Pasteur actually showed that if processed at temperatures 100° C, milk could be sterilized. Yet Pasteur's work on microorganisms as the causative agent of food spoilage was probably unknown to Gail Borden, who received a patent for the development of a process to can evaporated, tinned milk (Borden 1856).

The autoclave, although invented by Raymond Chevallier-Appert in 1853 (Goldblith 1972), only came into extensive use with the invention of a retort by Shriver in 1874 with steam provided by a separate boiler (Goldblith 1971, 1972).

Thus, in the century 1776-1876, the major developments of our extensive food processing armamentaria were yet to be developed. The biggest single achievement was that of Pasteur in developing the basic knowledge of spoilage of foods. The impetus to greater production of processed foods was growing urbanization and the Civil War.

ADVANCES IN FOOD PROCESSING
TECHNOLOGIES, 1876-1976

Pasteur's findings were practically unknown to most Americans. The result was that even though canneries were springing up all over America, they were organized around the cannery room presided over and run by a man who held mysterious secrets of the art of canning and vials of unknown material which enabled him to increase the throughput of the factory. *But* this did not work. The cannery superintendent merely increased the throughput without increasing the process temperature. It was Underwood and Prescott (Goldblith 1971, 1972)—a university scientist and a food processor—who were familiar with Pasteur's work (at least Prescott was initially) and teamed up to develop time-temperature requirements for different foods based on can sizes and the resistance of the spoilage microbes. This work done in 1895–1898 resulted in the basis of the modern canning industry.

Mechanical refrigeration was developed in 1874 by Linde (Woolrich 1968) and resulted in a cold-storage warehouse network, initially for cold-storage and later for frozen foods ("sharp" or slow freezing).

The great meat packers began about the time of the Civil War with the founding of Armour & Company in 1867 and Swift & Company in 1875. Gustavus Swift developed the first refrigerated railroad cars in 1877 (Fowler 1952).

The meat industry was flayed by Upton Sinclair in his famous book *The Jungle*, in 1906.

In 1906, Harvey Wiley pushed through his Pure Food Law—the foundation of our Food and Drug Administration today.

The use of dehydrated vegetables on a large scale and other dehydrated foods was probably a result of World War I (Prescott 1919; Prescott and Sweet 1919). These were air-dehydrated products, not always good—in fact, rarely so.

Frozen foods began to really develop when Clarence Birdseye developed his plate freezer and with it, quick-freezing of foods *in packages*. His first advertisement and sales were in Springfield, Massachusetts in 1930. However, it was not until after World War II that the frozen food industry began to grow at fantastic rates.

As a result of Pasteur's work, William T. Sedgwick cleaned up the water supply of the Merrimack River, using slow sand filters. Thus, typhoid fever so rampant in the 1880's was conquered.

The sanitary, clean milk crusade by William T. Sedgwick, his student, Samuel Cate Prescott, and Milton Rosenau of Harvard resulted in pasteurization of milk. This, together with cleanliness of production and refrigeration in ice boxes, resulted in markedly reducing milk-borne diseases such as typhoid fever, tuberculosis and enteric diseases.

World War II resulted in the development of boneless beef, clean plants and good quality dehydrated foods. The C and K rations in cans may well have been the forerunners of main courses in cans and in frozen foods.

With the invention of television, and with more women working as a result of World War II, demands for convenience in foods (reducing hours in the kitchen) began. T.V. dinners became popular, and these developed into the main courses and desserts of today.

Dehydration from air dryers to spray dryers to freeze dryers, was developed and today we have a whole armamentaria of processing technologies (Fig. 23.2) to produce good foods economically.

By 1976, we had developed a tremendous food industry and supplier industries of chemicals and ingredients as well as regulation and over-regulation. This century of 1876-1976 also witnessed the development of transportation and communication, both of which have had profound effects on the food industry. The corollary to this is that developments in food preservation and processing have made ships livable, aircraft meals more enjoyable—even feeding in space is nutritious—as well as providing gustatory delight.

Table 23.2 lists some developments in transportation from 1840 to the present. To these should be added the developments in shipping—refrigerated ships, containerization, etc. Transportation and societal changes have resulted in the food service industry growing to approximately 68 billion dollars per year in the United States.

TABLE 23.2

TRANSPORTATION DEVELOPMENTS, 1840-1976

1840	50,000 people traveled to West coast on covered wagons. Fifteen months to cross the continent.
1850	Stagecoach from Independence, Missouri to Sacramento, California in 25 days.
1860	Pony Express—St. Joseph, Missouri to Sacramento, California in 8 days.
1869	Completion of Transcontinental Railroad at Promentary, Utah.
1930	Wright brothers flew first airplane.
1909	Development of Model T car by Henry Ford.
1934	Development of DC3—approximately 180 mph.
1952	Development of DC7—approximately 300 mph.
1959	Commercial jets—approximately 550 mph.
1976	Supersonic jet travel—1500+ mph.

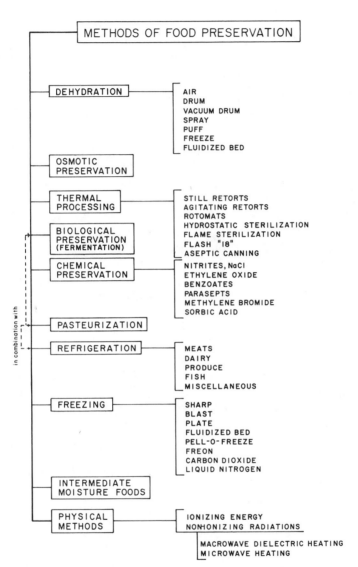

FIG. 23.2. ARMAMENTARIA OF FOOD PROCESSING TECH-
NIQUES AVAILABLE TODAY

I am sure that in this attempt to be brief and to highlight the developments of food processing methods, I have become obscure or skipped a number of developments which should have been expanded upon, for which I apologize.

THE FUTURE?

The population of the world has grown tremendously and so has America's capacity to grow food. With only 4% of the population on the farm, we have become highly mechanized and dependent on petrochemicals and agricultural equipment, and the achievements in plant and animal genetics. If we are to continue to lead the world in food production and processing, and keep costs at a minimum, let us recognize several factors in processing that are certain to become more important in the next decades.

Use of Radio-frequency Energy to Process Foods

This is now being used to reduce and level out moisture in biscuits, to defrost meat blocks, and to cook hamburgers prior to freezing for institutional use.

When control of energy can be achieved, when proper plastic or nonmetal containers are developed, sterilization using radio-frequency energy will become a reality.

Computer Technology

The development of the computer will offer many advantages to the processor. It will provide:

1. Optimization of process control.
2. Real inventory control and reduction of inventory costs by predicting needs more accurately.
3. Better engineering information for new process developments.
4. Better instrumentation.
5. A data bank for all types of information such as process parameters used in each batch, nutritional labeling, etc.
6. Instant retrieval of information for recall if needed.
7. Minimize losses in processing.
8. Reduce labor costs.
9. Better and more uniform products.

Spoilage Losses

I can see the need for reducing spoilage losses, and hopefully methods will be developed to achieve this in the field, on the farm, and in the factory.

Thermally Processed Pouch-Packed Foods

This is bound to happen in the next few years. Its success may be assured by demographic trends of smaller families.

Development of Equipment and Processes to Reduce Needs for Food Additives

While I, for one, do not agree that the food additives situation is nearly as bad as the vocal minority—consumer advocates—claims, I feel that the use of food additives can be cut down by design of alternative types of equipment and processes.

Productivity

This is an area where much more needs to be done to reduce food costs. We must increase productivity at all levels.

Continuous Automatically Controlled Factory

I have alluded to this earlier. I believe that the computer can make this possible with a minimum of labor; and this will result in more uniform products.

Soybean Technology Improvements

Developments in the knowledge of the basic chemistry of the soybean will lead to a host of thermally processed products with soybeans which will extend our protein supply.

Energy Conservation

We will see more efficient use of energy with wasted low-temperature Btu's being recycled and reused in processing and in heating of plants. I believe the next 50 years will witness the harnassing of solar energy to be used, in part at least, for food processing.

Utilization of Agricultural and Processing Wastes

I believe, now that we have finally learned that nature is not immortal, we will learn how to utilize wastes, process them, remove pesticides, heavy metals, etc., and convert them into animal feeds. In other words, we may learn how to make our factories into "ruminants," thus improving our environment.

Growth of Institutional Products

This is a societal trend that is growing, i.e., eating away from home. It will demand better products at cheaper prices in hospitals and restaurants. These are more achievable in the food factory than at the user establishment. Here there is room for all technologies.

Vertical Integration and Substitute Formulation

To achieve uniformly better quality and regular supply of raw materials, I foresee many more companies vertically integrating in some areas. To meet the rising costs of certain raw ingredients, the development of substitute formulations are a sine qua non.

FIG. 23.3. INTERRELATIONSHIP
AMONG THE CONSUMER, THE EN-
VIRONMENT, FOOD PRODUCTION,
MARKETING, AND RESEARCH

SUMMARY

I hope I have not set too many angels dancing on the head of a pin by my prognostications, but I feel the items I have listed are realizable in the lifetime of many in this room.

Food processing cannot be treated in extenso, it is part of the "troika" of marketing, research and production, all of which relate to the consumer and his environment (Fig. 23.3).

The consumers of tomorrow will demand convenience, quality, safety, economy and gustatory delight in their food supply. This will be achieved if we utilize the trinity of marketing, research and processing technology and produce products the consumer wants, products the supermarket will stock, and products which have gustatory delight.

BIBLIOGRAPHY

BORDEN, G. 1856. Process and improvement for the concentration and preservation of milk. U.S. Pat. 15,553.

FOWLER, B. B. 1952. Men, Meat and Miracles. Julian Meesner, New York.

GOLDBLITH, S. A. 1968. From Appert to Pasteur—A retrospect. Proc. Inst. Food Sci. Technol. *1*, No. 3, 7.

GOLDBLITH, S. A. 1972. Controversy over the autoclave. Food Technol. *26*, No. 12, 62–65.

GOLDBLITH, S. A. 1971, 1972. A condensed history of the science and technology of thermal processing. Part 1. Food Technol. *25*, No. 12, 44–46, 48–50, 1971. Part 2. Food Technol. *26*, No. 1, 64–69, 1972.

PRESCOTT, S. C. 1919. Relation of dehydration to agriculture. U.S. Dep. Agric. Circ. *126*, Jan. 25.

PRESCOTT, S. C., and SWEET, L. D. 1919. Commercial Dehydration: A Factor in the Solution of the International Food Problem. Annals, Am. Acad. Political Social Sci. Publ. *1294*, May.

WOOLRICH, W. R. 1968. The history of refrigeration, ice manufacture and cold storage. In: The Freezing Preservation of Foods, 4th Edition, Vol., 1, Refrigeration and Equipment. D. K. Tressler, W. B. Van Arsdel and M. J. Copley (Editors). AVI Publishing Co., Westport, Conn.

24

Food Acceptance

Stanley J. Kazeniac

The most important function of a food is its nutritive value. However, the safest, most nutritious and wholesome food will not necessarily be consumed unless "it tastes good." The decisive sensory properties in the selection, acceptance and ingestion of a food are: (1) texture and feel; (2) sight, appearance and color; and (3) flavor, including taste, odor or aroma and mouth satisfaction. Of these properties, flavor is the most important. Food flavors may be divided into these general classes: (1) natural flavors; (2) naturally processed flavors; and (3) artificial or imitation flavors. Many variations of these are possible. Natural flavors exist as such in the food and include direct metabolites that are produced in animal or vegetable organisms by extracellular biogenetic pathways. These are dependent on genetic characteristics, ripening, and in some cases on aging. Naturally processed flavors are produced either by extracellular biogenetic pathways (biochemically via enzymes) or by chemical reactions. Amino acids/proteins, carbohydrates, lipids, terpenes and even some vitamins are important flavor precursors. Artificial or imitation flavors may be synthetic compounds, either identical to or different from those in natural flavors, or mixtures of synthetic compounds and compounds from naturally-occurring foods.

Varying degrees of progress have been attained in improving these sensory properties during the last 100 years. Very good emulsifiers, surfactants, special starches, and plant gums as well as other ingredients, have been developed to provide processed foods with good texture, viscosity and other functional properties. Undoubtedly, better ingredients will be developed to improve the texture of the rapidly growing textured foods. Time does not permit a review of important developments in this area, though texture and feel can influence flavor.

Color is important to the appearance of a food and in many cases can affect its flavor acceptance. Many natural pigments such as the anthocyanins and chlorophylls have poor stabilities. Good progress has been made in developing carotenoid pigments but these have limited color ranges. Some artificial colorants are encountering difficulties in conforming to the food safety regulations which in some aspects may be unreasonable because of the testing procedures. Therefore, a review of the current status and the future outlook on food colorants have been included in this symposium.

Not only is food preservation against microorganisms necessary to ensure food safety, but the type of treatment used can also affect the flavor and color of a food. Degradation reactions of carbohydrates, amino acids and lipids, and interactions between carbonyl compounds and amino acids and proteins, can produce either desirable or objectionable flavors. Because these reactions are important to food flavors, existing methods of food preservation as well as future prospective processes will be reviewed.

Unquestionably, food safety is the paramount factor in the use of a food, and it must get first priority. Today, there is more concern about food safety than ever before, even though our foods are now safer than they have ever been (Morrison 1976). In a symposium of "Flavors—Safety and Regulations" (Wendt 1976), presented at a meeting of the Flavor Chemists Society, E. P. Grisanti, President-elect of the Flavor and Extract Manufacturers Association, stated, "Our food supply—thanks to this country's food industry and the flavour industry—is the best, the most nourishing and, yes, the tastiest in the world. It is also the safest. The public should know this, and be assured of the steps being taken to keep it that way." Commissioner A. M. Schmidt of the Food and Drug Administration is quoted as saying that the truth is that our food supply is safer now than it ever has been (Beers 1976). At some point the consumer must decide what risk he will assume to achieve a certain benefit. Food scientists must supply the information required to make such assessments. In this way, the food safety problem can be put in proper perspective and rational regulatory policies can be adopted (Beers 1976).

Without question, outstanding progress has been made in the flavor chemistry of foods. Greater advances have been achieved during the past two decades than in all previous history. Today, completely synthetic fruit flavors can be created with either fresh or cooked notes. Many pathways elucidating the flavor biogenesis of fruits and vegetables have been worked out. The importance of enzymes in the development of flavors in many fruits and vegetables and the effects of processing conditions have been established. Important flavor compounds present in foods at concentrations less than 0.1 ppm have been identified. Some examples of these are the alkylpyrazines in potatoes; 3-isobutyl-2-methoxypyrazine in green bell pepper; 2-methoxy-3-sec-butylpyrazine in carrots; and nootkatone in grapefruit. These results are encouraging, because they indicate that flavors can be modified and controlled naturally through processing and genetics.

Though considerable progress has been made, the complex nature of meat, poultry and fish flavors still remains to be solved. Only in the last few decades have the flavor enhancing and synergistic effects of 5'-ribonucleotides, such as inosinic acid (the latter was identified as a compound of beef extract in 1913) and monosodium glutamate been established. Though many good imitation meat-like flavorings have been developed, reproduction of the true delicate flavors of meats still requires more research.

Though over 500 compounds have now been identified in the volatiles of coffee, it has not yet been possible to formulate an imitation flavoring that has the true coffee notes, because of the extreme complexity of the composition of the total mixture of volatiles. Similarly, the flavors of cocoa, chocolate and tea have been difficult to define as will be evident from the review of research on these beverages (Chap. 28).

Undoubtedly, the breakthroughs in flavor chemistry were made possible with the elaborate techniques developed with the sophisticated instrumentation that became available. The newer improved analytical tools such as gas-liquid chromatography, liquid-liquid chromatography, thin layer chromatography, mass spectrometry, nuclear magnetic resonance spectrometry and computers to handle the data provided tremendous capabilities to identify compounds at concentrations below the 0.1 ppm range. This rapidly growing methodology shows great future promise for objective measurement of flavor.

The principal goal and result of flavor research should be improved nutrition by making nutritious foods more palatable. The consumer should benefit not only from improved nutrition but also at a relatively lower cost, since low cost nutritious foods become more acceptable to the consumer due to their improved flavor.

It is impossible to review all the research that has been done on food flavors. Just about every food has been examined to some degree, and many compounds have been identified. The topics indicated above illustrate some of the outstanding contributions made in flavor research during the last 100 years.

BIBLIOGRAPHY

BEERS, W. O. 1976. Openly confronting public anxieties over food safety. Food Prod. Dev. *10*, No. 6, 30, 33–34.
MORRISON, A. B. 1976. Food safety in the seventies. J. Milk Food Technol. *39*, 218–224.
WENDT, A. S. 1976. Flavour safety—fact or fantasy? Int. Flavours Food Additives *7*, 96–101.

25

Natural Food Colorants

F. J. Francis

It is generally conceded that consumers do prefer food with an attractive appearance. The physical factors, which contribute to an attractive appearance involve color, particle size, shape, etc. It may well be that if the appearance is not appealing, a consumer may never get to judge the other two major quality factors, namely flavor and texture. However, regardless of the weighting one may give to color, flavor and texture, it is obvious that consumers prefer food which has a desirable color. Much research has been devoted to maintaining the attractive color normally present in some foods. Much research has been devoted to the development of artificial colors to improve the appearance of formulated foods, or to use a term coined by Stephanie Crocca "civilized foods."

Artificial colors have had a rather difficult history because the list of permitted colors has been continually changing in the past 75 years (Fig. 25.1). This is due in part to refinements in the science of toxicology. The list is also growing shorter because of consumer demands for an ever-lowering risk/benefit ratio in all foods. The consuming public is willing to accept a well-documented risk/benefit ratio in, say cigarettes, but apparently will not accept any risk factor in food colors. Presumably this is because one can voluntarily stop smoking, but one can hardly stop eating. Whatever the consumer rationale may be, there is considerable public pressure to switch from artificial colorants to natural colorants. On the premise that humans have had thousands of years to develop physiological mechanisms to handle natural colorants, this hopefully will lower the risk/benefit ratio. There is little evidence to suggest that this is indeed true.

Natural colorants are usually confined to the yellow, orange, red and blue colors since no greens have been provided. Chlorophyll is an exception but it has never been used seriously as a food colorant apart from foods in which it occurs naturally. The yellow colors can be obtained from two groups of naturally occurring pigments: the

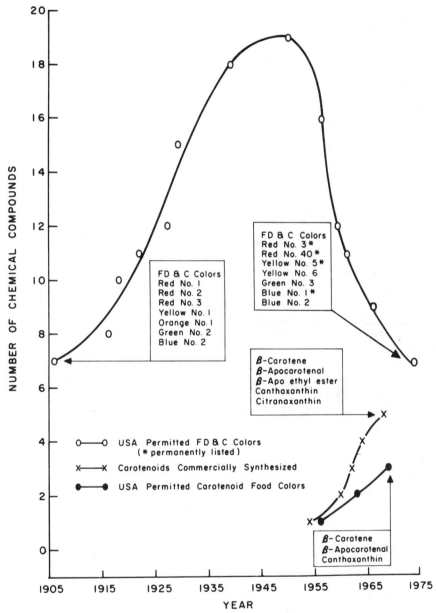

Courtesy of J. C. Bauernfeind, Hoffman LaRoche, Inc.

FIG. 25.1. PERMITTED FOOD COLORS IN THE UNITED STATES[1]

In addition to the seven FD&C colors permitted in 1976, three others are permitted in restricted amounts (Red #4[1] for maraschino cherries, 150 ppm; Citrus Red #2 for skins of oranges, 2 ppm; Orange B for casings and surfaces of frankfurters and sausages, 150 ppm.

[1] Since this paper was written, Red #4 and carbon black also have been banned.

261

carotenoids and the betazanthins. Orange colors can be obtained from three groups: carotenoids, betazanthins, and anthocyannis. The reds are confined to the carotenoids and the anthocyanins. Blue colors can only be obtained from anthocyanins.

CAROTENOIDS

Structure

The carotenoids are a group comprising about 300 compounds which occur very widespread in the plant kingdom (Isler 1971). They are of interest as food colorants since they are commercially available as natural colorants and as synthetic pigments. The general formula for a carotene is shown in Fig. 25.2, which shows the relationship between beta-carotene, the most widespread carotenoid, and lycopene, the major pigment in red tomatoes. All carotenoids are made up of isoprene units and differ only by the length of the middle chain and substitution on each end. The color is derived from the number of resonating double bonds in the molecule.

Hoffman-LaRoche has studied the potential of carotenoids as food colorants and are marketing synthetic beta-carotene, beta-8-carotenal, which occurs naturally as a minor pigment in many plants, and canthazanthin, which occurs in mushrooms, algae, some shellfish as well as in the feathers of flamingoes (Fig. 25.3A). Beta-

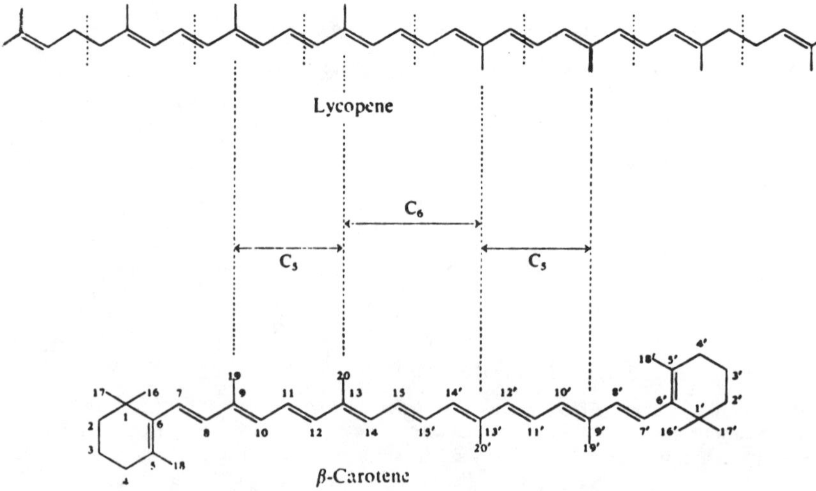

FIG. 25.2. STRUCTURE OF BETA-CAROTENE AND LYCOPENE

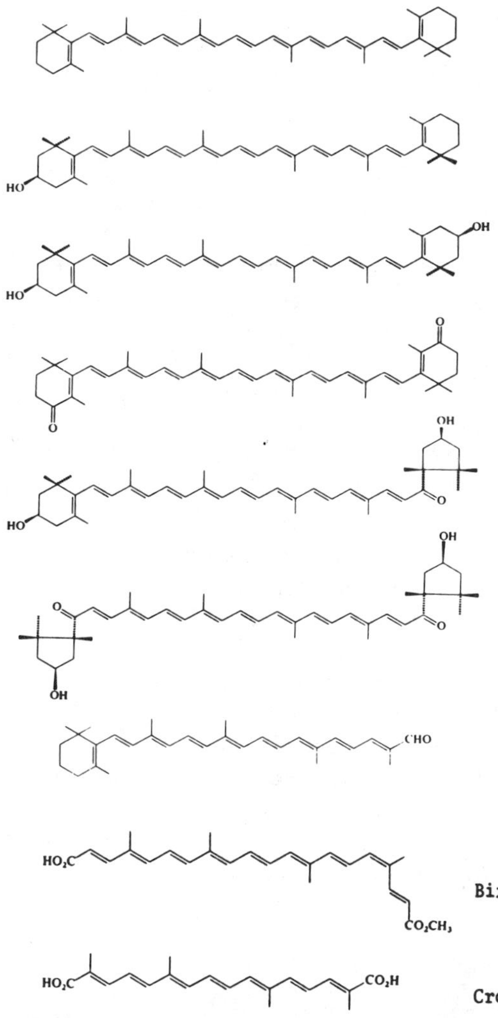

1 (A)
2
3
4
5
6
7
Bixin (B)
Crocetin

FIG. 25.3A. STRUCTURE OF SOME COMMON CAROTENOIDS: (TOP TO BOTTOM) BETA-CAROTENE, CRYPTOZANTHIN, ZEAZANTHIN, CANTHAZANTHIN, CAPSANTHIN, CAPSORUBIN, BETA-APO-8-CAROTENAL. (B) STRUCTURE OF BIXIN AND CROCETIN

carotene will supply yellow to orange colors, whereas beta-8-carotenal is orange-red and canthazanthin is red. All three are oil-soluble but are available in very finely dispersed particles which can be used in aqueous foods. They provide a spectrum of yellow to red colors. Canthazanthin is not quite as rosy red as would be desired for a raspberry shade but this could easily be corrected by adding a small quantity of Violet No. 1. However, the FDA recently barred the use of Violet No. 1 so this alternative is not available. Blue No. 2 is still permitted but the results are less satisfactory.

Sources

Carotenoids are present in many foods and contribute to the pleasing appearance of the food itself. They also have been used as sources of food colorants for addition to other foods. For example, lycopene in tomato paste or powder is often added to other foods both for color and flavor. Capsanthin and capsorubin (Fig. 25.3A) are the major pigments in red peppers (paprika) and form welcome additions to food for both color and flavor. Bixin (Fig. 25.3B) is the major colorant in annatto, a well-known food colorant obtained from the seeds of *Bixa orellana*. Crocetin and its digentiobioside, crocin (Fig. 25.3B), is the major colorant in saffron obtained from the petals of *Crocus sativus*. Carrot oil contains large quantities of beta-carotene and is used as a colorant. Many unrefined plant oils, such as palm oil contain large quantities of beta-carotene and are used as food colorants but obviously this is not their primary use. Green leaves contain beta-carotene, lutein, zeazanthin and cryptozanthin as major pigments (Fig. 25.3A). Lutein has the same formula as zeazanthin except that the former has a 5',6' double bond instead of a 4',5' bond. Crude carotenoid pigments are being prepared commercially from alfalfa meal and marketed as a poultry feed (The Pro-Xan process). It would seem to be a short step to make more highly purified products available for food colorants particularly with the current research emphasis on protein from leaves for human food.

Stability

The carotenoids in general are fairly stable as food colorants. They are stable to heat during food processing or storage. They are insensitive to pH changes over the ranges normally found in foods (pH 2-7). They are stable in the presence of vitamin C. They are susceptible to oxidation which can be accelerated in the presence of light, metals, peroxides and lipid-oxidizing enzymes. They interact with lipid components via the free radical mechanism to either promote or delay the onset of rancidity depending on the amount of lipid material and the degree of unsaturation. The susceptibility to oxidation can be a problem with powdered dehydrated products unless packed in vacuum or an inert atmosphere. However, the susceptibility to oxidation may be overcome by appropriate technology (Bauernfeind 1975).

For the above reasons, carotenoids are very successful as food colorants. In terms of natural replacements for synthetic colors, they are the most advanced. This statement ignores the fact that the three commercially available carotenoids are synthetic since they are identical in structure to naturally occurring compounds. One disadvantage of the synthetic carotenoids is that when compared with other red and yellow synthetic colorants, they are relatively expensive. One advantage is that they have a high tinctorial power.

Applications

Beta-carotene is used widely in its commercially available forms (liquid suspensions, semisolid suspensions, dry beadlets, gels and liquid emulsions). It has been used for margarine, cheese, shortening, ice cream, macaroni, breadings, vegetable oils, salad dressings, toppings, eggs, baked goods, cake mixes, soups, desserts and beverages. Beta-8-carotenal is used in the same types of products where a more reddish shade is desired. Canthazanthin can also be used in the same type of product to produce a redder shade. It also has application for frankfurters, simulated meats, gelatin desserts and beverages.

ANTHOCYANINS

Structure

Anthocyanins are a group comprising about 175 different compounds, which are very widespread in nature. They form the attractive red to blue colors in many fruits, vegetables and leaves. In terms of volume, the largest single contribution to the human diet is red wine. All anthocyanins have the flavylium cation basic structure shown in Fig. 25.4. The differences between the various pigments involve the substitution of hydroxy or methoxy groups at the $3'$, $4'$, $5'$, $3,5,7$ positions. The pigments may have sugar residues (glucose, rhamnose, galactose, xylose and arabinose in order of de-

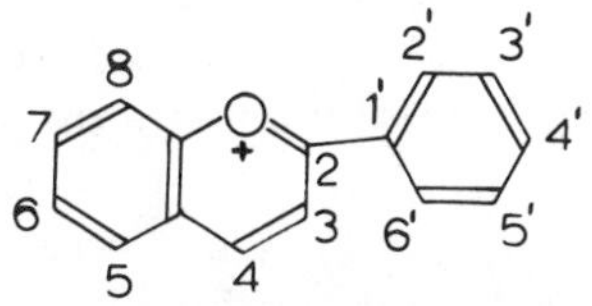

FIG. 25.4. THE FLAVYLIUM STRUCTURE COMMON TO ALL ANTHOCYANINS

creasing abundance), substituted on the 3, 5 and rarely in the 7 position. The sugars may also have acyl groups such as coumaric, ferulic, caffeic or acetic acids substituted on one or more of their hydroxy groups. The sugars and the acyl group contribute very little to the color of the anthocyanin molecule; the color is due to the resonance in the two ring structure. Substitution in the ring structure does influence the color (Table 25.1). Pelargonidin is orange in color, cyanidin is orange-red, delphinidin is blue, peonidin is red, petunidin is bluish-red and malvidin is reddish blue. The six compounds in Table 25.1 are called "anthocyanidins" since they do not have sugar substitution. With the addition of one or more sugar molecules, they become anthocyanins. Anthocyanidins are less stable than anthocyanins and are rarely found in nature in the free state.

The anthocyanins are usually accompanied in nature by yellow to colorless flavonoids. This group comprising about 2000 compounds in 15 classes is very widely distributed in the plant kingdom (Harborne *et al.* 1975). They are of little importance as colorants themselves but since crude anthocyanin preparations usually contain flavonoids, they modify the color slightly towards the yellow hues.

Sources

Anthocyanins are very widespread in nature but only a few sources have practical applications as potential food colorants. Several preparations from wine grapes *(Vitis vinifera)* are commercially available. They are usually marketed as concentrated solutions or powders derived from grape skins as a by-product of the wine industry. Philip (1974) published a method for recovering pigments from grape wastes. It involved a tartaric acid/methanol extraction followed by precipitation of excess tartaric acid as potassium hydrogen tartrate. Some commercial preparations (e.g., Enocyanin or Enocianine) are prepared by aqueous acid leaching with or without sulphur dioxide, or even by secondary fermentation on the skins (Garoglio 1965). Grape skins contain the monoglucosides of the six anthocyanidins in Table 25.1 in quantities varying with the type of grape. Consequently the mixture is complex to start with and usually contains an appreciable proportion of degradation compounds and other impurities. Most preparations currently being imported into the United States are considerably degraded and are difficult to define chemically. The degradation polymers are usually blue in color hence this type of product produces a reddish-blue colorant.

The Concord variety grape *(Vitis labrusca)*, which is produced in the United States in large quantities for juice, jam, jelly, etc., may be another source of grape pigments. The juice is usually collected from

TABLE 25.1

SUBSTITUTION ON THE FLAVYLIUM CATION STRUCTURE
TO PRODUCE THE MAJOR ANTHOCYANIDINS

	Substituent on Carbon Number[1]		
	$3'$	$4'$	$5'$
Pelargonidin (II)	H	OH	H
Cyanidin (III)	OH	OH	OH
Delphinidin (IV)	OMe	OH	H
Peonidin (V)	OMe	OH	OH
Petunidin (VI)	OMe	OH	OMe
Malvidin (VII)			

[1] Cpds (II) → (VII) have OH groups on carbons 3, 5 and 7 and hydrogens on all other carbon atoms.

mechanically harvested grapes and stored in bulk in large tanks. The "lees" (potassium acid tartrate and related compounds) precipitate from the juice during storage and carry considerable pigment down with them. This crude material remaining in the tanks after removal of the juice is a good source of tartrates and pigments. The pigment mixture is very complex, comprising the mono- and diglucoside derivatives of cyanidin, peonidin, malvidin, petunidin and delphinidin plus a number of acyl derivatives (Ingalsbe *et al.* 1963A). The pigment recovered from the precipitate can be fractionated into a red component containing essentially pure anthocyanins and a blue component containing polymers derived from the pigments and other components of the mixture. The chemical constitution of the blue components is not well understood. Preparations from both the red and blue components are suitable as beverage colorants (Ingalsbe *et al.* 1963B).

The two sources of grape pigments described above are both by-products. As such they should be less expensive than plant products grown for pigment content only. Yet, it may be economically feasible to grow varieties of grapes selected for high pigment content as a source of pigment only.

Roselle (*Hibiscus sabdariffa* L.) has been suggested as a source of pigment and the University of West Indies in Trinidad has grown and processed several acres of Roselle. Roselle is a tropical plant grown in many tropical areas as a colorant and a flavorant for a refreshing drink (Esselen and Sammy 1973). The calyces of this annual plant contain as much as 1.5% pigment. The pigment can easily be extracted by leaching with water to form a solution with few impurities other than citric acid. Apparently, little clean-up is required to produce a liquid concentrate or spray-dried powder. The major pigments in Roselle are delphinidin-3-sambubioside and cyanidin-3-sambubioside (Du and Francis 1973) which give a very desirable red shade associated with a raspberry color. If this plant is put into production for pigment purposes, it will probably require some horticultural research and FDA approval for the pigment concentrates.

Miracle fruit *(Synsepalum dulcificum)* has a red skin containing cyanidin-3-galactoside and cyanidin-3-glucoside as the major pigments (Buckmire and Francis 1976). These compounds would produce a red-orange color in beverages at pH 3.0 (Chiriboga and Francis 1970). Miracle fruit was being considered as a potential sweetener and since the pigments are present in low concentration (about 18 mg/100 g), its success as a colorant will depend on its success as a sweetener. The FDA recently rejected an application to market the compounds in miracle fruit as a taste modifier.

An anthocyanin recovery system from cranberry press cake was developed by Chiriboga and Francis (1970). It is rather complicated involving acid-alcohol extraction of the press cake, transference to a water solution for passage through an ion-exchange column, elution from the column with ethanol, and drying. The procedure is necessary because the pigments are difficult to extract and require purification before they can be added back to cranberry juice in appreciable quantities. The major pigments of cranberries, present in amounts up to 75 mg/100 g are the monogalactosides and arabinosides of peonidin and cyanidin (Zapsalis and Francis 1965). The pigment mixture produces an attractive pinkish-red color in beverages (Chiriboga and Francis 1973). Obviously the availability of cranberry anthocyanins depends on the availability of press cake from the production of cranberry juice cocktail since it is probably uneconomic to grow cranberries for pigment content alone.

Many plants contain high concentrations of anthocyanins. One example is the berries of *Viburnum dentatum* which contain about 1% of their fresh weight as pigment (Du and Francis 1975). The pigment mixture is very complex with cyanidin-3-glucoside, cyanidin-3-sambubioside and cyanidin-3-vicianoside as the major pigments. Unfortunately most members of the Viburnum family contain other phenolic bitter components and considerable "clean-up" is necessary. Other sources of anthocyanins such as strawberries, raspberries, blackberries, roses, apples, currents, blueberries, etc., probably command too high a price in their own right to be used as colorants.

Stability

Anthocyanins are not particularly stable in solution in food products. They are particularly susceptible to pH changes. The

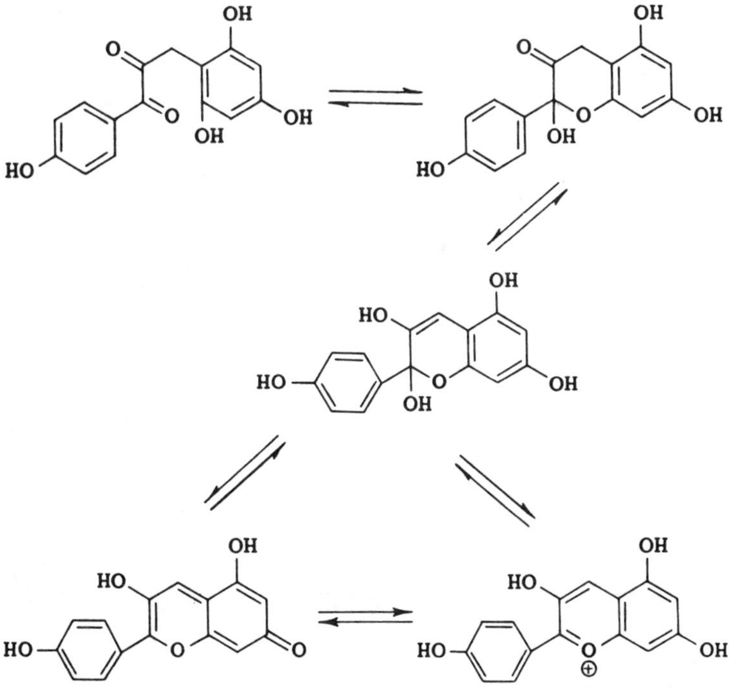

From Chichester (1972)

FIG. 25.5. EFFECT OF pH ON THE STRUCTURE OF PELARGONIDIN

At pH 1, the pigment exists primarily in the form at the upper left. At pH 3, the pigment is primarily in the quinoidal form (upper right). At pH 4, the carbinal base (middle) predominates. The bottom structures exist primarily at pH values above 4.

double ringed flavylium structure (Fig. 25.4) is very reactive chemically and the degree of ionization changes with pH (Fig. 25.5). At pH values of 1 or lower, the compound is almost completely unionized and exhibits the most intense color. At pH values of 4.5 and above they are almost colorless (Fig. 25.6, 25.7). This will obviously restrict their use to foods with pH values of 3.5 or lower. Anthocyanins are susceptible to degradation by hydrolysis, oxidation, (Fig. 25.8), reaction with vitamin C (Fig. 25.9), ring opening and subsequent polymerization. The chemistry of polymer formation is not well understood in spite of the fact that it has been a major problem with, for example, red wines and jams for centuries. Loss of pigment in foods is fairly rapid. Figure 25.10 shows overall loss of anthocyanins in cranberry juice cocktail as well as individual rates of degradation. The arabinosides degrade faster than the galactosides, but the difference is probably too small to be of practical importance. Robinson *et al.* (1966) reported that the diglucosides were

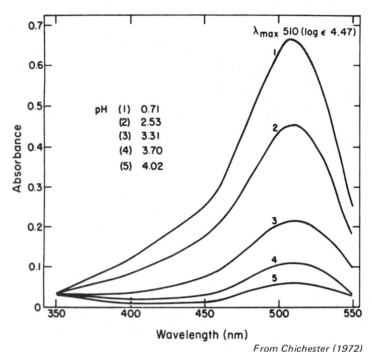

From Chichester (1972)

FIG. 25.6. ABSORPTION SPECTRA OF CYANIDIN-3-RHAMNOGLUCO-SIDE IN BUFFER SOLUTIONS AT pH 0.71–4.02. PIGMENT CONCENTRATION: 1.6×10^{-2} PER LITER

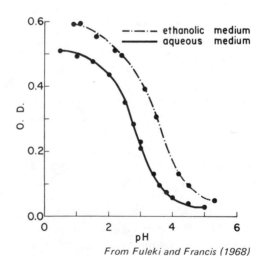

From Fuleki and Francis (1968)

FIG. 25.7. A PLOT OF ABSORBANCE VERSUS pH FOR CRANBERRY ANTHOCYANINS

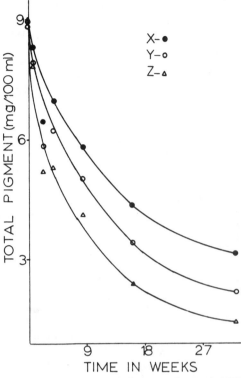

From Starr and Francis (1968)

FIG. 25.8. STABILITY OF ANTHOCYANINS IN CRANBERRY JUICE COCKTAIL WITH IN-CREASING OXYGEN IN THE HEADSPACE

X, Y, Z refers to 0.0, 0.3 and 2.0 ml O_2 in head-space of pint bottles.

less stable in wines than the monoglucosides. This difference could be important in wines since the sugars hydrolyzed from the pigments could contribute to increased browning reaction. The differences are probably not very important from a strictly colorant point of view.

The stability problems described above are probably general reactions for anthocyanins. This is shown by research on grapes by Palamidis and Markakis (1975), Van Buren et al. (1968) and Hrazdina et al. (1970); on strawberries by Sondheimer and Kertesz (1945) and Meschter (1953); and on cranberries by Francis and Servadio (1963), Starr and Francis (1973) and Chiriboga and Francis (1973). The rapid loss of anthocyanins on storage and their heat

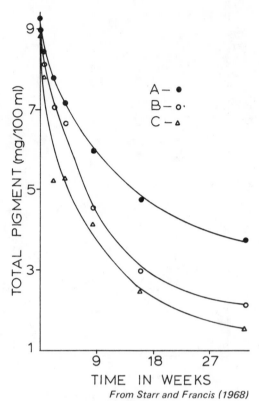

TIME IN WEEKS

From Starr and Francis (1968)

FIG. 25.9. STABILITY OF ANTHOCYANINS
IN CRANBERRY JUICE COCKTAIL WITH
INCREASING ASCORBIC ACID CONTENT

A, B, C refers to 40, 78, and 177 A.A. per ml
juice.

lability on thermal processing is not as important as one might think since some of the degradation products are brownish-red. For example, strawberry jam with zero content of anthocyanins (after 6 months' storage at room temperature) still retains a reasonably attractive color. So does red wine!

Applications

The pH sensitivity of anthocyanins will probably restrict them to foods with pH values of 3.5 or less. This means carbonated beverages, fruit drinks, toppings, jams, jellies, sauces, some ice creams, gelatin desserts, etc. Carbonated beverages are probably the biggest market.

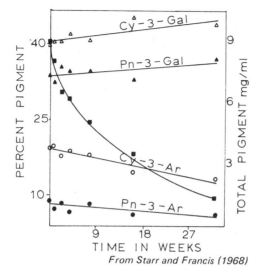

From Starr and Francis (1968)

FIG. 25.10. STABILITY OF INDIVIDUAL ANTHOCYANINS IN CRANBERRY JUICE COCKTAIL. THE LINE WITH SOLID SQUARES REPRESENTS THE TOTAL PIGMENT CONTENT

BETALAINS

Structure

The betalains are a group comprising about 70 compounds with the basic structure shown in Fig. 25.11 in which R and R^1 may be hydrogen or an aromatic substituent. The color is attributed to the resonating structures shown in Fig. 25.12 (Mabry and Dreiding 1968). If R or R^1 does not extend the resonance, the compound is yellow and is called a betazanthin. If R or R^1 does extend the resonance, the compound is red and is called a betacyanin. The major pigment in beets is betanin (Fig. 25.13) with the sugar-free portion (the aglycone) called betanidin. The only glycosides found in betalains contain glucose or glucuronic acid but the pigments may have acyl components in a manner similar to anthocyanins but possibly more complicated. Malonic, ferulic, p-coumaric sinapic, caffeic, and 3 hydroxy-3-methyl glutaric acids have been found attached to the sugar portions of the betalain molecule.

Betalains and anthocyanins, being both red, were confused in the literature in earlier years but they are actually chemically different. Their spectra are different and their electrophoretic behaviour is

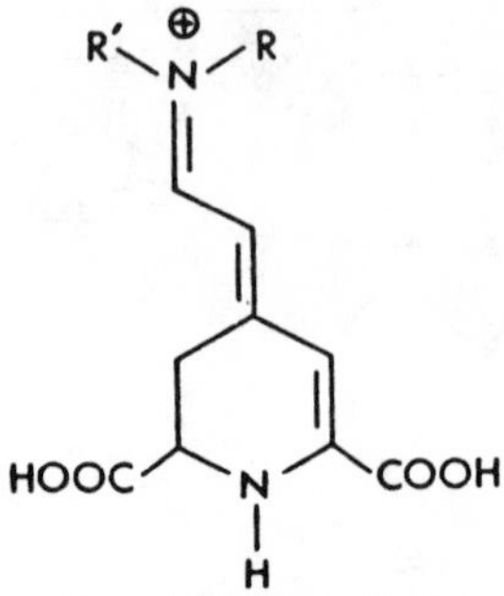

From Mabry and Dreiding (1968)

FIG. 25.11. THE 1,7-DIAZ-
OHEPTAMETHIN BASE
COMMON TO ALL BETA-
LAIN PIGMENTS

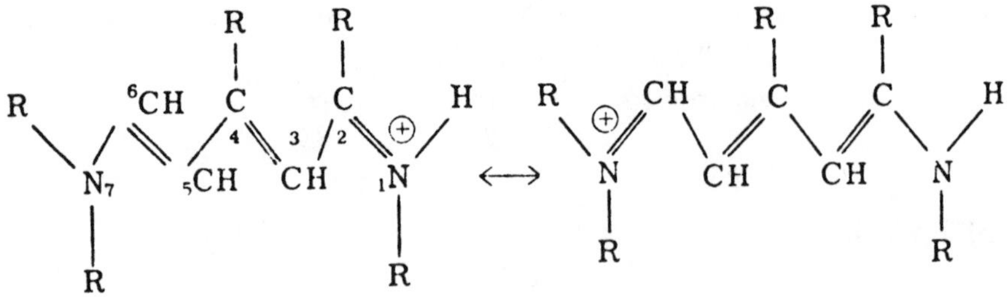

From Mabry and Dreiding (1968)

FIG. 25.12. BETACYANIN RESONATING STRUCTURES

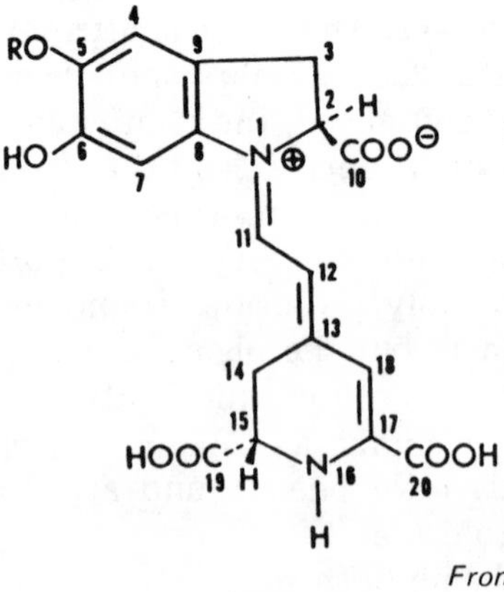

Betanidin, R = H

Betanin, R = Glucose

From Mabry and Dreiding (1968)

FIG. 25.13. THE STRUCTURES OF BETANIDIN AND BETA-
NIN

different. In weakly acidic buffers, anthocyanins migrate to the cathode and betalains to the anode. Betalains are more easily extractable with water whereas anthocyanins are more easily extractable with ethanol.

Sources

The betalains are confined to ten families of the Centrospermae in the plant kingdom. The major plant as a potential colorant is red beet *(Beta vulgaris)*. Betalains are also found in chard, cactus fruits, pokeberries, and a number of flowers such as bougainvillia and amaranthus.

There is an orange beet *(Beta vulgaris*, var. Burpee Golden Beet) which is identical with the normal reds except for color. The pigments are betaxanthins which produce an attractive orange color. Betazanthins are always present with betacyanins in red beets and do contribute to the color but their effect is masked by the betacyanins. Orange beets are a potential source of orange color.

Stability

Von Elbe and co-workers have studied the potential of red beet pigments as colorants (Von Elbe 1975). One advantage of the betalains over the anthocyanins is that they are less sensitive to pH changes (Fig. 25.14). Below pH 4, the hue shifts slightly towards the yellow and above pH 9, the hue shifts to the violet. The absorbance is at a maximum around pH 4-5. Betacyanins are more stable at the more acid pH ranges, and are very heat labile. They are also degraded by the presence of air and/or light. A lowered water activity of the system promotes greater pigment stability.

Applications

Potential applications for betalains include frankfurter and sausages, gelatin desserts, beverages, sherbets, ice cream, toppings, etc. With foods of pH 3.5 and below, anthocyanins may be the pigment of choice; whereas with products of pH 3.5-7.0, the betalains may be more suitable. One of the problems facing the development of the betalains is the preparation of pigment concentrates. Dehydrated beet juice or beet powder usually has a low content of pigment (1%). Adams *et al.* (1976) reported that the sugars could be fermented from beet juice to raise the concentration of pigment to 10%. They also explored gel chromatography (Hrazdina *et al.* 1970) as a means

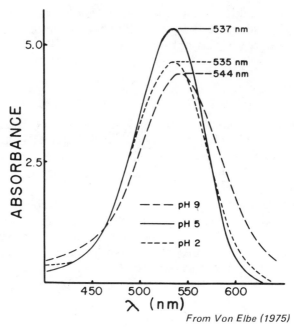

From Von Elbe (1975)

FIG. 25.14. CHANGES IN ABSORPTION SPECTRA OF
BETANIN WITH pH

of pigment assay and this approach may have some application for preparation of more concentrated extracts.

The pigments in pokeberries *(Phytolacca americana)* are identical to those in beets and the plant grows very vigorously in the United States. It is used as a source of greens ("poke") in the south. Pokeberries had a dubious claim to fame in France many years ago since they were grown in a corner of the vineyard as insurance against a grape crop with poor coloration. Their addition to wine was forbidden 100 years ago in France when it was discovered that they contained a toxic saponin. Driver and Francis (1977) worked out a method to remove the saponin and concluded that the purified pigment extracts were suitable as colorants for beverages and jellies (Driver and Francis 1977).

MISCELLANEOUS

A number of food colorants are permitted in the U.S. food supply which do not fall into the previous categories (Table 25.2). Also listed in Table 25.2 are permitted crude extracts mainly of plant

TABLE 25.2

APPROVED FOOD COLOR ADDITIVES EXEMPT FROM CERTIFICATION

Crude Organic Extracts	
Dried algae meal	(carotenoids)
Annatto	(bixin)
Paprika	(capsanthin and capsorubin)
Paprika oleoresin	(capsanthin and capsorubin)
Turmeric	(curcumin)
Turmeric oleoresin	(curcumin)
Saffron	(crocetin and crocin)
Fruit juice	(anthocyanins)
Vegetable juice	(anthocyanins)
Cochineal extract	(carmine)
Corn endosperm oil[1]	(carotenoids)
Grape skin extract[1]	(anthocyanins)
Dehydrated beets	(betacyanins)
Carrot oil	(carotenoids)
Tagetes (marigold) meal and extract[1]	(carotenoids)
Cottonseed flour	(gossypol and carotenoids

Miscellaneous	
Riboflavin	Ferrous gluconate[1]
Titanium dioxide	Synthetic iron oxide[1]
Caramel	Carbon black (channel process)
Ultramarine blue[1]	

SOURCE: Bauernfeind (1975).
[1] Restricted to specific uses.

origin but including one from insects (cochineal). The main colorant in the extracts is listed in parentheses. Many of these crude extracts have historical usage and recent toxicological clearance (Dineson 1975), but they pose a problem for regulatory officials. The composition of these extracts varies considerably depending on the processing, source of raw material, storage, etc. Regulations must necessarily be empirical and in some cases (e.g., anthocyanin preparation) the actual chemical makeup of the polymers produced by pigment degradation is unknown. The situation is further complicated by the classification of compounds by function, i.e., turmeric can be a flavor, a color or a seasoning. There probably will be pressure to define the chemical composition of crude organic extracts more accurately for regulatory purposes. There also will be pressure to develop more toxicological data on natural pigment preparations. For example, beet powder is presently permitted, but concentrated pigment preparations may also contain larger amounts of other undesirable compounds (e.g., saponins). Regulatory approval may be required for preparations that change the proportions of compounds which occur naturally. Probably the least one can ask is to have natural compounds subjected to the same safety requirements as synthetic compounds. It is not a question of the origin of a compound but whether or not the compound is in the food supply. There seems to be a schizophrenia at the present time with one standard of safety for natural compounds and another for synthetic compounds. This is illogical—there should be only one standard of safety for all foods and additives.

The Dynapol Corp. has developed a new concept for food colorants. They have attempted to synthesize compounds with a chromophore attached to a biologically inert polymer. This means that colors, flavors, preservatives, etc., could exert their desired technical effect and, after consumption, pass through the human body unchanged and unabsorbed. The critical proof in this concept is whether or not the additives are actually unabsorbed and unchanged. If the Dynapol concept is approved, I cannot resist the impulse to paraphrase William Shakespeare's saying, "What is past is prologue" to "What is passed may well be colored."

This chapter is contribution No. 2138 from the University of Massachusetts Agricultural Experiment Station. Appreciation is expressed to the Glass Container Manufacturers Institute, Washington, D.C. for partial financial support.

BIBLIOGRAPHY

ADAMS, J. P., and VON ELBE, J. H. 1976. Betanine separation and quantification by chromatography on gels. J. Food Sci. 42, 410.

ADAMS, J. P., VON ELBE, J. H., and AMUNDSEN, C. H. 1976. Production of a betacyanine concentrate by fermentation of red beet juice with Candida utilis. J. Food Sci. 41, 78-81.

BAUERNFEIND, J. C. 1975. Carotenoids as food colors. Food Technol. 29, No. 5, 48-49.

BUCKMIRE, R. M., and FRANCIS, F. J. 1976. Anthocyanins and flavonols of miracle fruit (Synsepalum dulcificum) J. Food Sci. 41, 1363.

BUCKMIRE, R. M., and FRANCIS, F. J. 1977. Anthocyanins of miracle fruit (Synsepalum dulcificum) as potential food colorants. J. Food Sci. (in press).

CALVI, P. J., and FRANCIS, F. J. 1977. Concord grape pigments as potential food colorants. J. Food Sci. (in prep.)

CHICHESTER, C. O. (Editor). 1972. The Chemistry of Plant Pigments. Academic Press, New York.

CHIRIBOGA, C. D. and FRANCIS, F. J. 1970. An anthocyanin recovery system from cranberry pomace. J. Am. Soc. Hortic. Sci. 95, 233.

CHIRIBOGA, C. D. and FRANCIS, F. J. 1973. Ion exchange purified anthocyanin pigments as a colorant for cranberry juice cocktail. J. Food Sci. 33, 464.

DINESEN, N. 1975. Toxicology and regulation of natural colors. Food Technol. 29, No. 5, 40.

DRIVER, M. G. 1977. Isolation and purification of phytolaccanin of Phytolacca americana. J. Food Sci. (in press).

DRIVER, M. G., and FRANCIS, F. J. 1977. Pigments of pokeberry (Phytolacca americana) as potential food colorants. J. Food Sci. (in press).

DU, C. T., and FRANCIS, F. J. 1973. Anthocyanins of Roselle (Hibiscus sabdariffa). J. Food Sci. *38*, 810-812.
DU, C. T., and FRANCIS, F. J. 1975. Anthocyanins of Viburnum. (unpublished).
ESSELEN, W. B., and SAMMY, G. M. 1973. Roselle—a natural red colorant for foods. Food Prod. Dev. 7, No. 1, 80.
FRANCIS, F. J., and SERVADIO, G. J. 1963. Relation between color of cranberries and stability of juice. Proc. Am. Soc. Hortic. Sci. *83*, 406-415.
FULEKI, T., and FRANCIS, F. J. 1968. Quantitative methods for anthocyanins. II. Determination of total anthocyanin and degradation index for cranberry juice. J. Food Sci. *33*, 78-83.
GAROGLIO, P. G. 1965. The new science of winemaking. Inst. Ind. Agrar. Florence, Italy. (Italian)
HARBORNE, J. B., MABRY, T. J., and MABRY, H. 1975. The Flavenoids. Chapman and Hall, London, England.
HRAZDINA, G., BORZELL, A. J. and ROBINSON, W. B. 1970. Studies on the stability of the anthocyanidin-3,5-diglucosides. Am. J. Enol. Vitic. *21*, 201-206.
INGALSBE, D. W., NEUBERT, A. M. and CARTER, G. H. 1963A. Colloidal blue pigments of juice. J. Agric. Food Chem. *11*, 263-266.
INGALSBE, D. W., NEUBERT, A. M. and CARTER, G. H. 1963B. Identification of the anthocyanins. J. Agric. Food Chem. *11*, 266-268.
ISLER, O. 1971. Carotenoids. Birkhauser Verlag, Basel, Switzerland.
JURD, L. 1972. Some advances in the chemistry of anthocyanin-type plant pigments. *In* The Chemistry of Plant Pigments. C. O. Chichester (Editor). Academic Press, New York.
MABRY, T. J., and DREIDING, A. S. 1968. The betalains. Recent Adv. Phytochem. *1*, 145-160.
MESCHTER, E. E. 1953. Effects of carbohydrates and other factors on strawberry products. J. Agric. Food Chem. *1*, 574-578.
PALAMIDIS, N., and MARKAKIS, P. 1975. Stability of grape anthocyanin in a carbonated beverage. J. Food Sci. *40*, 1047-1049.
PHILIP, T. 1974. An anthocyanin recovery system for grape wastes. J. Food Sci. *39*, 859.
ROBINSON, W. B., WEIRS, I. D., BERTINO, J. J. and MATTICK, L. R. 1966. The relationship of anthocyanin composition to color stability of New York State wines. Am. J. Enol. Vitic. *17*, 178-184.
SONDHEIMER, E., and KERTESZ, Z. I. 1948. Anthocyanin pigments. Colorimetric determination in strawberries and strawberry products. Anal. Chem. *20*, 245-248.
STARR, M. S., and FRANCIS, F. J. 1968. Oxygen and ascorbic acid effect on the relative stability of four anthocyanin pigments in cranberry juice. Food Technol. *22*, 1293-1295.
STARR, M. S., and FRANCIS, F. J. 1973. Effect of metallic ions on color and pigment content of cranberry juice cocktail. J. Food Sci. *38*, 1043-1045.
VAN BUREN, J. P., BERTINO, J. J. and ROBINSON, W. B. 1968. The stability of wine anthocyanins on exposure to heat and light. Am. J. Enol. Vitic. *19*, 147-151.
VON ELBE, J. H. 1975. Stability of betalains as food colors. Food Technol. *29*, No. 5, 42-44.
VON ELBE, J. H., MEING, IL-YOUNG, and AMUNDSEN, C. H. 1974. Color stability of betanin. J. Food Sci. *39*, 334-337.
WEISSLER, A. 1975. FDA regulation of food colors. Food Technol. *29*, No. 5, 38.
ZAPSALIS, C., and FRANCIS, F. J. 1965. Cranberry anthocyanins. J. Food Sci. *30*, 396.

26

Chemical and Physical Methods to Isolate and Identify Food Flavors

Roy Teranishi

Flavor has fascinated man from the dawn of history. Spices and essential oils are mentioned in the Old Testament (Exodus 30:22-25). Because the world faces a food crisis, we must utilize all the food that can be grown by making it acceptable. Thus, flavor research must be emphasized.

Acceptability includes sight, sound, and touch or texture, as well as flavor, and is guided by experience and cultural practices. Flavor is defined by Hall (1968) as "the sum of those characteristics of any material taken in the mouth, perceived principally by the senses of taste and smell and also by the general pain and tactile receptors in the mouth, as received and interpreted by the brain." Taste does make important contributions to flavor, such as bitter, salt, sweet, sour, etc., but it is well known that odor does play a predominant role in acceptance or rejection of foods. Because many researchers have devoted most of their efforts to studying volatile constituents in foods, this discussion will be limited to chemical and physical methods used in isolating and identifying odoriferous materials.

ISOLATION

Flavor compounds are present in extremely small amounts, as shown in Table 26.1, usually in water systems although some processed foods are predominantly lipids. Obviously, it is a formidable task to extract the characteristic flavors as concentrates. Unless the concentrate can be reconstituted to have the characteristic flavor, further fractionation and identification will not be very meaningful. Each step in isolation should be checked with sensory evaluations.

Methods for isolation and concentration of volatile food consti-

TABLE 26.1

FOOD COMPOSITION

	%
Water	up to 95
Carbohydrates	1–80
Lipids	1–40
Proteins	1–25
Minerals	1–5
Vitamins	ca 10^{-6}
Flavor compounds	10^{-6} –10^{-14}

tuents have been discussed in detail (Teranishi *et al.* 1971). Whether it is to obtain an isolate or to fractionate a complex mixture into its many components, any separation involves a redistribution between phases: as in crystallization, distillation, or partitioning; as in liquid-liquid extractions; and as in chromatography. This redistribution must be accomplished without loss of the desirable trace constituents and without introduction of contaminants or generation of artifacts which would overpower or mask the desirable odors.

A simultaneous distillation-extraction apparatus (Nickerson and Likens 1966) is proving to be very useful for research purposes if the material will survive steam distillation without degradation of flavor. A very high concentration is achieved in one operation with minimal introduction of contaminants because only small volumes of water and organic solvents are used. Temperature damage can be lessened by lowering the pressure of the simultaneous distillation-extraction system. Damaging off-flavor development can be lessened by preliminary separations from reactants by flash evaporation or solvent extractions. The degree of recovery of various compounds from an aqueous system by simultaneous steam distillation and extraction with an organic solvent under various conditions has been reported (Schultz *et al.* 1976).

Liquid carbon dioxide has been found to have good solvent properties for extracting essences from fruit juices and from solids such as coffee and shrimp (Schultz 1966), and subsequently from various other foods (Schultz *et al.* 1967; Schultz, 1969; Rey, 1972; Schultz *et al.* 1974). Liquid carbon dioxide has the advantage of low boiling so that concentrates can be obtained with little heat damage. Usually this solvent is free of contaminating odors or reactants, such as peroxides in ethyl ether. Because halogenated hydrocarbon solvents have been found to be injurious to health and environment, liquid carbon dioxide will have to be considered for its commercial possibilities (Rey 1972) since it is permitted in foods and beverages. As health considerations become more important, the cost of pressure equipment will become less important.

An added commercial possibility would be to adsorb volatiles on charcoal (Schultz 1976; Schultz *et al.* 1967). Such a procedure would have the advantage of reducing the size of pressure equipment necessary for liquid-liquid extractions with liquid CO_2. Both materials are permitted in food and beverage processing so that costly clearance would not be necessary.

It is even more difficult to isolate volatile flavor compounds from lipid systems than from aqueous systems. It has been shown that the air/solution partition coefficient of C_8 aldehyde is 1000 times less in a vegetable-oil system than in a water system (Buttery *et al.* 1973). If extrapolated to C_{10}, the difference is over 10,000. These data explain why it is so difficult to isolate volatiles from lipid systems by sweeping with an inert gas or by high vacuum distillation.

It is a well-known commercial practice to blow offending odors out of oils with steam. In order to study the odor qualities, the odoriferous materials must be collected and isolated. Standard steam distillations present problems of hot oil bumping and of extracting small amounts of volatiles from large amounts of steam distillate condensates. Use of steam for isolating volatile flavor compounds from oil systems has been described (Chang 1976; Teranishi *et al.* 1976).

In all isolation procedures, great care must be exercised to obtain isolates with desired sensory properties. In some cases, as in milk products (Moinas 1973), the source material must be kept at room temperature or lower to prevent off-flavors from developing. In the case of red dried beans (Buttery *et al.* 1975), the amount of oil obtained at atmospheric pressure and 100°C was about 10 times that obtained at 100 mm Hg vacuum and 50°C. In every case, good judgment must be used for proper isolation conditions.

Since gas chromatography started being applied in flavor research (Dimick and Corse 1956), sample size requirements have become considerably less than before this powerful method was available. Various methods have been utilized to trap trace amounts of organics from foods (MacKay 1960; Hornstein and Crowe 1962; Morgan and Day 1965; Issenberg 1971). Organic substances, down to concentrations of 1 part in 10^{13} (w/w), can be adsorbed on, then desorbed from, charcoal; separated by capillary gas chromatography; then identified by mass spectrometry (Grob 1973). Time, sample preparation, deterioration, etc., can be kept to a minimum with such vapor condensation and trapping systems.

SEPARATION

The purpose in isolating odoriferous material is to separate it into its constituents for identification of chemical and sensory properties. Usually, unless one is working with essential oils which are available in gram quantities, the flavor researcher has only milligrams of a very complex material. Under ideal conditions, one would have enough material to divide the complex mixture into narrow boiling range fractions, then each fraction can be divided into different functional groups with preparative liquid chromatography, or the sequence can be reversed for convenience of attacking a specific problem.

If one has sufficient material and if this material will survive prolonged periods at elevated temperatures, vacuum fractional distillation can be used for preparation of pure materials, especially with a Teflon spinning band (Yost 1974). With adequate temperature and pressure controls, very good separations can be achieved. In some cases, better separations can be obtained with vacuum fractional distillation than with preparative gas chromatography. Therefore, in cases in which it is applicable, fractional distillation should not be overlooked.

Before the advent of gas chromatography and mass spectrometry, the problems that could be studied were only those in which gram quantities were available. Therefore, types of problems studied were severely limited. Now only milligram quantities are necessary for surveys, especially with programmed temperature control of open tubular columns (Teranishi et al. 1960; Harris and Habgood 1966). Figure 26.1 shows a chromatogram of strawberry oil (Teranishi et al. 1960). This is an early example of what can be done with this technique. Although the range of molecules and amounts of sample have been increased, not much improvement in resolution of separation has been accomplished since this early work. Certainly the technique of gas chromatography (James and Martin 1952) has greatly facilitated research in aroma chemistry.

It is not sufficient in flavor research to merely separate. We must know what compounds or groups of compounds are important to various characteristic odors; i.e., we must correlate sensory evaluations to fractions separated. Word description identifications from human sensors (Fuller et al. 1964; Ryder 1966) of effluents from gas chromatographs can be very useful in determining what fractions are important. Packed columns have sufficient capacity but poor separation resolution. Small bore open tubular columns have good resolution but insufficient capacity. The large bore, 0.75 mm id, open

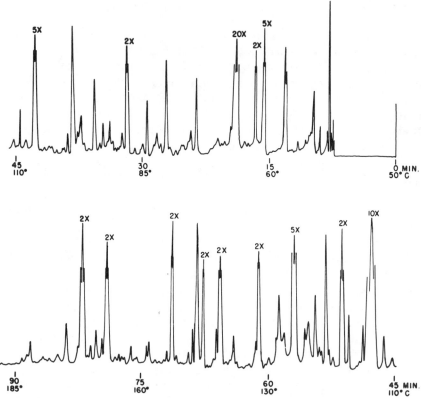

FIG. 26.1. CHROMATOGRAM OBTAINED WITH PROGRAMMED TEMPERATURE CONTROL OF 60 M X 0.25 MM ID OPEN TUBULAR COLUMN COATED WITH TWEEN-20

tubular columns have good resolution with large enough capacity for the effluents to be characterized by sniffing (Mon 1971). Such columns also can be used for direct vapor analyses (Mon 1971) for studying vapors from various food products (see Issenberg, 1971, for papers given at the Symposium on Direct Vapor Analyses).

Some compounds, such as 2, 5-dimethyl-4-hydroxy-2, 3-dihydro-3-furanone (Flath and Forrey 1970), in very small amounts, do not come through stainless steel open tubular columns. Therefore, there has been considerable interest in glass capillary columns (see Anon. 1975 for the papers presented at the First International Symposium on Glass Capillary Chromatography). Use of glass capillaries has not yet been widely applied in aroma research in the United States, but as more stable glass columns become available, such columns undoubtedly will be used in many applications in flavor research.

IDENTIFICATION AND CHARACTERIZATION

Physical and Chemical

Gas chromatography and mass spectrometry have been combined (Teranishi *et al.* 1971; McFadden 1973) to greatly facilitate aroma research. Material separated and purified by high resolution open tubular columns in microgram quantities or less can be identified by mass spectral patterns if the chemical structures are simple enough. Even if effluents follow each other closely, if the separation is good and if the pumping capacity of the mass spectrometer vacuum system is high enough, clean mass spectra can be obtained (Teranishi *et al.* 1963); see Fig. 26.2. Since such work was done, there have been many improvements in extending the range of molecules to be analyzed, different applications, sample manipulations, data handling, etc. (McFadden 1973), but the greatest impetus in flavor research was in the ability to obtain results such as those shown in Fig. 26.2.

The structure of a complex and closely related compounds cannot be completely elucidated or differentiated by mass spectra alone. In such situations, enough material must be isolated by batch methods for nuclear magnetic resonance, infrared, and Raman spectra (Freeman 1965; Ettre and McFadden 1969; Teranishi *et al.* 1967A, B, 1971). Figure 26.3 shows an example of a nuclear magnetic resonance analysis of the protons in the humulene molecule, sample size 0.45 mg in 34μl carbon tetrachloride. The frequencies cannot be separated and identified so well in every case, but this example does illustrate what a powerful tool nuclear magnetic resonance is in obtaining information about the intimate structure of molecules. If the purified material is crystalline, or if a crystalline derivative can be made, X-ray crystallographic analysis will yield the absolute configuration (Nyburg 1961; Stout and Jensen 1968).

Although computer techniques have lowered sample size requirements for nuclear magnetic resonance spectrometry (Gray 1975), this method still requires the batch method. The most promising computer application is in infrared spectrometry (Freeman 1969; Koenig 1974). Some gas chromatography-infrared combinations are already in operation. Since the infrared-mass spectrometry combination is possible, effluents from a gas chromatograph could be passed into an infrared cell then into a mass spectrometer. Such a system would yield much valuable information in a short time. Hundreds of compounds could be scanned in a few hours. A greater number of compounds is likely to be identifiable by the combination of MS and

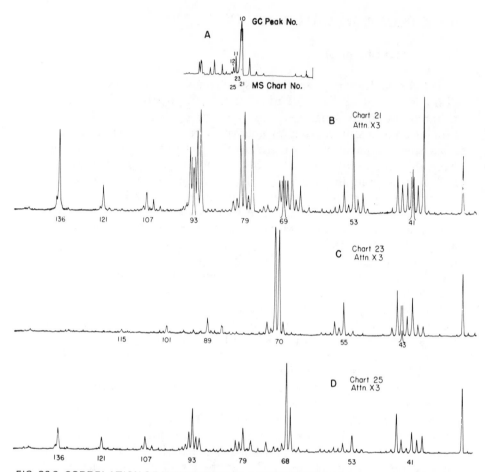

FIG. 26.2. CORRELATION OF MASS SPECTRAL PATTERNS TO GAS CHROMATOGRAM PEAKS

A—Chromatogram showing GC peak numbers and MS chart numbers. B—MS chart number 21, myrcene. C—MS chart number 23, 2-methylbutyl isobutyrate. D—MS chart number 25, limonene.

IR data than each alone. Certainly, functional groups can be reliably identified by infrared spectra. To accomplish such a task of handling such small samples and such large numbers of samples would be very difficult by the batch method. Also, such a system would provide spectra of compounds sensitive to light and/or oxygen immediately as they are separated.

The characterization of 2-methoxy-3-isobutylpyrazine, an important component of green bell peppers (Buttery et al. 1969), is an excellent example of a complete characterization. The volatile oil

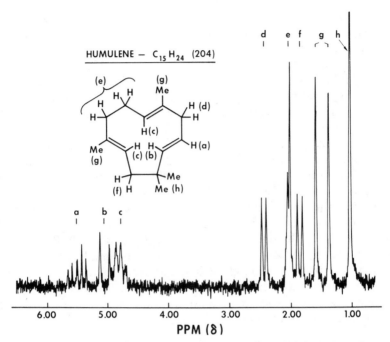

FIG. 26.3. NUCLEAR MAGNETIC RESONANCE ANALYSIS OF HUMULENE,
0.45 MG IN 34 μ LITER CARBON TETRACHLORIDE, HA-100

from green bell peppers *(Capsicum annuum)* was isolated by vacuum steam distillation, and the important aroma constituent was separated by gas chromatography. The chemical characterization utilized infrared, ultraviolet, nuclear magnetic resonance, and mass spectrometric methods. For further proof, the material was synthesized. The odor threshold was established as 2 parts per 10^{12} parts of water by a trained panel of 20 judges. The aroma character of this material was described as typical of freshly chopped green bell peppers. This accomplishment is one of the milestones in flavor research and illustrates the utilization and application of modern technology that has been developed in this last century.

Sensory

In flavor research it is not sufficient to merely identify the physical and chemical properties. Even though lists of constituents (van Straten and de Vrijer 1973) are very valuable, compositional studies must be correlated to sensory evaluation as was done in the case of 2-methoxy-3-isobutylpyrazine (Buttery *et al.* 1969). Much valuable information can be obtained from single observers, especially if they are experts like perfumers (Fuller *et al.* 1964), but statistically reliable panel studies (Guilford 1936; Amerine *et al.* 1965) are desirable. Concepts and techniques important in olfactory research in general (Moulton *et al.* 1975) are also applicable in flavor studies; such as, olfactometry (Dravnieks 1975), purity of substances (Turk and Turk 1975), pharmacological aspects of olfaction (Beets 1975), etc.

A systematic approach in coordinating instrumental and sensory techniques has been proposed (Guadagni 1968). Amounts of material present and their threshold values (Stahl 1973) are utilized to calculate odor units (Guadagni 1968). This method has been used to determine the compounds important in imparting the characteristic odor to various materials, such as bell peppers (Buttery *et al.* 1969), apple essence (Guadagni *et al.* 1966; Guadagni 1968), bay oils (Buttery *et al.* 1974), etc. It cannot be over emphasized that this odor unit approach is merely a first order approximation of the important contributors to a characteristic odor. The complexities and nuances are still in the creative minds and hands of the experts, the flavorists and compounders.

SUMMARY AND PREDICTIONS

Although flavor research involves complex systems, progress is being made because remarkable instruments and techniques have been developed and applied. Special attention is called to Chap. 27 by van den Ouweland, Olsman and Peer who cover "Challenges in Meat Flavor Research," to Chap. 28 by Reymond covering "Flavor Chemistry of Tea, Cocoa and Coffee," and to Chap. 29, "Flavor Chemistry of Fruits and Vegetables" authored by Nursten. However, only the surface has been scratched: only a few definitive characterizations have been accomplished compared to what can be done and what needs to be done. It is discouraging to note that progress in flavor research seems to be slowing down because of shift of funding to other areas, such as food safety and nutrition. Appropriations for flavor research should be increased again. With application of computer methods to analyze massive data, complex problems heretofore considered impractical for study can now be solved. If we are to feed the world and new foods are to be consumed, flavors and appearances of new foods must meet familiar and customary standards. Our quality of living, the quality of food we ingest, will be lowered if flavor research is not re-emphasized in the near future.

BIBLIOGRAPHY

AMERINE, M. A., PANGBORN, R. M., and ROESSLER, E. B. 1965. Principles of Sensory Evaluation of Food. Academic Press, New York.

ANON. 1974. Odour and Taste. 100 Years H & R. Haarmann and Reimer GmbH., Holzminden, Germany.

ANON. 1975. 1st Int. Symp. Glass Capillary Chromatography, Hindeland, Germany, May 4–7, 1975. Chromatographia 8, No. 9, Sept.

ARCTANDER, S. 1960. Perfume and Flavor Materials of Natural Origin. Arctander, Elizabeth, New Jersey.

ARCTANDER, S. 1969. Perfume and Flavor Chemicals, Vols. 1 and 2. Arctander, Montclair, New Jersey.

BEETS, M. G. J. 1975. Pharmacological aspects of olfaction. In Methods in Olfactory Research. D. G. Moulton, A. Turk, and J. W. Johnston, Jr. (Editors). Academic Press, New York.

BUTTERY, R. G. et al. 1969. Characterization of an important aroma component of bell peppers. Chem. Ind. (London) 490–491.

BUTTERY, R. G., GUADAGNI, D. G. and LING, L. C. 1973. Flavor compounds: Volatiles in vegetable oil and oil-water mixtures. Estimation of odor thresholds. J. Agric. Food Chem. 21, No. 2, 198–201.

BUTTERY et al. 1974. California bay oil. I. Constituents, odor, properties. J. Agric. Food Chem. 22, No. 5, 773–776.

BUTTERY, R. G., SEIFERT, R. M. and LING, L. C. 1975. Characterization of some volatile constituents of dry red beans. J. Agric. Food Chem. 23, No. 3, 516–519.

CHANG, S. S. 1976. Apparatus for the isolation of trace volatile constituents from foods. Abstracts of Papers, 172nd ACS National Meeting, San Francisco, California. American Chemical Society. AGFD-132 (Abstr.).

DIMICK, K. P., and CORSE, J. 1956. Gas chromatography—A new method for the separation and identification of volatile materials in foods. Food Technol. 10, 360–364.

DRAVNIEKS, A. 1975. Instrumental aspects of olfactometry. In Methods in Olfactory Research. D. G. Moulton, A. Turk, and J. W. Johnston, Jr. (Editors). Academic Press, New York.

ETTRE, L. S., and McFADDEN, W. H. (Editors). 1969. Ancillary Techniques of Gas Chromatography, Wiley-Interscience, New York.

FLATH, R. A., and FORREY, R. R. 1970. Volatile components of smooth Cayenne pineapple J. Agric. Food Chem. 18, No. 2, 306–309.

FREEMAN, S. K. 1965. Interpretive Spectroscopy. Van Nostrand Reinhold Co., New York.

FREEMAN, S. K. 1969. Gas chromatography and infrared and Ramon spectrometry. In Ancillary Techniques of Gas Chromartography. L. S. Ettre and W. H. McFadden (Editors). Wiley-Interscience, New York.

FULLER, G. H., STELTENKAMP, R., and TISSERLAND, G. A. 1964. The gas chromatography with human sensor: perfumer model. In Recent Advances in Odor: Theory, Measurement, and Control. Ann. N.Y. Acad. Sci. 116, part 2, 711–724.

GRAY, G. A. 1975. Nuclear magnetic resonance spectroscopy. Anal. Chem. 47, No. 6, 546A–564A.

GROB, K. 1973. Organic substances in potable water and in its precursor. Part I. Methods for their determination by gas-liquid chromatography. J. Chromatogr. 84, 255–273.

GUADAGNI, D. G., OKANO, S., BUTTERY, R. G., and BURR, H. K. 1966. Correlation of sensory and gas-liquid chromatographic measurement of apple volatiles. Food Technol. 20, No. 4, 518–521.

GUADAGNI, D. G. 1968. Requirements for coordination of instrumental and sensory techniques. *In* Correlation of Subjective-Objective Methods in the Study of Odors and Taste. ASTM *440*. Am. Soc. Test. Mater. Philadelphia.

GUILFORD, J. P. 1936. Psychometric Methods. McGraw-Hill Book Co., New York.

HALL, R. L. 1968. Food flavors: benefits and problems. Food Technol. *22*, 1388-1392.

HARRIS, W. E., and HABGOOD, H. W. 1966. Programmed Temperature Gas Chromatography. John Wiley & Sons, New York.

HORNSTEIN, I., and CROWE, P. F. 1962. Gas chromatography of food volatiles—an improved collection system. Anal. Chem. *34*, 1354-1356.

ISSENBERG, P. 1971. Symposium on direct vapor analysis. J. Agric. Food Chem. *19*, No. 6, 1045-1073.

JAMES, A. T., and MARTIN, A. J. P. 1952. Gas-liquid partition chromatography: the separation and micro-estimation of volatile fatty acids from formic to dodecanoic acid. Biochem. J. *50*, 679-690.

KOENIG, J. L. 1974. Computerized infrared spectroscopy via Fourier transform techniques. Am. Lab. Sept., 9-16.

MACKAY, D. A. M. 1960. General discussion. *In* Gas Chromatography— Edinburgh 1960. R. P. W. Scott (Editor). Butterworths, London, England.

McFADDEN, W. H. 1973. Techniques of Combined Gas Chromatography/Mass Spectrometry: Applications in Organic Analysis. Wiley-Interscience, New York.

MOINAS, M. 1973. Volatile aromas of milk products—extraction and identification. Trav. Chim. Aliment. Hyg. *64*, No. 1, 60-65.

MON, T. R. 1971. Preparation of large bore open tubular columns for GC. Res. Dev. *22*, No. 12, 14-17.

MORGAN, M. E., and DAY, E. A. 1965. Simple on-column trapping procedure for gas chromatographic analysis of flavor volatiles. J. Dairy Sci. *48*, 1382-1384.

MOULTON, D. G., TURK, A., and JOHNSTON, J. E. JR. 1975. Methods in Olfactory Research. Academic Press, New York.

NICKERSON, G. B., and LIKENS, S. T. 1966. Gas chromatographic evidence for the occurence of hop oil components in beer. J. Chromatogr. *21*, 1-5.

NYBURG, S. C. 1961. X-ray Analysis of Organic Structures. Academic Press, New York.

REY, L. 1972. Recent developments in freeze-drying and cryogenics. Nestlé Res. News, 48-53.

RYDER, W. S. 1966. Progress and limitations in the identification of flavor components. *In* Flavor Chemistry. Advances in Chemistry Series *56*. I. Hornstein (Editor). American Chemical Society, Washington, D.C.

SCHULTZ, T. H. *et al.* 1967. Volatiles from Delicious apple essence—extraction methods. J. Food Sci. *32*, 279-283.

SCHULTZ, T. H., FLATH, R. A., MON., T. R., and TERANISHI, R. 1976. Isolation of volatile components from a model system. Abstract of papers, 172nd ACS National Meeting, San Francisco, Calif. *AGFD-131* (Abstr.).

SCHULTZ, W. G. 1966. Liquid carbon dioxide for selective aroma extraction. Paper presented at the 26th Annu. Meet. Inst. Food Technologists, Portland, Oregon, May.

SCHULTZ, W. G. 1969. Process for extraction of flavors. U.S. Pat. 3,477,856, Nov. 11.

SCHULTZ, W. G. *et al.* 1974. Pilot-plant extraction with liquid CO_2. Food Technol. June, 32-36, 88.

STAHL, W. H. 1973. Compilation of Odor and Taste Threshold Value Data. ASTM Data Ser. *DS48*. Am. Soc. Test Mater., Philadelphia.

STOUT, G. H., and JENSEN, L. H. 1968. X-ray Structure Determination. A Practical Guide. Macmillan Co., New York.

TERANISHI, R., NIMMO, C., and CORSE, J. 1960. Gas-liquid chromatography.

Programmed temperature control of the capillary column. Anal. Chem. *32*, 1384-1386.

TERANISHI, R., *et al.* 1963. Role of gas chromatography in aroma research. Am. Soc. Brew. Chem. Proc. 52-57.

TERANISHI, R., HORNSTEIN, I., ISSENBERG, P., and WICK, E. L. 1971. Flavor Research: Principles and Techniques. M. Dekker, New York.

TERANISHI, R., LUNDIN, R. E., and SCHERER, J. R. 1967A. Analytical technique. *In* Symposium on Foods: The Chemistry and Physiology of Flavors. H. W. Schultz, E. A. Day, and L. M. Libbey (Editors). AVI Publishing Co., Westport, Conn.

TERANISHI, R., LUNDIN, R. E., McFADDEN, W. H., and SCHERER, J. R. 1967B. Ancillary systems. *In* The Practice of Gas Chromatography. L. S. Ettre and A. Zlatkis (Editors). Wiley-Interscience Publishers, New York.

TERANISHI, R. MURPHY, E. L., and MON, T. R. 1976. Steam distillation-solvent extraction recovery of volatiles from fats and oils. Abstracts of Papers, 172nd ACS National Meeting, San Francisco, Calif. American Chemical Society, *AGFD-155* (Abstr.).

TURK, A., and TURK, V. 1975. The purity of odorant substances. *In* Methods in Olfactory Research. D. G. Moulton, A. Turk, and J. W. Johnston, Jr., (Editors). Academic Press, New York.

VAN STRATEN, S., and DE VRIJER, F. 1973. Lists of volatile compounds in foods. Cent. Inst. Nutr. Food Res. TNO. Zeist, Netherlands.

YOST, R. W. 1974. Distillation primer—a survey of distillation systems. Am. Lab. Jan. 63-71.

27

Challenges in
Meat Flavor Research

Godefridus A. M. van den Ouweland
Herman Olsman
Hein G. Peer

MEAT FLAVOR AND THE CONSUMER

Meat is definitely one of the most popular items of the human diet and it is consumed all over the world as a nutritious and flavorful food. It is regarded by many as a daily necessity and as the center piece of the main meal. Meat is an excellent source of good protein; its consumption, though, is primarily a matter of gustatory satisfaction and, to this, flavor makes a very important contribution. The intriguing point is that not only does each type of meat (e.g., beef, pork, chicken) possess its own characteristic flavor but different methods of cooking, such as stewing, simmering, frying and roasting, also give rise to quite distinct flavor impressions, each with its own special appeal to the consumer (Fig. 27.1).

This is hardly surprising when one realises that the heat treatments to which the meat is subjected differ widely for the various forms of preparation. For example, a steak will be heated for only several minutes during grilling or frying, with temperatures varying from no more than 50°C in the center to about 150°C on the outside, while a stew is simmered at 100°C for several hours.

The heat treatments to which canned meats are subjected, on the other hand, must be sufficient to prevent spoilage and deterioration during their storage and distribution. Sterilization, but also other steps in a processed food chain, can lead to loss of flavor and to the development of flavors not recognized as being characteristic for the home-cooked equivalent.

The increasing popularity of convenience foods and other processed foods, such as soups, sauces and T.V. dinners, has created a

FIG. 27.1. EACH TYPE OF MEAT POSSESSES ITS OWN FLAVOR, AND DIFFERENT METHODS OF COOKING GIVE RISE TO DISTINCT FLAVOR IMPRESSIONS

demand for attractive flavors, capable of conveying a better boiled, fried or roasted meat impression to these foods. This demand for a range of characteristic meat flavors had been boosted by the development and marketing of a new generation of products in which part or all of the meat has been replaced by other protein sources, e.g., vegetable protein materials.

The need for meaty flavors undoubtedly presents a formidable challenge to the flavor chemist. But before going on to discuss this challenge, let us look in turn at the origins of meat flavor research and the progress made in understanding and imitating meat flavor formation. In doing so, we will focus on beef flavor, but data on the flavor of other types of meat will be touched upon where relevant.

ORIGIN OF MEAT FLAVOR RESEARCH

The occasion of another Centenary, that of the Liebig Meat Extract Company, was celebrated, in 1965. Events leading to the birth of the company can be considered as the origin of all meat flavor work, and it is therefore interesting to recall this story here.

We all know that Von Liebig's contribution to organic chemistry is very impressive. But, although he definitely made greater scientific achievements, he is remembered most of all for developing a method for producing meat extract. He recognized the possibility of bringing meat in "concentrated form" to Europe from the distant plains of South America where the carcasses of cattle, slaughtered for their hides and fat, were left to rot by the thousands. His "Extractum Carnis," which he showed at Munich in 1847 (Fig. 27.2), became a successful commercial product, one which is still selling well today.

Acting on Von Liebig's instructions, Giebert erected the company which we now know as the Liebig Meat Extract Company. Initially, Liebig's meat extract was applied commercially in soups, sauces and other products. By 1875, one million pounds per year of the extract were being produced. Shortly after that, canning of beef was commercialized, while the export of corned beef started around 1900, thus adding impulse to the meat extract market. Later on, Liebig's subsidiary in the United Kingdom started the production of protein hydrolysates and in 1911, a product which was to be a big commercial success came onto the market, namely "OXO" cubes, made from meat extract, dried beef powder, protein hydrolysate, spices, caramel and salt. From then onwards, as more and more fabricated foods came onto the market, a growing need was felt to understand more about meat flavor, particularly as regards the: (1) chemical components of meat flavor; (2) precursors of these flavor components and the reactions involved in the formation of these components during cooking; and (3) the origin of the flavor precursors and factors affecting their formation, both ante- and postmortem.

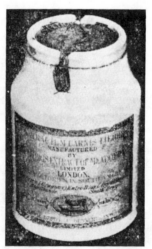

FIG. 27.2. VON LIEBIG'S
"EXTRACTUM CARNIS,"
THE MEAT EXTRACT DE-
VELOPED IN 1847

We will review the present state of knowledge of those aspects starting with the latter.

FORMATION OF MEAT FLAVOR

Influence of Ante- and Postmortem Factors on Precursor Formation

The origin of meat flavor lies in processes occurring both in the antemortem and the postmortem stages; these have been discussed thoroughly by Dwivedi (1975). Nature and concentration of meat flavor precursors depend to a large extent on metabolic activities in the animal body and the postmortem conditions. The precursors are developed under the influence of enzymes; the enzymatic reactions concerned, however, have not been studied in detail.

Some important antemortem factors are as follows.

Age of the Animal at the Time of Slaughter.—The composition of muscle changes with age; the older the animal, the more intense the odor of the cooked meat.

Sex of the Animal.

Type of Joint.—Differences in the flavor of different joints are widely recognized. Thus rump and top-side yield roasts with markedly more flavor but greater toughness, than sirloin (Lawrie 1970).

Breeding Conditions.—Harris *et al.* (1968) reported that meat from conventionally raised chicken had a more characteristic flavor than meat from the chicken raised in a germ-free environment.

Biochemical State of the Muscle.—After slaughter, muscle glycogen is converted into lactic acid, causing the pH to fall. The lower the glycogen content, the higher the final pH. The relationship between the final pH of the muscle and meat flavor has been reported by Lawrie *et al.* (1957).

Postmortem factors also have a big influence on flavor. There are considerable differences in the flavor of meat cooked before and after rigor mortis. It may be concluded that the drop in pH and other chemical and physicochemical changes taking place during rigor are essential to the formation of meat flavor precursors. Enzymes present in raw meat remain active during aging. ATP is decomposed enzymatically to IMP. Most of the nucleotides decompose to sugar phosphates and then to sugars (mainly glucose, fructose and ribose). Proteins are slowly hydrolyzed and the structural elements, such as cell walls and the z-lines of the contractile apparatus are changed. The latter changes are also responsible for making the meat more tender. Free fatty acid content rises as a result of phospholipid hydrolysis. Deep freezing can also affect flavor. Firstly, undue delay in freezing can cause rancidity of the fat. Secondly, even during storage processes leading to off-flavor development can start after about three months.

Flavor Precursors and Their Interactions

What is the nature of the flavor precursors formed in the enzymatic processes referred to and how are they responsible for formation of the multitude of compounds contributing to meat flavor? For, whereas raw meat has little odor and contains few volatile compounds, more than 200 volatile odorous components have been separated and identified from roasted, fried and boiled meat or from the headspace of cooked meat; and many more remain to be identified. All skeletal muscles are built up from the same basic constitutents, i.e., proteins, lipids, carbohydrates and mineral salts; and all have similar biochemical functions. Consequently, interspecial differences in muscle composition, and hence in the nature of the flavor precursors, are small. If very lean beef, lamb and pork are cooked identically, it is very difficult to distinguish between the resultant flavors. As we shall see later, the components of the fatty tissue are responsible for the formation of flavors characteristic for the various species.

Sugars, Amino Acids, Peptides

Flavor precursors have been successfully extracted with ice-cold water from fresh, lean beef, pork, lamb and chicken by Hornstein and Crowe (1960, 1963), Batzer et al. (1960, 1962), Wasserman and Gray (1965), Pippen et al. (1954), and many others. It was concluded that the precursors are water-soluble and of relatively low molecular weight. Hornstein and Crowe (1960) described a method of obtaining a flavor concentrate powder by lyophilizing a cold-water extract of raw lean beef. On heating the dry powder an odor reminiscent of roast beef evolved; heating an aqueous solution of the powder produced the aroma of boiled beef. They concluded that the flavor was produced by reactions between amino acids and carbohydrates. Similar work was done by Batzer et al. (1960, 1962) who fractionated an aqueous extract of beef muscle tissue and found a precursor fraction containing a glycoprotein; inosinic acid; the amino acids asparagine, glutamine, and taurine; and the peptides, anserine and carnosine. Wasserman and Gray (1965), working along similar lines, found precursor fractions which contained only amino acids, nucleotides, inosine and hypoxanthine. Macy *et al.* (1964) included lamb and pork in their investigations; amounts of taurine, alanine, anserine and carnosine which are abundant in all three species, were considerably reduced during heating. Of the carbohydrates initially present in beef, lamb and pork—primarily glucose with lower amounts of fructose and ribose—only ribose was in all cases completely absent after heating.

These results indicate that amino acids, peptides and carbohydrates, particularly ribose, are important precursors of meat aroma. The results above were confirmed and extended by other workers such as May (1974) and Heath (1970). They extracted raw beef with cold water and found that only those fractions containing compounds with a molecular weight of less than 200 produce any beef flavor on heating. These fractions are composed of amino acids (cysteine, β-alanine, glutamic acid and tryphophan), peptides and carbohydrates (glucose, glucosamine, fructose and ribose). May found that ribose and cysteine disappeared completely on cooking and concluded that it is mainly these two compounds which are involved in the production of meat flavor. Heath showed that the composition of amino acids and reducing sugars in beef, pork and lamb are indeed quite similar.

Other workers such as Pepper and Pearson (1969, 1971) and later Wasserman and Spinelli (1972) identified a number of amino acids and sugars in adipose tissue. With the aid of thin-layer chromatography, Pepper and Pearson (1974) demonstrated the presence of creatine, creatinine, cytosine and uracil in a diffusate of beef adipose tissue. Macy *et al.* (1964, 1970) found that the level of amino acids increases during the roasting of beef. The liberation of amino acids during cooking is considered to be important in the development of meat flavor. Whenever cysteine was released, recognizable beef aromas resulted.

Let us now turn our attention to the types of reactions involved in flavor formation from these precursors. During heating, both amino acids and carbohydrates can be degraded as they can interact with each other. The chemistry of sugar degradation (caramelization) and of sugar-amino acid interactions, and their importance in flavor formation, have been investigated and reviewed fairly thoroughly elsewhere. An interesting example of the way in which the amino acids can act as flavor precursors has been described by Schutte and Koenders (1972). They proposed a system of reactions for the formation of 1-methylthioethanethiol, a volatile isolated from the headspace of beef broth. Alanine, methionine and cysteine react in the broth with a diketone such as pyruvic aldehyde to form the immediate precursors of the flavor compound (Fig. 27.3).

Ribonucleotides, Ribose-5-phosphate, Methylfuranolone

Apart from the degradation of sugars and amino acids and their mutual interaction, there are other routes, possibly even more important, by which meat flavor is formed. As mentioned earlier, ribonucleotides decompose enzymatically to ribose-5-phosphate during rigor and, from the work of Tonsbeek et al. (1969), we learned that ribose-5-phosphate is present in fractions giving a meat flavor when heated. He found that ribose-5-phosphate is initially converted into 4-hydroxy-5-methyl-2,3-dihydrofuran-3-one (methylfuranolone) on heating. This conversion is represented schematically in Fig. 27.4. A milestone in the development of meat flavor was the subsequent recognition at our laboratory that methylfuranolone is not only a flavor component but, probably, itself a precursor of other characteristic flavor components. This represents one of the most important approaches in the development of meat flavor. We will return to this later on.

Lipids, Fatty Acids

The following important group of precursors are the lipids. Hornstein and Crowe (1960, 1963) drew attention to these as a source of flavor differences between the various species. Lipids in meat can be

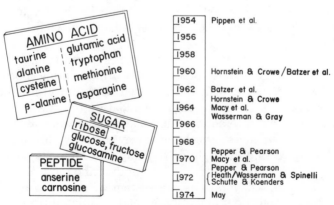

FIG. 27.3. AMINO ACIDS, SUGARS AND PEPTIDES AS PRECURSORS OF FLAVOR COMPOUNDS

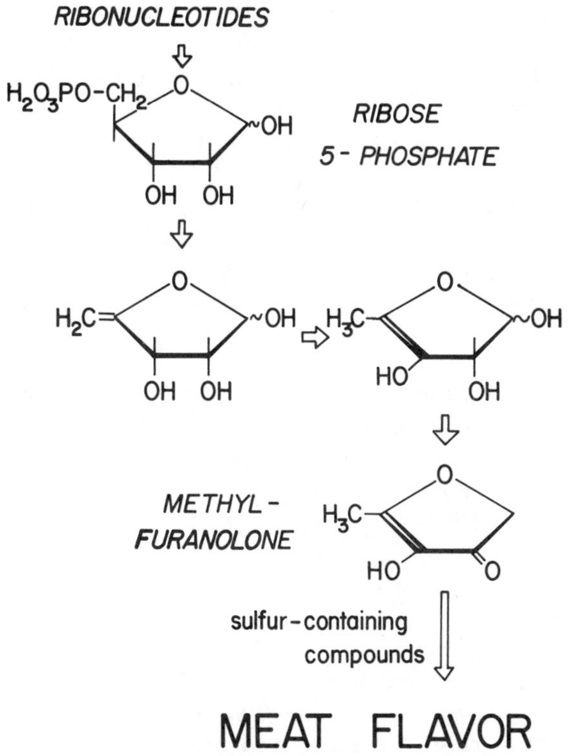

FIG. 27.4. CONVERSION OF RIBOSE 5-PHOSPHATE INTO MEAT FLAVOR

classified as intermuscular or depot and as intramuscular or tissue lipids.

The depot lipids are generally stored in relatively large deposits in connective tissues. They are able to dissolve and retain aroma originating outside the fat. Pippen *et al.* (1969) found that fat isolated from cooked poultry has a characteristic poultry flavor, but this is not the case for fat isolated from raw poultry and then heated. Evidently, flavor compounds, derived from the lean portions of the meat dissolve in fat during cooking. Fat from cooked poultry, particularly roasted poultry, contains sulphur compounds such as methional and methylmercaptan.

The tissue lipids on the other hand are distributed throughout the muscle tissues. They exist in close association with proteins and contain a large proportion of the total phospholipids. As a result of their high content of unsaturated, fatty acids (particularly linoleic acid and arachidonic acid) the phospholipids are highly susceptible to oxidation. This oxidation may lead to flavor deterioration (rancidity) and to the formation of species-specific flavors. Hornstein and Crowe (1960) found that characteristic pork and beef odors are produced when fat of these meats is heated in air. Fatty acid oxidation leads to the formation of carbonyl compounds. Hornstein and Crowe (1963) analyzed pork and beef fat for free fatty acids. Not only was the total level of fatty acids in pork fat twice as high as in beef fat, but pork fat also contains a much higher proportion of poly unsaturated fatty acids. The relatively high content of linoleic acid and arachidonic acid in the muscle phospholipid of pork belly explains the susceptibility of these cuts to rancidity (Love and Pearson 1971).

As for the formation of species-specific flavor, the work of Harkes and Begemann (1974) is worth mentioning. They found that arachidonic acid from the phospholipid fraction of chicken meat is oxidized to mono- and polyunsaturated aldehydes, which play an important role in cooked chicken flavor (Fig. 27.5).

Hydrogen Sulfide

The next type of precursors we mention are the labile, low molecular weight type of compounds, comprising mercaptane, ammonia, and particularly, hydrogen sulfide. The latter is evolved continuously during the simmering of chicken and beef. Apart from contributing itself to flavor it is without doubt an important precursor of other flavor compounds.

Pippen and Mecchi (1969) have described the role of hydrogen

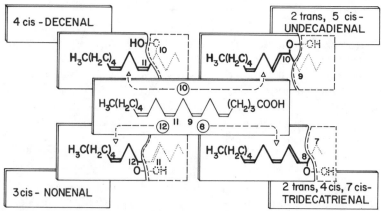

FIG. 27.5. AUTOXIDATION OF ARACHIDONIC ACID TO ALDEHYDES WHICH ARE COMPONENTS OF CHICKEN FLAVOR

sulfide in chicken flavor. It occurs in the broth of freshly simmered chicken and in the freshly cooked ready-to-eat meat of simmered, roasted and fried chicken. Pepper and Pearson (1969) studied the changes in the hydrogen sulfide content of beef adipose tissue upon heating. They concluded that the hydrogen sulfide evolved on heating may contribute indirectly to the meaty aroma of cooked beef by acting as a precursor to components of this aroma.

When hydrogen sulfide is passed through molten chicken fat containing acetaldehyde, sulfur-containing volatiles are formed. Further work is required to characterize the reaction products and to determine whether reactions of this type do indeed occur during cooking.

Thiamin

Another precursor worthy of note is thiamin, which, when heated, produces a number of sulfur compounds with meatlike flavors. Degradation products of thiamin have been shown to contribute to meat odor (Arnold et al. 1969).

Summing up, we can list the main groups of precursors and types of reactions as being responsible for meat flavor formation (see Fig. 27.6).

MEAT FLAVORING AND MEAT FLAVOR COMPONENTS

Flavoring in the Early Days

So far, little has been said about the identity of the meat flavor components themselves. Because of their close association with the flavoring of meat and other products, let us first take a look at the progress made in the development of meat flavor compositions.

Beef extract has been used for many years to introduce the aroma and taste of beef into many convenience foods. It is, however, expensive and its price tends to fluctuate. This spurred the flavor industries to produce a variety of alternative flavor compositions prepared from readily available raw materials. The first "meat flavors" were skillfully blended spices that were and are still used in processed meats. Once the importance of monosodium glutamate and 5'-nucleotides as flavor potentiators was recognized and these materials became commercially available, they were combined with spices to enhance the flavor of natural meat. The introduction of protein hydrolysates and autolysates, which were combined with the nucleotide and spice blends, culminated in fairly satisfactory replacement of meat extract in certain areas of the food industry.

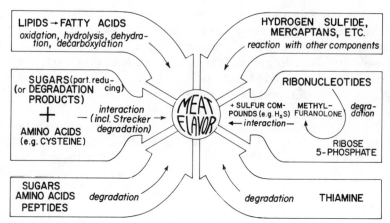

FIG. 27.6. MAIN GROUPS OF FLAVOR PRECURSORS AND TYPES OF
REACTION LEADING TO MEAT FLAVOR

Thus in the United States, protein hydrolysates have almost completely eliminated the use of beef extract in dry soups, bouillon, and canned soups and gravy (Wilson and Katz 1974). Until 1955, the recognized way of creating or enhancing a meat flavor involved some or all of the following ingredients: meat extract, spice blends, flavor potentiators (e.g., monosodium glutamate, 5'-ribonucleotides), protein hydrolysates and autolysates.

These early meat flavors gave a savory rather than a meaty impression. The need was felt for more realistic, meaty flavors which would enable the many characteristic notes of meat flavor to be imitated more closely. Two approaches have been followed in the development of such flavors, namely: (a) the "reaction flavor" approach, in which meat flavor precursors are allowed to react with each other; (b) the compounding of flavor components, isolated from and identified in cooked meat. Combination of the two approaches is also possible. In fact, in meat flavor work the two cannot always be clearly distinguished. As we shall see, studies of reaction flavors or "model systems" may help in the identification of interesting meat flavor components.

Reaction Flavors

Since the late 1950's, the knowledge built up on flavor precursors has been applied in patents describing the manufacture of meat flavors by the reaction of reducing sugars with amino acids in general and with cysteine in particular. Gradually, "artificial" mixtures of amino acids were replaced by protein hydrolysates which, possibly on account of the presence of aldehydes, ketones and sulfur compounds, give an even better flavor effect. The first patents on such reaction flavors, granted to May et al. (1960) and May and Akroyd (1960), represented an important breakthrough. They were followed by many others, too numerous to mention here. These patented flavors are now used worldwide for the flavoring of a variety of food products.

Years later, in 1973, Mussinan and Katz demonstrated that volatiles identified in natural beef flavor were indeed formed when xylose and cysteine are heated together in water.

The first generation of reaction flavors was followed in the 1970's by ones based on the reaction of methylfuranolone with sulfur sources such as cysteine and H_2S (van den Ouweland and Peer 1970; Katz et al. 1971). These resemble much more closely the authentic flavor of roasted meat. In this context it is interesting to note that some of the earlier patents describe meat flavors produced by reaction of cysteine with sugar degradation products such as furans (May 1961) and with alifatic aldehydes (May and Morton 1961; May and Soeters 1968).

Another series of reaction flavor patents are based on the knowledge of the precursor activity of thiamin (Giacino 1968, 1970).

Identification and Blending of Meat Flavor Components

The "reaction flavors" described simulate only partially natural processes. At the most they can be seen as basic ingredients of meat flavors. It is obvious that even closer matches of and more subtle variations of natural flavors can be obtained if "key" individual components, characteristic for a particular flavor, are introduced. Thus, during the last decade, many investigators, particularly Chang, Wick, Watanabe and Sato, and Liebich and their co-workers, and the groups at Unilever and IFF, have concentrated on the analysis of cooked meat, meat broth, and similar sources of natural taste and odor components. This work has resulted in the identification of a range of components for potential application in flavoring compositions. Other key components have been isolated from model systems while a number of other flavor chemicals with interesting "meaty" properties have been identified in studies other than those on natural and model systems. In our discussion of the important compounds, we will distinguish between taste and odor components.

Taste Compounds

Recognition of the importance of taste compounds came with the finding that 5'-ribonucleotides, besides acting in combination with monosodium glutamate as flavor potentiators, had themselves interesting taste properties. Work on the isolation of nonvolatile flavor components started in the 1960's, in our laboratory in 1964.

Commercial meat extract and natural beef broth were selected as starting materials. After removal of the proteins and other high molecular weight compounds, the aqueous concentrate was fractionated by ion-exchange chromatography. After neutralization to the original pH of the concentrate, the fraction containing the organic acids had almost the same taste character as the concentrate, albeit much weaker. None of the other fractions as such had an interesting taste. Upon addition of a rather bitter-tasting fraction containing, among others, the amino acids to the acid fraction, however, the taste of the latter was enhanced to a level comparable to that of the original concentrate. This enhancing effect could be explained only partially by the secondary synergistic action[1] of glutamic acid (in the amino acid fraction) and 5'-IMP (in the organic acid fraction). None of the other individual amino acids or peptides contribute significantly to the taste-enhancing effect of the fraction; together however, they contribute to the meaty taste by way of ternary synergism.[1] As for the meaty taste of the organic acid fraction, we established that, besides 5'-IMP, the most important contributors are lactic acid, succinic acid, pyrrolidonecarboxylic acid and orthophosphoric acid. Their combined taste effect depends strongly on the ratio in which they are mixed.

Some of these findings have been patented; for example, a taste mixture containing pyrrolidonecarboxylic acid, glutamine, succinic acid and 5'-ribonucleotides (Tonsbeek 1968); taste mixtures of amino acids, ribonucleotides, succinic acid and lactic acid (Tonsbeek 1970); and similar mixtures from Ajinomoto (Anon. 1967).

---

1. Only those compounds which are nonvolatile or possess a negligible vapor pressure are considered as taste compounds. Their sensoric effect can only be perceived by the taste buds of the tongue. Besides direct basic taste effects such as acid, sweet, salt and bitter, some taste compounds also demonstrate taste modifying or enhancing effects. Especially in savory foods, such as meat products, the taste enhancement by salt, monosodium glutamate and the 5'-ribonucleotides is of great practical (and economical) importance. These compounds not only enhance the natural taste of many foods (primary synergism) but they also work synergistic on each other (secondary synergism). The term ternary synergism refers to the sensoric effect of nonvolatiles, particularly amino acids and peptides, which individually demonstrate no specific taste or taste-enhancing effect but, if present in substantial amounts, are together responsible for a general taste background.

Odor Compounds

Let us now take a look at the odor compounds. Volatiles isolated from a variety of sources of beef and pork have been recently reviewed by Dwivedi (1975). Some characteristic compounds are listed in Table 27.1

The precursors and reactions responsible for the formation of some of the individual flavor compounds have already been mentioned. For example methylfuranolone is formed from 5'-ribonucleotides, higher aldehydes from fatty acids. Another example is the formation of lower aldehydes and pyrazines via Strecker degradation of amino acids.

It would take us beyond the context of this lecture to deal with all the work on the isolation and identification of odor compounds from model meat flavor systems. The valuable information obtained on the formation of sulfur compounds in the reaction of glucose with sulfur-containing amino acids (Arroyo and Lillard 1970; Linsday and Lau 1972; Mussinan and Katz 1973) and of pyrazines in similar systems (Dawes and Edwards 1966; Koehler et al. 1969; Kato et al. 1972; Shigematsu et al. 1972; Mussinan 1973) is worth mentioning, as are the contributions of Ferretti et al. (1970, 1971, 1973) in this field of nonenzymic browning. We have learned that many of the compounds present in cooked meat flavor are formed in the model systems as well. An important example which we ourselves experienced and which we mentioned earlier is the reaction of methylfuranolone or its thio analog with hydrogen sulfide (van den Ouweland and Peer 1975). Complex mixtures of compounds were produced with an overall odor resembling that of roasted meat. The major components of these mixtures have been isolated by gas-liquid chromatography (GLC) and many of them have been identified. Those identified include mercapto-substituted furan and thiophene derivatives, some of which are listed in Fig. 27.7 with their respective odors.

We established that the initial stage in the formation of the thiophene derivatives involves a partial substitution of the ring oxygen by sulfur to give the thio analog (Fig. 27.8).

It may be postulated that formation of the odorous components proceeds by the pathways illustrated in Fig. 27.9

The main point about these particular model systems is that many of the individual reaction products, in contrast to most components isolated from meat itself, have a real meaty odor. Moreover, various components have a GLC retention time and odor strikingly similar to those of trace components of a natural beef broth which, as yet, have defied identification. Thus, methylfuranolone, besides con-

TABLE 27.1

SOME VOLATILE COMPONENTS ISOLATED FROM
HEAT-TREATED MEAT

Volatile Component	Type of Meat	References
Alcohols, Aldehydes, Ketones		
1-Octanol	Roast Beef	Liebich *et al.* (1972)
2-Heptenal	(drippings)	
2,4-Decadienal		
3-Hydroxy-2-butanone		
Aldehydes		
3 *cis*-Nonenal	Cooked chicken	Harkes and Begemann
4 *cis*-Decenal		(1974)
2t,5c-Undecadienal		
2t,6c-Dodecadienal		
2t,4c,7c-Tridecatrienal		
Alifatic Sulfur Compounds		
	Beef:	
Dimethyl sulfide	Boiled beef	Liebich *et al.* (1972)
1-Methylthio-ethanethiol	Simmered beef	Brinkman *et al.* (1972)
	(head space)	
3-Methylthiopropanal	Shallow fried	Watanabe and Sato
	beef	(1972)
Oxygen-containing Heterocyclic Compounds		

4-butanolide (and
other 1,4/1,5- lactones)

Roast beef — Liebich *et al.* (1972)

2- acetylfuran

Shallow fried beef — Watanabe and Sato (1972)

4-hydroxy-5-methyl-
3(2H)- furanone

Beef broth — Tonsbeek *et al.* (1968)

4-hydroxy-2,5-dimethyl-
3(2H)- furanone

Beef broth — Tonsbeek *et al.* (1968)

5-(methylthio)-
2-furaldehyde

Cooked beef — Herz and Chang (1970)

TABLE 27.1 (Continued)

Volatile Component	Type of Meat	References
Sulfur-containing Heterocyclic Compounds		

2-Thiophene aldehyde

Boiled beef — Herz (1968)

2-Acetylthiophene

Pressure-cooked beef — Mussinan *et al.* (1973)

2,5-Dimethyl-
-1,3,4-trithiolane

Boiled beef — Herz (1968)
Chang *et al.* (1968)
Brinkman *et al.* (1972)
Hirai *et al.* (1973)
Persson and
    von Sydow (1973)
Mussinan *et al.* (1973)

2-Methyl-
-3(2,4,5H)-thiophenone

Pressure-cooked beef — Mussinan *et al.* (1973)

Nitrogen-containing Heterocyclic Compounds

2-acetylthiazole

Pressure-cooked beef — Mussinan *et al.* (1973)

2-acetylthiazoline

Beef broth — Tonsbeek *et al.* (1971)

2,4,5-trimethyloxazoline

Boiled beef — Chang *et al.* (1968)

Continued

TABLE 27.1 (Continued)

Volatile Component	Type of Meat	References
Nitrogen-containing Heterocyclic Compounds (Continued)		

CH₃ structure
1-methyl-2-acetylpyrrole — Shallow fried beef — Watanabe and Sato (1972)

acetylpyrazine structure — Pressure-cooked beef — Mussinan *et al.* (1973)

alkylpyrazines and alkylpyridines — Cooked beef, Shallow fried beef — Koehler *et al.* (1969), Watanabe and Sato (1971)

tributing as such to cooked meat flavor, is almost certainly involved in the formation of other components responsible for meaty flavor notes. The detection of these compounds in cooked meat remains a challenge to the flavor chemist with modern, highly sensitive, analytical tools at his disposal.

We may conclude by observing that of the 250 or more substances which have been reported to contribute to the flavor of cooked meat, only a few have a typically meaty aroma, e.g., the mercapto-substituted furan and thiophene derivatives just mentioned. Evidently, the creation of attractive meat flavors is not merely a question of identifying characteristic meat flavor compounds but also calls for subtle blending of all potential ingredients in just the right proportions for the desired flavor impression.

Much progress has been made in the imitation of natural meat flavors. What challenges still face the flavor chemist, besides those of detection of key trace components and of opportune blending?

CHALLENGES FOR FUTURE RESEARCH

Undoubtedly, considerable progress has been made since Herz and Chang (1970) compiled their survey on meat flavor. Nevertheless there are still areas where our knowledge is inadequate and which are deemed fruitful for further research.

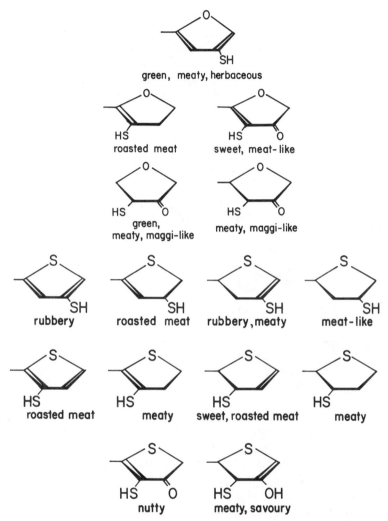

FIG. 27.7. SOME MERCAPTO-SUBSTITUTED FURAN AND THIOPHENE
DERIVATIVES (REACTION PRODUCTS OF Me-FURANOLONE OR ITS
THIO-ANALOG + H₂S)

We have seen that meat quality, including meat flavor, is sensitive
to postmortem conditions, particularly those during rigor. This is
especially true for pork. Yet the differences in metabolic pathways
responsible for flavor modification are poorly understood. Improved
knowledge could be beneficial in determining optimal conditions for
slaughter and postmortem storage.

Another aspect that needs further investigation is the difference in

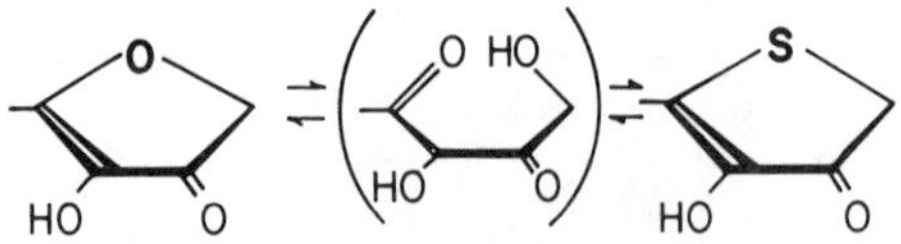

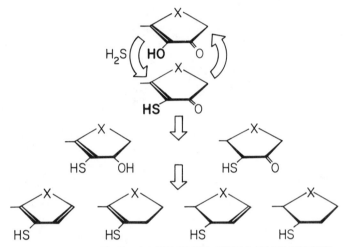

FIG. 27.8. INITIAL STAGE IN THE REACTION
OF METHYLFURANOLONE WITH H₂S

FIG. 27.9. FORMATION OF MEATY FLAVORS FROM METHYL-
FURANOLONE OR ITS THIO-ANALOG AND H₂S

composition between beef, pork, lamb and chicken volatiles. We may wonder whether we know enough about the reactions giving rise to the compounds present in flavor concentrates of cooked meat. In 1970, Herz and Chang thought we needed to know more about this aspect. This still holds today, particularly for roasted meat flavor.

An important challenge to those involved in meat flavor work is presented by the canned foods market. For such products it is important to develop flavors with increased thermal stability or precursor systems which liberate the desired flavor during processing.

An even bigger challenge, though, is posed by the trend towards the extension or replacement of meat by other protein sources and the requirement that, at least for most products, the characteristic meat flavor should be preserved.

The reasons for the challenge are obvious:

(a) First, there is the huge *quantity* of meat flavor expected to be needed for the flavoring of meat extenders and analogs. As the world food problem becomes more critical, meat will become a luxury in the diet of most people. It does not seem possible for beef production, for instance, to keep pace with worldwide demand. Partial replacement of meat (e.g., by vegetable protein) can help to provide the growing world population with high-quality protein foods at reasonable cost. Already there is a growing market for processed protein-rich foods (Wilson and Katz 1974). Heath (1972) estimates that by 1980, 20% of the total amount of manufactured "beef" in the United Kingdom will be accounted for by vegetable protein.

(b) Even more important than the quantity will be the *quality* of meat flavors needed. It will be a challenge to find flavoring materials capable of imitating the appetizing or satisfying aroma of meat in a protein base material, which, at best, has a bland flavor. The meat flavors developed so far, albeit of reasonable quality, are far from meeting the demands of the food industry for today's expanding meat replacement market. There is a clear need for a new generation of meat flavors. This will almost certainly involve blending of reaction flavors with key flavor compounds isolated from meat itself.

(c) Last but not least, more and more attention will have to be paid to improved *perception* of flavors, particularly in "extended" meat products, in which texture, tenderness, succulence and flavor are strongly interrelated. In this respect one may wonder exactly what parameters make meat so highly appreciated. Without going into details of texturing technology, it is obvious that there is a great difference in texture and consequently in the perception of flavor, between meat and textured protein materials.

We started this talk by recalling how meat flavor research owed its origins to local surpluses of meat. Paradoxically, more than a century later, it seems that results of considerable research effort will be put to the greatest benefit in the meat replacement area, which has been stimulated by the threat of meat supply shortages.

BIBLIOGRAPHY

ANON. 1967. Seasoning Compositions. (Assigned to Ajinomoto Co., Inc.) Brit. Pat. 1076948, July 25.

ARNOLD, R. G., LIBBEY, L. M. and LINDSAY, R. C. 1969. Volatile flavor compounds produced by heat degradation of thiamine (vitamin B_1). J. Agric. Food Chem. *17*, 390.

ARROYO, P. T. and LILLARD, D. A. 1970. Identification of carbonyl and sulfur compounds from nonenzymatic browning reactions of glucose and sulfur-containing amino acids. J. Food Sci. *35*, 769.

BATZER, O. F. *et al.* 1960. Precursors of beef flavor. J. Agric. Food Chem. *8*, 498.

BATZER, O. F., SANTORO, A. T., and LANDMANN, W. A. 1962. Identification of some beef flavor precursors. J. Agric. Food Chem. *10*, 94.

BRINKMAN, H. W., COPIER, H., DE LEUW, J. J. M., and TJAN, S. B. 1972. Components contributing to beef flavor. Analysis of the headspace volatiles of beef broth. J. Agric. Food Chem. 20, 177.
CHANG, S. S. et al. 1968. Isolation and identification of 2,4,5-trimethyl-3-oxazoline and 3,5,-dimethyl-1,2,4-trithiolane in the volatile flavor compounds of boiled beef. Chem. Ind. 1639.
DAWES, I. W. and EDWARDS, R. A. 1966. Methyl substituted pyrazines as volatile reaction products of heated aqueous aldose/amino acid mixtures. Chem. Ind. 2203.
DWIVEDI, B. K. 1975. Meat flavor. Crit. Rev. Food Technol. 5, 487.
FERRETTI, A., FLANAGAN, V. P. and RUTH, J. M. 1970. Nonenzymatic browning in a lactose-casein model system. J. Agric. Food Chem. 18, 13.
FERRETTI, A. and FLANAGAN, V. P. 1971. The lactose-casein (Maillard) browning system: volatile components. J. Agric. Food Chem. 19, 245.
FERRETTI, A. and FLANAGAN, V. P. 1973. Characterization of volatile constituents of an N^α-formyl-L-lysine-D-lactose browning system. J. Agric. Food Chem. 21, 35.
GIACINO, C. 1968. Product and process of reacting a proteinaceous substance with a sulfur-containing compound to provide a meat-like flavor. (Assigned to IFF.) U.S. Pat. 3,394,015, July 23.
GIACINO, C. 1970. Meat flavor compositions. (Assigned to IFF.) U.S. Pat. 3,519,437, July 7.
HARKES, P. D. and BEGEMANN, W. J. 1974. Identification of some previously unknown aldehydes in cooked chicken. J. Am. Oil Chemists' Soc. 51, 356.
HARRIS, N. D., STRONG, D. H. and SUNDE, M. L. 1968. Intestinal flora and chicken flavor. J. Food Sci. 33, 543.
HERZ, K. O. 1968. A study of the nature of boiled beef flavor. Diss. Abstr. B29, 1398-B.
HERZ, K. O. and CHANG, S. S. 1970. Meat flavor. Adv. Food Res. 18, 1.
HEATH, H. B. 1970. Flavours: a brief consideration of the chemistry and technology. Flavour Ind. 1, 586.
HEATH, H. B. 1972. Flavours in novel foods. Food Manuf. 47, No. 1, 21.
HIRAI, C., HERZ, K. O., POKORNY, J. and CHANG, S. S. 1973. Isolation and identification of volatile flavor compounds in boiled beef. J. Food Sci. 38, 393.
HORNSTEIN, I. and CROWE, P. F. 1960. Flavor studies on beef and pork. J. Agric. Food Chem. 8, 494.
HORNSTEIN, I. and CROWE, P. F. 1963. Food flavors and odors. Meat flavor: lamb. J. Agric. Food Chem. 11, 147.
KATO, H., SHIGEMATSU, H., KURATA, T. and FUJIMAKI, M. 1972. Maillard reaction products formed from L-rhamnose and ethylamine. Agric. Biol. Chem. 36, 1639.
KATZ, I., PITTET, A. O., WILSON, R. A. and EVERS, W. J. 1971. Procédé et compositions from aromatiser les denrées alimentaires (Method and compositions for flavouring foodstuffs.) (Assigned to IFF.) French Pat. 2,075,449, Sept. 13.
KOEHLER, P. E., MASON, M. E., and NEWELL, J. A. 1969. Formation of pyrazine compounds in sugar-amino acid model systems. J. Agric. Food Chem. 17, 393.
LAWRIE, R. A. 1970. Variation of flavour in meat. Flavour Ind. 1, 591.
LAWRIE, R. A., BOUTON, P. E. and HOWARD, A. 1957. Studies on beef quality. Part 6. Food Invest. Board (London), Spec. Rep. 66.
LIEBICH, H. M. et al. 1972. Volatile components in roast beef. J. Agric. Food Chem. 20, 96.
LINDSAY, R. C. and LAU, V. K. 1972. Some reaction products from nonenzymatic browning of glucose and methionine. J. Food Sci. 37, 787.

LOVE, J. D. and PEARSON, A. M. 1971. Lipid oxidation in meat and meat products. A review. J. Am. Oil Chemists' Soc. 48, 547.
MACY, R. L., JR., NAUMANN, H. D. and BAILEY, M. E. 1964. Water-soluble flavor and odor precursors of meat. 1. Qualitative study of certain amino acids, carbohydrates, non-amino acid nitrogen compounds, and phosphoric acid esters of beef, pork, and lamb. 2. Effects of heating on amino nitrogen constituents and carbohydrates in lyophilized diffusates from aqueous extracts of beef, pork, and lamb. J. Food Sci. 29, 136 (Part 1), 142 (Part 2).
MACY, R. L., JR., NAUMANN, H. D. and BAILEY, M. E. 1970. Water-soluble flavor and odor precursors of meat. 3. Changes in nucleotides, total nucleosides and bases of beef, pork and lamb during heating. 4. Influence of cooking on nucleosides and bases of beef steaks and roasts and their relationship to flavor, aroma and juiciness. 5. Influence of heating on acid-extractable nonnucleotide chemical constituents of beef, lamb and pork. J. Food Sci. 35, 78 (Part 3), 81 (Part 4), 83 (Part 5).
MAY, C. G. 1961. Flavouring substances and their preparation. Brit. Pat. 858,333, Jan. 11.
MAY, C. G. 1974. An introduction to synthetic meat flavour. Food Trade Rev. 44, 7.
MAY, C. G. and AKROYD, P. 1960. Flavouring substances and their preparation. Brit. Pat. 855,350, Nov. 30.
MAY, C. G. and MORTON, J. D. 1961 Flavouring substances and their preparation. Brit. Pat. 858,660, Jan. 11.
MAY, C. G., MORTON, I. D. and AKROYD, P. 1960. Flavouring substances and their preparation. Brit. Pat. 836,694, June 9.
MAY, C. G. and SOETERS, C. J. 1968. Edible products. Brit. Pat. 1,130,631, Oct. 16.
MUSSINAN, C. J. and KATZ, I. 1973. Isolation and identification of some sulfur chemicals present in two model systems approximating cooked meat. J. Agric. Food Chem. 21, 43.
MUSSINAN, C. J., WILSON, R. A. and KATZ, I. 1973. Isolation and identification of pyrazines in pressure-cooked beef. J. Agric. Food Chem. 21, 871. Also Isolation and identification of some sulfur chemicals present in pressure-cooked beef. J. Agric. Food Chem. 21, 873.
PEPPER, F. H. and PEARSON, A. M. 1969. Changes in hydrogen sulfide and sulfhydryl content of heated beef adipose tissue. J. Food Sci. 34, 10.
PEPPER, F. H. and PEARSON, A. M. 1971. Possible role of adipose tissue in meat flavor. The nondialyzable aqueous extract. J. Agric. Food Chem. 19, 964.
PEPPER, F. H. and PEARSON, A. M. 1974. Identification of components of the dialyzable fraction of beef adipose tissue. J. Agric. Food. Chem. 22, 49.
PERSSON, T. and VON SYDOW, E. 1973. Aroma of canned beef: gas chromatographic and mass spectrometric analysis of the volatiles. J. Food Sci. 38, 377.
PIPPEN, E. L., CAMPBELL, A. A. and STREETER, I. V. 1954. Flavour studies. Origin of chicken flavour. J. Agric. Food Chem. 2, 364.
PIPPEN, E. L. and MECCHI, E. P. 1969. Hydrogen sulfide, a direct and potentially indirect contributor to cooked chicken aroma. J. Food Sci. 34, 443.
PIPPEN, E. L., MECCHI, E. P. and NONAKA, M. 1969. Origin and nature of aroma in fat of cooked poultry. J. Food Sci. 34, 436.
SCHUTTE, L. and KOENDERS, E. B. 1972. Components contributing to beef flavor. Natural precursors of 1-methylthio-ethanethiol. J. Agric. Food Chem. 20, 181.
SHIGEMATSU, H., KURATA, T., KATO, H. and FUJIMAKI, M. 1972. Volatile compounds formed on roasting DL-α-alanine with D-glucose. Agric. Biol. Chem. 36, 1631.
TONSBEEK, C. H. T. 1968. Werkwijze voor het bereiden van kunstmatige

smaakstofmengsels met vleessmaak (Procedure for the preparation of artificial aroma mixtures with meat flavour.) Neth. Pat. 6,707,232, Nov. 26.

TONSBEEK, C. H. T. 1970. Flavour composition. Brit. Pat. 1,205,882, Sept. 23.

TONSBEEK, C. H. T., COPIER, H. and PLANCKEN, A. J. 1971. Components contributing to beef flavor. Isolation of 2-acetyl-2-thiazoline from beef broth. J. Agric. Food Chem. *19*, 1014.

TONSBEEK, C. H. T., KOENDERS, E. B., VAN DER ZIJDEN, A. S. M. and LOSEKOOT, J. A. 1969. Components contributing to beef flavor. Natural precursors of 4-hydroxy-5-methyl-3[2H]-furanone in beef broth. J. Agric. Food Chem. *17*, 397.

TONSBEEK, C. H. T., PLANCKEN, A. J. and VAN DE WEERDHOF, T. 1968. Components contributing to beef flavor. Isolation of 4-hydroxy-5-methyl-3[2H]-furanone and its, 2,5-dimethyl homolog from beef broth. J. Agric. Food. Chem. *16*, 1016.

VAN DEN OUWELAND, G. A. M. and PEER, H. G. 1970. Aromasubstanzen und Verfahren zu ihrer Herstellung (Aroma substances and procedure for their preparation.) (Assigned to Unilever N.V.) West German Pat. 1,932,800, Jan. 8.

VAN DEN OUWELAND, G. A. M. and PEER, H. G. 1975. Components contributing to beef flavor. Volatile compounds produced by the reaction of 4-hydroxy-5-methyl-3[2H]-furanone and its thio analog with hydrogen sulfide. J. Agric. Food Chem. *23*, 501.

WASSERMAN, A. E. and GRAY, N. 1965. Meat flavor. 1. Fractionation of water-soluble flavor precursors of beef. J. Food Sci. *30*, 801.

WASSERMAN, A. E. and SPINELLI, A. M. 1972. Effect of some water-soluble components on aroma of heated adipose tissue. J. Agric. Food Chem. *20*, 171.

WATANABE, K. and SATO, Y. 1971. Some alkyl-substituted pyrazines and pyridines in the flavor components of shallow-fried beef. J. Agric. Food Chem. *19*, 1017.

WATANABE, K. and SATO, Y. 1972. Shallow-fried beef: additional flavor components. J. Agric. Food Chem. *20*, 174.

WILSON, R. A. and KATZ, I. 1974. Synthetic meat flavours. Flavour Ind. *5*, 30.

28

Flavor Chemistry of
Tea, Cocoa and Coffee

D. Reymond

Tea, coffee and cocoa beverages contain xanthines. Caffeine is the main xanthine constituent of coffee and tea whereas theobromine is present in relatively large amounts in cocoa. It is beyond the scope of this presentation to review in detail the constituents which were identified up to now in these three commodities. We intend to show how the detailed knowledge of their composition brought scientific explanations for the development of cup flavor during the processing of these stimulant items.

The technology involved in processing is summarized in Table 28.1. It involves a fermentation step during which the tissues of the tea leaf and the cocoa bean cotyledons are submitted to enzymatic degradations. A short wet fermentation step is also performed on

TABLE 28.1

STEPS IN STIMULANT BEVERAGES TECHNOLOGY

Tea (*Camelia sinensis*)	Cocoa (*Theobroma cacao*)	Coffee (*Coffea arabica*)	(*Coffea robusta*)
Fresh Tea Leaves	Cocoa Pods	Coffee Cherries	
Fermentation	Fermentation 3 to 10 days after breaking the pods.	Fermentation 24 hr after breaking cherries	Drying, then breaking the cherries.
Firing in drier	Drying 2–3 weeks in the sun or more rapidly in driers.	Drying 2–3 weeks in the sun or more rapidly in driers.	
Optional scenting			
Black Tea	Cocoa Beans	Coffee Beans	
	Roasting at temperature not exceeding 150°C during 40 min.	Roasting at more than 180°C.	
	Defatting, scenting.		
Infusion	Suspension	Infusion	

high quality *Arabica* beans such as Columbian coffee. Cocoa beans are generally submitted to external fermentations which involve the degradation of mucilage into ethanol by yeasts and a subsequent oxidative fermentation by lactic acid and acetic acid bacteria. This fermentation step is followed by a mild drying operation in the case of tea. The firing operation produces further degradations which participate in the flavor generation in black coffee beans. The temperature/time treatment is particularly pronounced for coffee. The genuine volatile flavor of coffee is therefore due to pyrolytic degradations.

During fermentation, flavor precursors are formed. Since we will see that the composition of volatiles is a very complex one, the identification of precursors open interesting ways to ascertain the origin of certain classes of substances. Varietal and seasonal characteristics contribute to the characteristic flavor notes which are used for blending the commodities in order to ensure an appealing beverage. Future aspects include the establishment of correlative evidences between sensory analyses data and objective determination of quality.

IDENTITY OF VOLATILE AND NONVOLATILE CONSTITUENTS

Flavor impressions are due to the interaction of active constituents with our senses of taste and smell. Volatile and nonvolatile constituents have been submitted to intense investigation since the 1930's. Recent reviews on volatiles show the presence of 296 substances in tea (Yamanishi 1975), with the additional publication of data on 56 substances (Vitzthum *et al.* 1975A). Coffee volatiles groups now show 438 constituents (Vitzthum 1975). A review shows the presence of 225 constituents in cocoa (Van der Wal *et al.* 1971), to which were added 59 substances (Vitzthum *et al.* 1975B). Nonvolatiles constitute also an impressive list of substances; their role in flavor was reviewed by Feldman *et al.* (1969) on coffee, by Millin *et al.* (1969) on tea and by Fincke *et al.* (1965) on cocoa. From these data, the influence of the important contributors to flavor can be established by comparisons between the compositions of tea, cocoa and coffee. These important contributors are:

1. Alcohols related to monoterpenes, oxidation products of carotenes, lactones and oxidation products of flavanols in tea.

2. Aliphatic esters, unsaturated aromatic carbonyls, pyrazines, polyphenols and peptides in cocoa.

3. Furan derivatives including alcohols, carbonyls and sulfur-containing compounds, pyrazines, pyrroles, oxazoles and acids in coffee.

It is, however, impossible to select flavor-active substances from the direct examination of the composition data. We will see that mathematical methodology is required for localizing these important constituents on differential chromatograms.

CHANGES INDUCED BY FERMENTATION

These changes were reviewed in cocoa by Roelofsen (1958) and in tea by Sanderson (1972). Less information is presently available on coffee fermentation. Main changes involve proteolytic formation of free amino acids and small peptides as well as the formation of reducing sugars.

In tea, a model of oxidative fermentation was established by Sanderson and Graham (1973). Flavanols, such as (-)epicatechin and (-)epigallocatechin, undergo an oxidative polymerization to give a theaflavin (Takino *et al.* 1965) as shown in Fig. 28.1. Theaflavin ex-

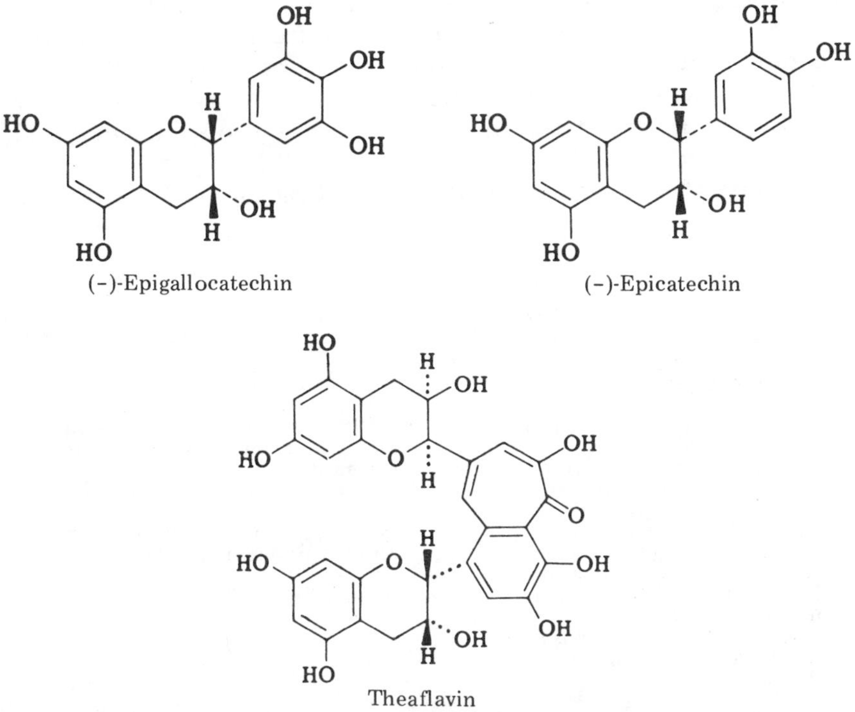

(-)-Epigallocatechin   (-)-Epicatechin

Theaflavin

FIG. 28.1. THEAFLAVIN STRUCTURE

During the oxidative fermentation of tea leaves, flavanols such as (-)epicatechin and (-)epigallocatechin undergo a polymerization. From these precursors theaflavin is formed.

hibits a benztropolone structure. The enzyme-catalized oxidative coupling seems rather unspecific towards substrates so that other theaflavins are also formed from other flavanols and flavanol-gallates (Coxon *et al.* 1970B; Bryce *et al.* 1970). This polymerization continues further and thearubigins exhibiting complex structures are formed (Brown *et al.* 1969). These oxidative patterns can be followed by means of gel permeation chromatography. Millin *et al.* (1969) confirmed earlier findings of E. H. Roberts in establishing that polymerization products of flavonols play an outstanding role in the mouth feel of black tea infusions as mentioned in Table 28.2. Biswas *et al.* (1973) evaluated statistically the factors affecting the valuation of North East Indian plain teas; three constituents, theaflavins, thearubigins and (–)epicatechin gallate, are considered to play an important role in the briskness quality of the samples. Similar conclusions were reached by Takeo (1974) on characterizing the quality of Japanese tea samples.

The oxidative fermentation of black tea reduces the astringency of the beverage; it is accompanied by changes in essential oil composition (Yamanishi *et al.* 1966), particularly by an increase in *trans*-2-hexenal, the formation of linalol-oxides from linalool, and the generation of dimethylsufide from a salt of methylmethionine-sulfonium (Kiribuchi and Yamanishi 1963). *Trans*-2-hexenal was recently shown to arise from the oxidation of linolenic acid (Gonzalez *et al.* 1972).

The model proposed by Sanderson and Graham (1973) gives also

TABLE 28.2

MOUTH FEEL OF TEA NONVOLATILES

Nonvolatiles Present in the Fraction	Contribution to Mouth Feel
Low molecular weight flavonols and their 3-glycosides	Slightly astringent, metallic, woody; play minor role in the taste of tea beverage.
Low molecular weight polysaccharides combined with nitrogenous constituents	Malty taste.
Intermediate molecular weight fraction including theaflavins, leucoanthocyanins, theogallin	Briskness, astringency, strength; maybe unpleasant and harsh when in excess.
High molecular weight fraction including polysaccharides (34%), proteins (6%), nucleic acids (1%), thearubigins and phenolic substances (32%)	Softness, flatness.

NOTE: Fractions were first separated by means of gel permeation chromatography and by dialysis; they were then freeze-dried. Tasting was performed on preparations dissolved in water at a suitable concentration.

an explanation for the presence of carotene oxidation products in black tea aroma. This group of substances includes ionones, damascones, damascenones as well as derived condensation products (see Fig. 28.2). The important role of these substances in flavor contribution is due to the fact that some as β-ionone exhibit very low thresholds of perception. Theaspirone and dihydroactinidiolide play also a preponderant role in black tea aroma (Marx 1973). Carotene oxidation products arise from the partial oxidation of the 14 carotenes described in tea (Tirimanna and Wickremasinghe 1965; Blanc 1972).

The oxidative changes induced by the oxidation of tea flavonols explain also the presence of carbonyls which arise from the oxidative Strecker degradation of free amino acids. The conversion of labelled amino acids to aldehydes in tea leaf homogenates was followed (Co and Sanderson 1970); a model fermentation system was even realized showing that some amino acids, among which leucine, isoleucine, valine and phenylalanine, are partially converted to corresponding aldehydes. These experiments confirm the findings of earlier authors (Finot *et al.* 1967; Saijo and Takeo 1970). The complex fermentation of cocoa produces important changes in the structure and the composition of the cotyledons. The review of Roelofsen (1958) shows the important contributions of W. G. Forsyth in elucidating polyphenol modifications. The external alcoholic and lactic acid

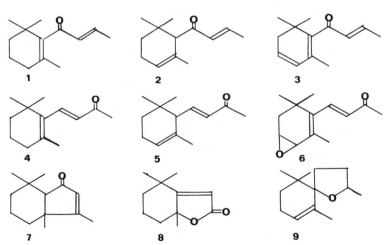

FIG. 28.2. VOLATILES RESULTING FROM CAROTENE'S OXIDATION IN BLACK TEA

1—β-Damascone. 2—α-Damascone. 3—β-Damascenone. 4—β-Ionone. 5—α-Ionone. 6—5,6-Epoxy-β-ionone. 7—1,5,5,9-Tetramethylbicyclo(4.3.0)non-8-en-7-one. 8—Dihydroactinidiolide. 9—Theaspirone.

fermentation steps induce autolytic changes in reserve cells of cocoa cotyledons; at the end of these fermentations, the cytoplasmic material is surrounded by a lipidic phase formed by the diffusion of vacuoles (Biehl 1973). During this fermentation, anthocyanins diffuse from reserve cells and they are submitted to an oxidative polymerization under the contact with polyphenoloxidase. The astringent taste of cocoa is thereby reduced. Changes in polyphenols can be followed by means of chromatographic methods. A dimeric leucocyanidin was characterized by Forsyth and Roberts (1960). During fermentation, flavor precursors are formed; cocoa flavor constitutes a particular field in which the importance of biodegradations for the formation of flavor has been especially well studied (Mohr et al. 1971; Rohan 1970). Important enzymatic reactions are related to the formation of amino acids and peptides from proteins, the formation of reducing sugars by the hydrolysis of anthocyanins, and to the oxidation of polyphenols. Separation into chemical classes of constituents was performed by means of gel chromatography. Carbohydrates, flavonoids, flavylogens, phenolic acids and amino acids are necessary to produce cocoa flavor when this mixture is submitted to roasting conditions (Table 28.3). A semiartificial mixture reconstituted from these data was also shown to create cocoa flavor on heating.

Volatile aldehydes are generated from free amino acids by Strecker degradations. Ethylesters of aliphatic and aromatic acids are relatively important in cocoa volatiles. Their presence can be attributed to the fact that the cocoa beans undergo first an anaerobic alcoholic fermentation. Less information is presently available on the changes induced in the green coffee beans by the fermentation step. Among the 96 volatiles that were recently identified (Vitzthum et al. 1975C), 2-methoxy-3-methylpyrazine certainly plays an important role in the pungent odor of fermented and dried green coffee beans. These volatiles are present at a level which is 50 times less abundant than the volatiles level in roasted coffee. Russwurm (1969) also performed

TABLE 28.3

PRECURSORS FOR VOLATILE FLAVOR OF COCOA

Carbohydrates: glucose, fructose
Flavonoids: epi-catechin, catechin
gallocatechin, epi-gallocatechin
Flavylogens: quercetin, quercetrin
Phenolic acids: p-coumaric acid
chlorogenic acids
Amino acids: 22 constituents including
valine and leucine

a preliminary fractionation of coffee flavor precursors by means of gel permeation chromatography. Earlier work in this field was mainly based on a comparison of compositions between green and roasted coffee samples. The most important flavor precursors are trigonelline, sucrose, fructose, glucose, free amino acids and peptides. Nonvolatile acids also play a very important role, which will be covered in the discussion below of flavor generation during the roasting of coffee beans.

CHANGES INDUCED BY THERMAL TREATMENTS

We have seen that the flavor of stimulant beverages is developed by thermal treatments such as firing in the case of tea and roasting for cocoa and coffee. The thermal generation of flavors has been extensively studied.

The firing step used in black tea manufacture produces a loss in volatile constituents (Yamanishi *et al.* 1966). The recent publication on the occurrence of pyrazines, pyridines and quinolines in the basic fraction of tea (Vitzthum *et al.* 1975A) led us to presume that firing can also contribute to the formation of these substances from precursors. Free sugars and amino acids are known, indeed, to generate such substances on heating. This complex reaction was recently reviewed by Adrian (1975). Heterocyclic constituents are created such as thiazoles, furanic compounds, pyrazines and oxazoles. The formation of pyrazines (Vitzthum 1975) can be explained as follows: a-diketones are initially formed from carbohydrates; they then react with a nitrogen source such as ammonia or amino acids to form aminoketones; these compounds easily form dihydropyrazines which are then oxidized into pyrazines (see Fig. 28.3). Pyrazines can also be formed by the cyclization of a-hydroxyamino acids such a serin or threonin (Odell 1973). An acylation step explains also the formation of oxazoles from a-aminoketones. The rate of a-diketones formation can be followed during coffee roasting by means of head-space gas chromatography (Reymond *et al.* 1965); it shows a maximum rate at the initial phase of the thermal treatment. The pyrazine content in cocoa beans is temperature-dependent (Keeney 1972); therefore, it can be used as an analytical index for following the roasting operation.

Trigonelline, the inner salt of 3-carboxy-1-methylpyridinium hydroxide, is present at the 1% level in green coffee. This substance undergoes a pyrolytic degradation during coffee roasting; in this manner, pyridines and pyrroles are generated and their origin can be guessed from model reactions (Viani and Horman 1974) as outlined

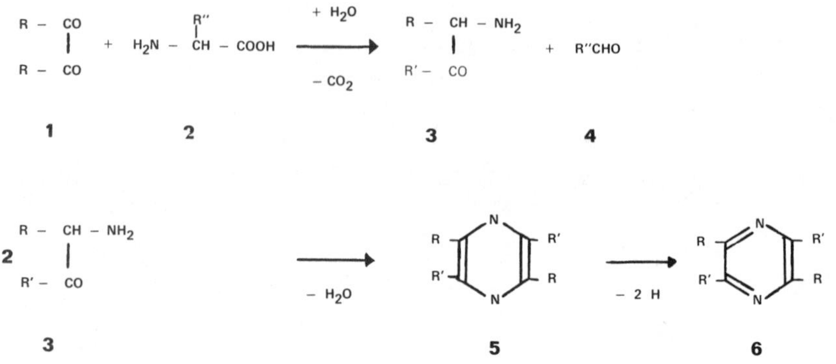

FIG. 28.3. FORMATION OF PYRAZINES FROM α-DIKETONES

α-diketones (1) are formed during the browning reaction by the degradation of carbohydrates. They can react with amino-acids (2) to form α-aminoketones (3) and aldehydes (4). Aminoketones react together to give cyclization products, dihydropyrazines (5), which are further dehydrogenated to give pyrazines (6).

in Fig. 28.4. Worth mentioning is the fact that nicotinic acid is also generated in trace amounts during coffee roasting; its concentration in coffee beverage is, however, sufficient for recommanding coffee consumption as a complementary source of vitamin PP (Adrian *et al.* 1971) for curing endemic pellagra.

Volatile aldehydes are generated from free amino acids by Strecker degradations; the presence of 5-methyl-2-phenyl-2-hexenal in cocoa (van Praag *et al.* 1968) can be attributed to an aldol condensation between 3-methylbutanal and phenylethanal as explained in Fig. 28.5. Similar aldol condensation products were also recently identified in tea aroma, as outlined in the review by Yamanishi (1975). Several furan derivatives were identified in coffee volatiles; their origin can be attributed to the pyrolytic decomposition of reducing sugars (Flament *et al.* 1967). The occurrence of important furan-sulfur derived substances in coffee volatiles (see Fig. 28.6) cannot yet be attributed to given precursors; they certainly arise from interactions between carbohydrates and sulfur-containing substances. Worth mentioning is the early description of furfurylmercaptan by Reichstein and Staudinger in 1925. This substance is an important contributor to coffee flavor.

The pleasant bitterness of cocoa was recently shown to be due to the interaction of cyclic dipeptides with the xanthine theobromine (Pickenhagen *et al.* 1975); these cyclic dipeptides are formed in trace amounts during cocoa roasting.

Acidity plays an important role in the flavor acceptance of coffee beverages. During the roasting operation, citric and malic acids are

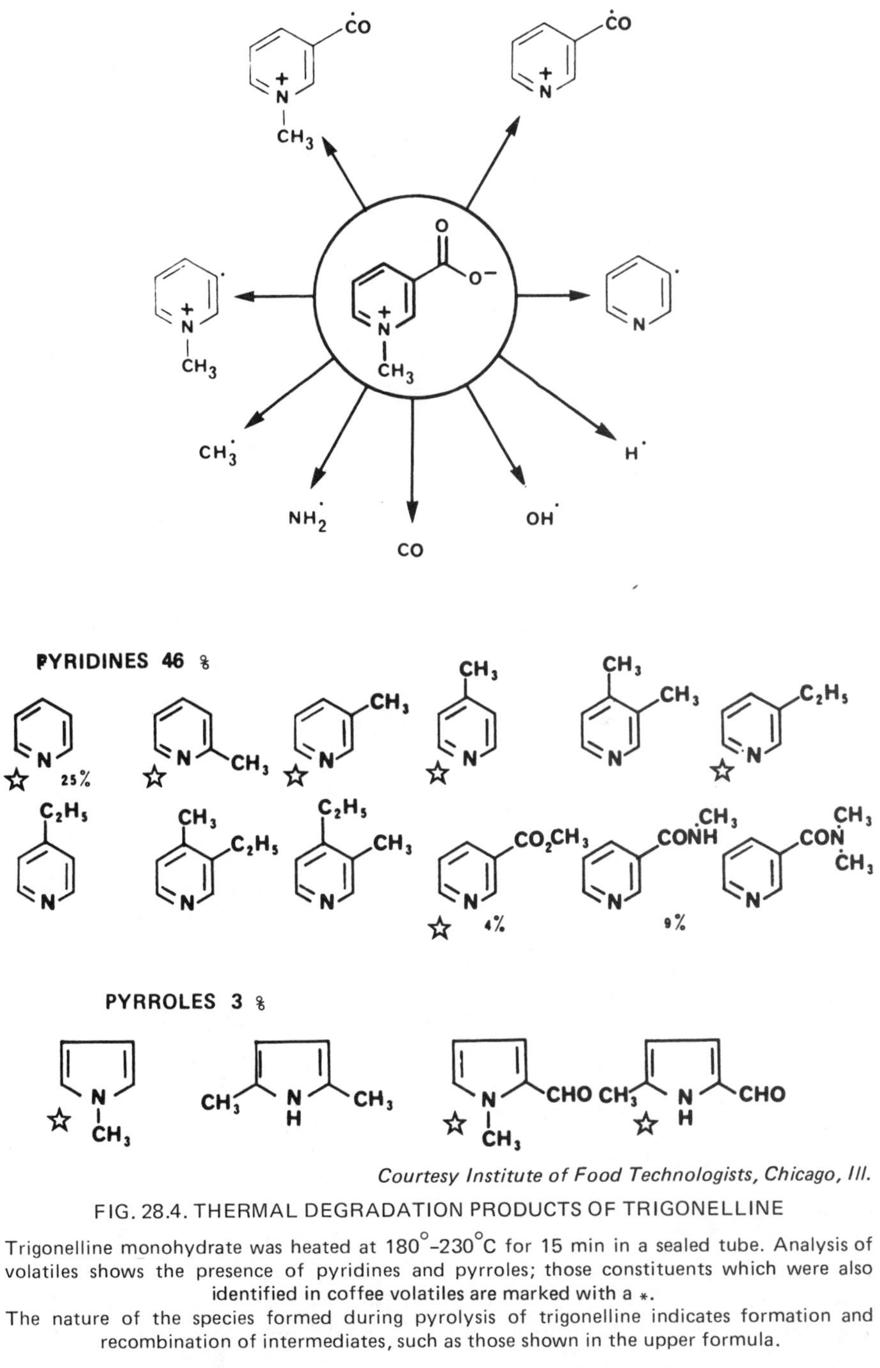

PYRIDINES 46 %

PYRROLES 3 %

Courtesy Institute of Food Technologists, Chicago, Ill.

FIG. 28.4. THERMAL DEGRADATION PRODUCTS OF TRIGONELLINE

Trigonelline monohydrate was heated at 180°–230°C for 15 min in a sealed tube. Analysis of volatiles shows the presence of pyridines and pyrroles; those constituents which were also identified in coffee volatiles are marked with a *.

The nature of the species formed during pyrolysis of trigonelline indicates formation and recombination of intermediates, such as those shown in the upper formula.

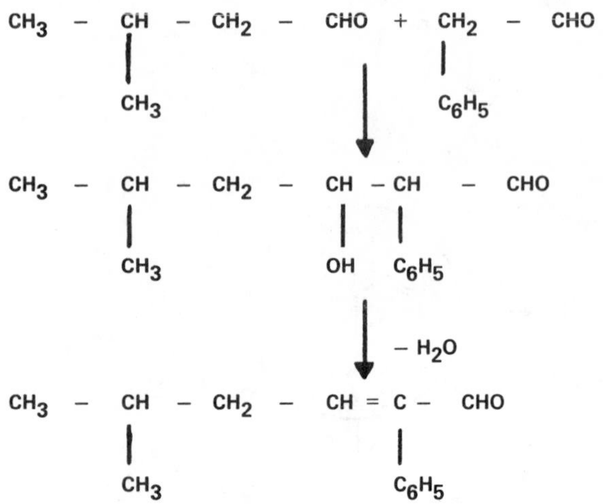
CH3 – CH – CH2 – CHO + CH2 – CHO
CH3
C6H5
CH3 – CH – CH2 – CH – CH – CHO
CH3
OH
C6H5
– H2O
CH3 – CH – CH2 – CH = C – CHO
CH3
C6H5

FIG. 28.5. THERMAL ALDOLIZATION REACTIONS

By Strecker degradations, L-leucine gives 3-methylbutanal and L-phenylalanine gives phenylethanal. These two aldehydes are certainly the precursors of an important flavor contributor in roasted cocoa, 5-methyl-2-phenyl-2-hexenal via an intermediary aldolization product.

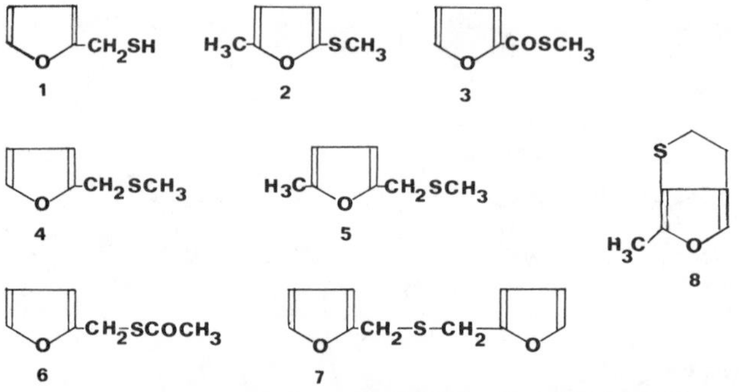
CH2SH
1
H3C
SCH3
2
COSCH3
3
CH2SCH3
4
H3C
CH2SCH3
5
S
H3C
8
CH2SCOCH3
6
CH2–S–CH2
7

FIG. 28.6. FURAN SULFUR COMPOUNDS IN COFFEE

1—Furfurylmercaptan. 2—2-Methylthio-5-methylfuran. 3—Methylthiofuroate. 4—Furfurylmethylsulfide. 5—5-Methylfurfuryl-2-methyl sulfide. 6—Furfurylthioacetate. 7—Difurfuryl sulfide. 8—Kahweofuran.

partly degraded into aconitic, itaconic, *cis*-citraconic, *trans*-mesaconic, *cis*-maleic and *trans*-fumaric acids as shown by quantitative separations by means of silica-gel chromatography (Woodman *et al.* 1967) (see Fig. 28.7). The contributions of various classes of acids to the flavor of coffee beverages was established; as shown in Table 28.4, these previously-mentioned acids, as well as phenolic acids, such as chlorogenic acids, and volatile acids contribute to the

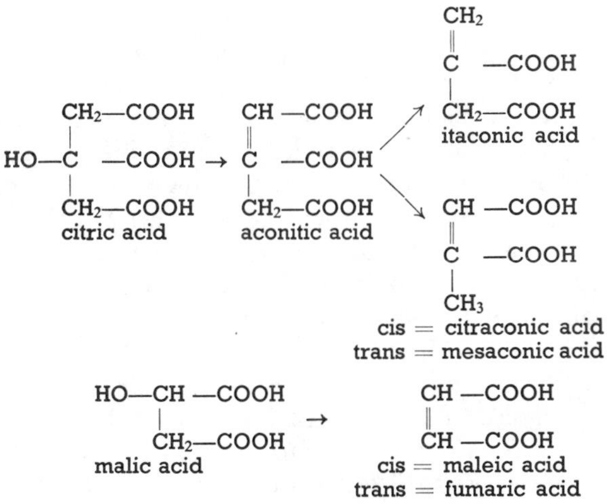

$$CH_2$$
$$\parallel$$
$$C\ —COOH$$
$$\mid$$
$$CH_2—COOH$$
itaconic acid

$$CH_2—COOH \qquad CH\ —COOH$$
$$\mid \hspace{2cm} \parallel$$
$$HO—C\ —COOH \rightarrow C\ —COOH$$
$$\mid \hspace{3cm} \mid$$
$$CH_2—COOH \qquad CH_2—COOH$$
citric acid   aconitic acid

$$CH\ —COOH$$
$$\parallel$$
$$C\ —COOH$$
$$\mid$$
$$CH_3$$
cis = citraconic acid
trans = mesaconic acid

$$HO—CH\ —COOH \qquad\qquad CH\ —COOH$$
$$\mid \hspace{3cm} \rightarrow \hspace{2cm} \parallel$$
$$CH_2—COOH \hspace{3cm} CH\ —COOH$$
malic acid   cis = maleic acid
                          trans = fumaric acid

Courtesy ASIC, Paris

FIG. 28.7. DEGRADATIONS OF ACIDS DURING COFFEE ROASTING

Silicagel chromatographic analyses were performed on coffee samples which were roasted at various intensities. Partial degradations were observed for citric acid and malic acid. These reactions involve dehydrations and decarboxylations.

balance in acidity (Feldman *et al.* 1969). These acids are determined by means of gel permeation and gas chromatographic techniques.

INFLUENCE OF VARIETAL AND SEASONAL FACTORS ON THE COMPOSITION

Previous knowledge gained on the chemical composition of stimulant beverages allow us to determine the influence of the blend on

TABLE 28.4

DISTRIBUTION OF ACIDS IN COFFEE BEVERAGES
IN MILLIEQUIVALENT VALUES ON 100 GRAM D.W.

| | Colombian | | Santos | |
	Dark Roast	Medium Roast	Dark Roast	Medium Roast
C_1 to C_{10} volatile acids	7.34	7.41	6.72	7.68
Hydroxyacids diacids	4.20	4.55	4.50	4.13
Phenolic acids	19.25	16.18	15.50	20.82
pH of beverages	4.62	4.82	5.35	5.18

the overall characteristics of the beverage. It is important here to stress the complexity of the analytical information due to the refinements of separation techniques. Relevant information must be selected from numerous data. It is therefore necessary to use mathematical methods for obtaining this selection. Biggers *et al.* (1969) determined, for instance, chromatographic regions which can be used for ascertaining the quality of coffee samples which are used in blends. In this manner, regions of gas chromatograms were determined for characterizing Arabica and Robusta varieties. By submitting blends of various compositions to a taste panel, a good agreement was found between the sensory ranking and quality indices derived from chromatographic information as mentioned in Table 28.5. In some cases, simpler methodology can also be used for obtaining the selection of relevant information. In differential chromatographic analysis (Figure 28.8), quantitative data obtained for a sample considered as standard are compared with data presented by a sample of another variety or other origin. This technique was applied for determining the influence of the origins of the beans on the composition of cocoa volatiles (Mueggler-Chavan and Reymond 1967). Trimethylethylpyrazine, an important flavor contributor, was found to be much more important in beans from Ecuador than in beans from Accra. Similar comparisons also allow characterization of typical flavors. Bahia beans, for instance, present a relative excess in gaiacol, phenol and 2-formylpyrrole; this composition explains the typical "smoked" note of the sample (see Table 28.6). Tannin contents were also used for cocoa beans differentiation (Gryúner *et al.* 1971).

Mathematical methodology can also be used for determining

TABLE 28.5

QUALITY INDICES OF COFFEE BLENDS

Composition of Blend		GLC Quality Index	Rank of Increasing Robusta Content by Quality Index
Arabica (%)	Robusta (%)		
100	0	1081.1	2
95	5	1108.4	1
90	10	978.7	3
85	15	957.5	5
80	20	962.0	4
75	25	934.9	6
70	30	815.7	7
65	35	793.2	8
60	40	689.8	9
55	45	678.7	10
50	50	656.4	11

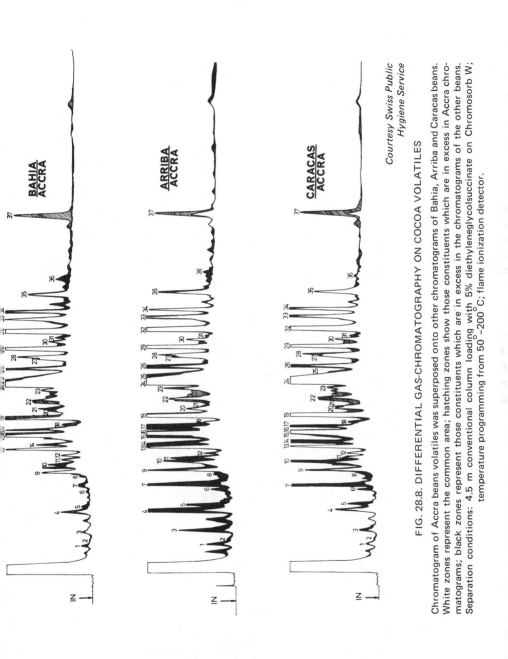

FIG. 28.8. DIFFERENTIAL GAS-CHROMATOGRAPHY ON COCOA VOLATILES

Chromatogram of Accra beans volatiles was superposed onto other chromatograms of Bahia, Arriba and Caracas beans. White zones represent the common area; hatching zones show those constituents which are in excess in Accra chromatograms; black zones represent those constituents which are in excess in the chromatograms of the other beans. Separation conditions: 4.5 m conventional column loading with 5% diethyleneglycolsuccinate on Chromosorb W; temperature programming from 50–200°C; flame ionization detector.

Courtesy Swiss Public Hygiene Service

TABLE 28.6

INFLUENCE OF VARIETAL FACTORS ON COCOA VOLATILES

Balance index (I) was calculated from gas chromatographic data as:

$$I = \frac{\text{substance area}}{\text{surface of 2-phenylethanol}} \times \frac{\text{surface of 2-phenylethanol in Accra beans}}{\text{substance area in Accra beans}}$$

	Balance Index I	
	Bahia	Arriba
Trimethylethylpyrazine	3.1	14.7
Gaiacol	2.6	0.8
Phenol	2.0	0.8
Formyl-2-pyrrole	2.8	0.9
Hexanol	1.2	6.1
Dimethylsulfide	0.1	0.1
Tetramethylpyrazine	0.5	0.3

chromatographic regions which show variations in intensity related to seasonal variations. In applications to tea volatile analyses, important flavor contributors were selected in this manner (Vuataz and Reymond 1971; Gianturco et al. 1974). Table 28.7 shows the importance of trans-2-hexenal, linalool and its oxides, pentenols, hexenols, and β-ionone to various odor notes which play an important role in the variations of tea quality. The importance of balance between odor notes in tea was also established by using the differential chromatographic analysis (Wickremasinghe et al. 1973; Kozhin and Treiger 1973).

CORRELATIONS BETWEEN BEVERAGE QUALITY AND COMPOSITION

The knowledge gained up to now shows the usefulness of composition data for following changes induced by the action of technolog-

TABLE 28.7

IMPORTANT FLAVOR CONTRIBUTORS IN SEASONAL VARIATION OF TEA

Constituents	Odor Note
trans-2-hexenal	Coarse vegetable aroma
Linalool	
Linalool-5-membered oxides	Light floral aroma
Hexanal	
1-Penten-3-ol	
cis-2-penten-1-ol	
cis-3-hexen-1-ol	Fresh leafy aroma
trans-2-hexen-1-ol	
β-Ionone	Strong floral aroma

ical treatments on tea, cocoa and coffee. Furthermore, the influences of origins and seasonal variations can also be followed by applying quantitative techniques. For building up these contributions, refined analytical techniques as those outlined by Teranishi (1976) were used.

Difficulties are encountered, however, when correlations have to be established between beverage sensory quality assessments and composition data. Modern sensory descriptions use many terms for assessing the quality of a beverage. Descriptive multidimensional models were recently used for the description of tea (Palmer 1974) and cocoa (Vuataz et al. 1974). In evaluating the quality of chocolate samples, a sensory evaluation taste panel was first trained to perform with satisfactory reliability on the profile analysis of an unknown sample in comparison with a standard. Factor analysis was then applied to check the correlations between samples and standard. The description of flavor was expressed as quality attributes. The configurations of samples was then represented in a space of minimum dimensions by using multidimensional scaling and other suitable mathematical techniques. Principal component analysis shows which quality attributes exhibit the largest loadings on each of the first canonical axis. By using a multidimensional scaling, a classification of the samples is obtained.

This new way of expressing sensory assessments of beverage quality opens a future field in which the selection of important analytical indices could perhaps be related to sensory data. We see, therefore, that one important future aspect of flavor research will be the establishment of such correlations as outlined on preliminary studies relating coffee volatile profiles with quality assessments (Tassan et al. 1974).

SUMMARY

Present information gained on the chemistry of tea, coffee and cocoa flavors is due to the publications of many laboratories in which scientists of several disciplines have brought their contributions. The accumulated scientific knowledge affords a better understanding of the technological processes of fermentation and roasting. During fermentation, flavor precursors are formed during various enzymatic degradations. These flavor precursors react during the roasting processes and generate important flavor notes. Refined analytical techniques follow the development of volatile and non-volatile constituents during the processing operations. Mathematical

methodology affords a convenient tool for selecting flavor contributors; the balance between these contributors plays an important
role in ascertaining characteristic changes induced by seasonal variations and for analyzing the typical flavor due to the origin of the
commodities.

The objective characterization of flavors requires more basic research to permit correlation of sensory analyses data with the
selected information obtained by physicochemical techniques.

Another fascinating aspect is the assistance afforded food technologists in their work to improve processing—and thereby quality—of
tea, cocoa and coffee through application of the biotechnological
sciences.

BIBLIOGRAPHY

ADRIAN, J. 1975. The Maillard reaction as a source of food aromas. Labo.
Pharma. Prob. Tech. *244*, 614-619. (French)
ADRIAN, J. *et al.* 1971. Curing pellagra by means of coffee consumption due to
its PP vitamin content. 4th Intern. Colloquium on Coffee Chemistry, Amsterdam. ASIC, Paris, France. (French)
BIEHL, B. 1973. Changes in subcellular structure of cocoa (*Theobroma cacoa*
L). cotyledons during fermentation and drying. Z. Lebensm. Unters. Forsch
153, 137-150. (German)
BIGGERS, R. E., HILTON, J. J. and GIANTURCO, M. A. 1969. Differentiation
between *Coffea arabica* and *Coffea robusta* by computer evaluation of gas
chromatographic profiles—comparison of numerically derived quality predic-
tions with organoleptic evaluations. J. Chromatogr. Sci. *7*, 453-472.
BISWAS, A. K., SARKAR, A. R. and BISWAS, A. K. 1973. Biological and
chemical factors affecting the valuation of North Indian plain teas. III. Statistical evaluation of the biochemical constituents and their effect on colour,
brightness and strength of black teas. J. Sci. Food Agric. *24*, 1457-1477.
BLANC, M. 1972. Distributions and variations of tea carotinoids during technological treatments. Lebensm. Wiss. Technol. *5*, No. 3, 95-97. (French)
BROWN, A. G., EYTON, W. B., HOLMES, A. and OLLIS, W. D. 1969. Identification of the thearubigins as polymeric proanthocyanidins. Nature (London)
221, 742-744.
BRYCE, T. *et al.* 1970. The structures of the theaflavins of black tea. Tetrahedron Lett. *32*, 2789-2792.
CO, H. and SANDERSON, G. W. 1970. Biochemistry of tea fermentation: conversion of amino acids to black tea aroma constituents. J. Food Sci. *35*,
160-164.
COXON, D. T., HOLMES, A., OLLIS, W. D. and VORA, V. C. 1970A. The
constitution and configuration of theaflavin pigments of black tea. Tetrahedron Lett. *60*, 5237-5240.
COXON, D. T., HOLMES, A. and OLLIS, W. D. 1970B. Isotheaflavin. A new
black tea pigment. Tetrahedron Lett. *60*, 5241-5246.
FELDMAN, J. R., RYDER, W. S. and KUNG, J. T. 1969. Importance of non-
volatile compounds to the flavor of coffee. J. Agric. Food Chem. *17*, 733-739.
FINCKE, A., LANGE, H. and KLEINERT, J. 1965. Handbook on cocoa
products. Springer Verlag, Berlin, Heidelberg, New York. (German)

FINOT, P. A. MÜGGLER-CHAVAN, F. and VUATAZ, L. 1967. Phenylalanine as precursor of phenylethanal in black tea aroma. Chimia *21*, 26-27 (French)

FLAMENT, I. *et al.* 1967. Chemical and spectroscopical aspects of furanic compounds in coffee aroma. 3rd Int. Colloquium on Coffee Chemistry, Trieste, ASIC, Paris, France. (French)

FORSYTH, W. G. and ROBERTS, J. B 1960. Cacao polyphenolic substances. 5. The structure of cacao "leucocyanidin 1." Biochem. J. *74*, 374-378.

GIANTURCO, M. A., BIGGERS, R. E. and RIDLING, B. H. 1974. Seasonal variations in the composition of the volatile constituents of black tea. A numerical approach to the correlation between composition and quality of tea aroma. J. Agric. Food Chem. *22*, 758-764.

GONZALEZ, J. G., COGGON, P. and SANDERSON, G. W. 1972. Biochemistry of tea fermentation: formation of t-2-hexenal from linoleic acid. J. Food Sci. *37*, 797-798.

GRYUNER, V. S., SELEZNEVA, G. D. and ZAKHAROVA, N. V. 1971. Quantitative determination of polyphenols in cocoa beans. Izvestyia Vysshikh Uchebnykh Zvadenii, Pishch. Tekhnol. *4*, 141-142.

KEENEY, P. G. 1972. Various interactions in chocolate flavor. J. Am. Oil Chemists' Soc. *49*, 567-572.

KIRIBUCHI, T. and YAMANISHI, T. 1963. Studies on the flavor of green tea. IV. Dimethylsulfide and its precursor. Agric. Biol. Chem. *27*, 56-59.

KOZHIN, S. A. and TREIGER, N. D. 1973. Gas liquid chromatography and computer aided evaluation of the influence of components on the essential oil of Georgian black tea on its aroma. Prikl. Biokhim. i Mikrobiol. *9*, 895-900.

MARX, J. N. 1973. Synthesis, absolute stereochemistry and odor properties of theaspirone, an odoriferous principle of tea. Am. Chem. Soc. Abstr. Papers *166*, AGFD 43.

MILLIN, D. J., CRISPIN, D. J. and SWAINE, D. 1969. Non-volatile components of black tea and their contribution to the character of the beverage. J. Agric. Food Chem. *17*, 717-722.

MOHR, W., RÖHRLE, M. and SEVERIN, T. 1971. On the formation of cocoa aroma from its precursors. Fette, Seif., Anstr. Mittel *73*, 515-521. (German)

MÜGGLER-CHAVAN, F. and REYMOND, D. 1967. Aroma constituents of cocoa samples from various countries. Mitt. Geb. Lebensm. Unters. Hyg. *58*, 466-472. (French)

ODELL, G. 1973. The role of pyrazine compounds in the flavor of coffee. 6th Intern. Colloquium on the Chemistry of Coffee, Bogota. ASIC, Paris, France.

PALMER, D. H. 1974. Multivariate analysis of flavour terms used by experts and non-experts for describing teas. J. Sci. Food Agric. *25*, 153-164.

PICKENHAGEN, W. *et al.* 1975. Identification of the bitter principle of cocoa. Helv. Chim. Acta *58*, 1078-1086.

REYMOND, D., PICTET, G. and EGLI, R. H. 1965. Analytical features of coffee aroma. 2nd Intern. Colloquium on Coffee chemistry, Paris, ASIC, Paris, France. (French)

ROELOFSEN, P. A. 1958. Fermentation, drying and storage of cocoa beans. Adv. Food Res. *8*, 225-296.

ROHAN, T. A. 1970. Chocolate flavor, its precursors and their reactions. Gordian, *1*, 5-8, 53-62, 111-115. (German)

RUSSWURM, H. 1969. Fractionation and analysis of aroma precursors in green coffee. 4th Intern. Colloquium on Coffee Chemistry, Amsterdam. ASIC, Paris, France.

SAIJO, R. and TAKEO, T. 1970. The formation of aldehydes from amino acids by tea leaves extracts. Agric. Biol. Chem. *34*, 227-233.

SANDERSON, G. W. 1972. The chemistry of tea and tea manufacturing. *In* Symposia Phytochemical Society of N. America: Structural and Functional Aspects of Phytochemistry, Recent advances in Phytochemistry. V. C.

Runeckles (Editor). Academic Press, New York, San Francisco, and London, England.

SANDERSON, G. W. and GRAHAM, H. N. 1973. On the formation of black tea aroma. J. Agric. Food Chem. *21*, 576–585.

TAKEO, T. 1974. Photometric evaluation and statistical analysis of tea infusion. Japan. Agric. Res. Q. *8*, No. 3, 159–164.

TAKINO, Y. *et al.* 1965. The structure of theaflavin, a polyphenol of black tea. Tetrahedron Lett. *45*, 4019–4025.

TASSAN, C. G. and RUSSELL, G. F. 1974. Sensory and gas chromatographic profiles of coffee beverage headspace volatiles entrained on porous polymers. J. Food Sci. *39*, 64–68.

TERANISHI, R. 1976. Chemical and physical methods used to isolate and identify food flavors. Centennial Am. Chem. Soc. Meeting, N.Y., April 4–9, AGFD *21*.

TIRIMANNA, A. S. L. and WICKREMASINGHE, R. L. 1965. The quality and flavor of tea. II. The carotenoids. Tea Q. *36*, No. 3, 115–121.

VAN DER WAL, B., KETTENES, D. K., STOFFELSMA, J. and SEMPER, A. T. J. 1971. New volatile components of roasted cocoa. J. Agric. Food Chem. *19*, 276–280.

VAN PRAAG, M., STEIN, H. S. and TIBBETTS, M. S. 1968. Steam volatile constituents of roasted cocoa beans. J. Agric. Food Chem. *16*, 1005–1008.

VIANI, R. and HORMAN, I. 1974. Thermal behavior of trigonelline. J. Food Sci. *39*, 1216–1217.

VITZTHUM, O. G., WERKHOFF, P. and ABLANQUE, E. 1975C. Volatile constituents of green coffee. 7th Intern. Colloquium on Coffee Science, Hamburg. ASIC, Paris, France. (German)

VITZTHUM, O. G. 1975. Chemistry and technology of coffee. *In* Kaffee und Koffein. O. Eichler (Editor). Springer Verlag, Berlin, Heidelberg, New York. German)

VITZTHUM, O. G., WERKHOFF, P. and HUBERT, P. 1975A. New volatile constituents of black tea aroma. J. Agric. Food Chem. *23*, 999–1003.

VITZTHUM, O. G., WERKHOFF, P. and HUBERT, P. 1975B. Volatile components of roasted cocoa: basic fraction. J. Food Sci. *40*, 911–916.

VUATAZ, L. and REYMOND, D. 1971. Mathematical treatment of GC data: Application to tea quality evaluation. J. Chromatogr. Sci. *9*, 168–172.

VUATAZ, L., SOTEK, J. and RAHIM, H. M. 1974. Profile analysis and classification. 4th Intern. Congress Food Sci. Technol. Madrid, Spain. (in press).

WICKREMASINGHE, R. L., WICK, E. L. and YAMANISHI, T. 1973. Gas chromatographic-mass spectrometric analyses of "flavory" and "non-flavory" Ceylon black tea aroma concentrates prepared by two different methods. J. Chromatogr. *79*, 75–80.

WOODMAN, J. S., GIDDEY, A. and EGLI, R. H. 1967. The carboxylic acids of brewed coffee. 3rd Intern. Colloquium on Coffee Chemistry, Trieste, Italy. ASIC, Paris, France.

YAMANISHI, T. *et al.* 1966. Flavor of black tea. IV. Changes in flavor constituents during the manufacture of black tea. Agric. Biol. Chem. *30*, 784–792.

YAMANISHI, T. 1975. Tea aroma. Nippon Nogei Kagaku Kaishi, *49*, No. 9, R1–R9.

29

Flavor Chemistry of
Fruits and Vegetables

H. E. Nursten, Ph.D.

In this centenary year of the American Chemical Society, it is appropriate to note that a great number of the advances in flavor chemistry have, in fact, originated in the United States, but significant contributions have come from all quarters of the globe, as Fig. 29.1 of some important recently discovered natural food odorants shows. Every one of the compounds is a character-impact compound, is of considerable industrial importance, and has given rise to a great deal of research.

Here the intention is to concentrate on the most recent advances, i.e., those which in general have taken place since the latest reviews (MacLeod 1976; Nursten 1972). In this period no character-impact compounds of comparable significance have been discovered. The only exception is geosmin (Murray *et al.* 1975; Acree *et al.* 1976; see below), but Buttery *et al.* (1975) have isolated interesting thiazoles from dry beans and potato products (Buttery and Ling 1974; *cf.* Buttery *et al.* 1976B; Maga 1975A) and two 2,4,6-nonatrienals from blended dry beans (Buttery 1975).

Several of the more exotic fruits and vegetables have been examined by modern methods: arctic bramble (Kallio and Linko 1973; Kallio and Honkanen 1974; see also Hiirsalmi *et al.* 1974), asparagus (Ney and Freytag 1972), blackberry (Gulan *et al.* 1973), blueberry (Parliment and Kolor 1975), loganberry (Miller *et al.* 1973), musk-mellon (Kemp *et al.* 1972A, 1972B), passion fruit (Kadota and Nakamura 1972; Murray *et al.* 1972; Parliment 1972; Winter and Klöti 1972), plum (Forrey and Flath 1974), rose-apple (Lee *et al.* 1975A), tamarind (Lee *et al.* 1975B), water melon (Kemp 1975B; Katayama and Kaneko 1969).

Many constituents have been identified, but overall no new general principle of flavor chemistry has emerged. The compounds found

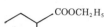

COOCH₂H₅

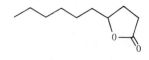

COOC₂H₅

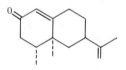

ETHYL 2-METHYLBUTYRATE

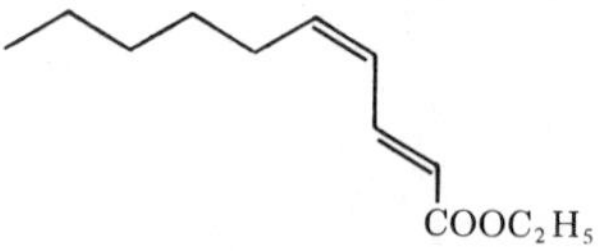

ETHYL *TRANS-2,CIS-*
4-DECADIENOATE

γ-DECALACTONE

APPLE:
FLATH *ET AL.* (1967) USA

BILBERRY:
VON SYDOW *ET AL.* (1970)
SWEDEN AND UK

PEAR:
JENNINGS AND
SEVENANTS (1964A)
USA

PEACH:
JENNINGS AND
SEVENANTS (1964B)
USA

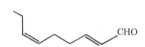

CHO

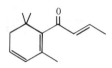

TRANS-2,CIS-6-NONADIENAL

CUCUMBER:
TAKEI AND ONO (1939) JAPAN
FORSS *ET AL.* (1962)
AUSTRALIA

DAMASCENONE

RASPBERRY:
WINTER AND ENGGIST
(1971) SWITZERLAND

APPLE:
NURSTEN AND WOOLFE (1972) UK

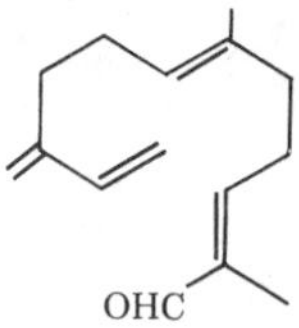

OHC

β-SINENSAL

ORANGE:
STEVENS *ET AL.* (1965) USA

NOOTKATONE

GRAPEFRUIT:
MACLEOD AND BUIGUES (1964)
USA

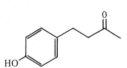

HO

SH

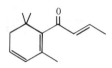

RASPBERRY KETONE

RASPBERRY:
SCHINZ AND SEIDEL (1961)
SWITZERLAND

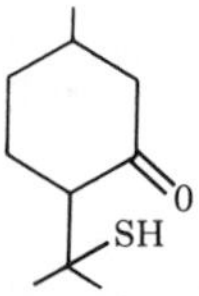

P-MENTHANE-8-
THIOL-3-ONE

OIL OF BUCHU:
SUNDT *ET AL.* (1971)
SWITZERLAND
LAMPARSKY AND
SCHUDEL (1971)
SWITZERLAND

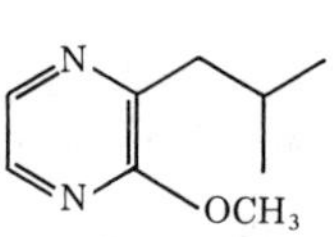

2-ISOBUTYL-3-METH-
OXYPYRAZINE

BELL PEPPER:
BUTTERY *ET AL.*
(1969) USA

PEAS:
MURRAY *ET AL.*
(1970) AUSTRALIA

FIG. 29.1. SOME IMPORTANT RECENTLY DISCOVERED FOOD ODORANTS

are, or will shortly be, listed in Weurman's invaluable compilation (van Straten and de Vrijer 1973).

All these advances are welcome, but of the wealth of topics currently under investigation, four appear to be the most important: (1) The pyrazines. (2) Lipid-derived volatiles. (3) Onions and related matters. (4) Sensory aspects.

Each of these will now be considered in turn.

The Pyrazines

The last compound in Fig. 29.1, 2-isobutyl-3-methoxypyrazine, is a very powerful and characteristic odorant and it continues to give rise to much synthetic activity (see for example Takken et al. 1975). Pyrazines in foods have been comprehensively reviewed by Maga and Sizer (1973).

A most valuable paper from Australia is concerned with the occurrence of 3-alkyl-2-methoxypyrazines in raw vegetables (Murray and Whitfield 1975). They have developed a method for collecting large volumes of headspace volatiles from the expressed, salt-saturated juice, which is almost as efficient as the widely-used concurrent steam distillation/solvent extraction. Twenty-seven vegetables were examined. The occurrence of compounds belonging to this important group of flavorants was found to be widespread, with many orders represented. The amounts determined covered the range from < 1 to 20,000 parts in 10^{13}. The isobutyl compound dominates in green and red peppers; the s-butyl isomer in carrot, parsnip, beetroot, and silverbeet; and the isopropyl homolog in peas, pea shells, broad beans, cucumber, and asparagus. Independently, Cronin and Stanton (1976) identified the s-butyl compound in boiled carrots. It would be interesting to see where Alabran et al. (1975) would place this stimulus in their multidimensional scaling diagram (see Chap. 28). The isobutyl compound has just been identified in an aroma fraction characteristic of Cabernet Sauvignon grapes (Bayonove et al. 1975).

In raw potato, substantial amounts of the isobutyl compound were found as well as the isopropyl (Murray and Whitfield 1975). No ethyl homolog was present. This is in agreement with the results obtained by Buttery and Ling (1973) and by Sevenants (1972), but is not in accord with the work on boiled potatoes (Meigh et al. 1973; Nursten and Sheen 1974). Examination of the differences in experimental conditions used is of immediate interest.

Alkylpyrazines, without a 2-methoxy group, are important volatiles in many heated foods. Thus, the 2-ethyl-3,6-dimethyl compound has been confirmed to be one of the most important odorants in baked potatoes, though other pyrazines also play a role (Pareles and Chang 1974).

Beetroot not only contains all the above three methoxypyrazines, but yet other odorants with strong earthy aromas, one of which is geosmin (Murray *et al.* 1975; Acree *et al.* 1976). This compound, *trans*-1,10-dimethyl-*trans*-9-decalol, is a metabolite of many actinomyces and of several blue-green algae. It is thought to be responsible for the odor of freshly-ploughed soil and earthy-musty taints in water supplies. It can also cause a musty off-flavor in dry beans (Buttery *et al.* 1976B).

Lipid-derived Volatiles

The area which is currently being opened up most effectively is that of the biogenesis of important flavor volatiles (see, for example, Nursten 1970; Tressl and Drawert 1973; Drawert 1975; Tressl *et al.* 1975). Flavor enzymes have also been reviewed recently (Chase 1974).

Three of the character-impact compounds in Fig. 29.1 clearly are lipid-derived products: the pear, peach, and cucumber compounds. Since on the whole the similarities in cell composition are more remarkable than the differences, why does lipid metabolism not proceed always in the same manner?

Careful work is clarifying the complexities. On the one hand, these are due to the existence of isoenzymes, on the other to further reactions of the products.

Some plants contain more than one lipoxygenase. Thus, Grosch and Laskawy (1975) have shown that the isoenzymes of soy differ in pH_{opt} (L2, L3, neutral:pH 6.5; L1, alkaline:pH 8.5–9.0) in their ability to co-oxidize carotenoids in substrate specificity, in specificity of position of hydroperoxidation, and in the volatiles produced.

With isoenzyme L1, linolenate was 85% converted into dienes and 1·5 μmol carbonyls were formed, whereas L2 under the same conditions except for pH led to only about 30% dienes, but about 7 μmol carbonyls. The carbonyls in the former case consisted primarily of *trans*-2-hexenal, whereas in the latter they were mainly of propanal, with noticeable amounts of *trans*-2,*cis*-4-heptadienal, *trans*-2-pentenal, *trans*-2-hexenal, 3,5-octadien-2-one, and 2,4,6-nonatrienal (see Table 29.1). The pea isoenzymes show similar differences.

Because L1 oxidises more substrate, but yields less volatiles, Grosch and Laskawy conclude that hydroperoxides are not their

TABLE 29.1

VOLATILE CARBONYLS FORMED BY 3 SOYBEAN LIPOXYGENASES

Enzyme: pH	L1 8.5	L2 6.5	L3 6.5
		Carbonyls, mol %	
Acetaldehyde		1.5	
Propanal	18	46	41
trans-2-Pentenal	5	8	11
trans-2-Hexenal	77	9	9
trans-2,cis-6-Nonadienal			2.5
trans-2,cis-4-Heptadienal		20.5	20
3,5-Octadien-2-one		8	8
2,4,6-Nonatrienal		7	8.5

SOURCE: Grosch and Laskawy (1975)

precursors. Morita and Fujimaki (1973) had already shown that the precursors were more polar peroxides of unknown structure.

Sekiya *et al.* (1976) provide evidence that the conversion of linolenic acid to 3-hexenal occurs in the chloroplasts of tea leaves, but not in those of spinach.

In many fruits the carbonyls are not the only products of oxidation of fats. Often they are reduced to the corresponding alcohols and the occurrence of this in tomatoes has been looked at by Stone *et al.* (1975), using U-^{14}C-labeled linolenic and linoleic acids. Considerably less label appeared in *cis*-3-hexenol when the tomatoes were blended rather than crushed and this was attributed to extensive cell rupture with probably some lysosome breakage and to greater access of air. Thus, the method of fruit preparation greatly influences the enzyme systems and the volatiles that are ultimately developed. Isolation of volatiles by distillation *in vacuo* rather than at atmospheric pressure led to recovery of much more *cis*-3- rather than *trans*-2-hexenal, *i.e.*, the method of isolation is also significant. Some evidence was obtained for an enzymic isomerization of *cis*-3-hexenal, rather slower than the chemical one occurring at 100°C. The reduction of *cis*-3-hexenal by crushed tomatoes at 25°C was shown to be faster than that of the *trans*-2-isomer, which was thought to fit in with *trans*-2-hexenol's not being a natural tomato constituent (but see Seck and Crouzet 1973). The reduction of hexenal paralleled that of *cis*-3-hexenal; and added benzaldehyde and phenylacetaldehyde also gave rise to the corresponding alcohols. No evidence was found of the reverse reaction, the oxidation of alcohol to aldehyde.

The situation in cucumber appears to be parallel, but different. Hatanaka *et al.* (1975), though not using radio-carbon, have shown that blending increases markedly the amounts formed of *trans*-2-nonenal and *trans*-2,*cis*-6-nonadienal. Addition of linoleic acid to blended cucumber greatly increases the amount of the monoene, and of linolenic acid that of the diene. Scission of the latter takes place at the 9–10 bond, with formation of *cis*-3,*cis*-6-nonadienal and 9-oxononanoic acid, both of which were conclusively identified. Some recent work by Kemp (1975A) has shown that cucumbers can give rise to unsaturated C_{17} and C_{16} aldehydes also, possibly by one and two α-oxidations, respectively (see also Kemp *et al.* 1974). β-Oxidation, among other reactions, was thought by Jennings and Tressl (1974) to account for the formation of many of the esters of *trans*-2,*cis*-4-decanoic and other unsaturated acids found in the Bartlett pear (Creveling and Jennings 1970).

Tomatoes and cucumbers provide a situation in which two fruits degrade the *same* precursors in different ways; but, in fact, things are more subtle still. Different varieties of fruit may show considerable contrasts in lipid metabolisms (Gholap and Bandyopadhyay 1975); and Tressl and Drawert (1973) have reported that the *same* fruit at different stages of ripeness uses different pathways. Thus homogenates of green bananas, not treated with ethylene, produce *trans*-2-nonenal, *trans*-2,*cis*-6-nonadienal, and 9-oxononanoic acid; whereas after treatment with ethylene and storage for 4 days at 15°C, hexanal, hexanol, and 12-oxo-*trans*-10-dodecenoic acid are produced, as in the climacteric and postclimacteric fruit.

In tomatoes, aldehydes were reduced to alcohols and not *vice versa*. From the work of Singleton *et al.* (1976), the situation is clearly different in peanuts. Here, added alcohols are converted to aldehydes.

In other kinds of tissue, added alcohols are converted to esters. Thus, whole strawberries are capable of converting virtually any added alcohol into its acetate, propionate, butyrate, isovalerate, and caproate (Yamashita *et al.* 1975). In fact, addition of isoamyl alcohol gave strawberries smelling of banana. For formate, isobutyrate, and valerate formation, addition of the corresponding acid was also required. Cutting the strawberries into small pieces remarkably depressed ester formation.

The banana can similarly convert alcohols and acids into esters, acyl-CoA being more effective than the acid (Gilliver and Nursten 1976).

It should be clear from the above that lipid oxidation is only part of a complex network of reactions, many of which are mediated by enzymes, and that there still is much left for further study.

Onion Flavor and Related Matters

The flavor of onions and related vegetables and flavoring materials also continues to receive a great deal of attention, 67 volatiles having recently been identified in a steam-distilled oil from leeks (Schreyen *et al.* 1976). About 5 years ago, it was finally settled that thiopropanal *S*-oxide is the lachrymator of onion, formed as shown in the well-known scheme (Fig. 29.2). Propyl propanethiosulfonate makes an important contribution to freshly cut onions, propyl propenyl di- and trisulfides to boiled onions, and dimethylthiophenes to fried onions (Boelens *et al.* 1971; *cf.* Ledl 1975; Maga 1975B; Schutte 1974).

Methods for determining the lachrymator have been devised (Freeman and Whenham 1975B; Tewari and Bandyopadhyay 1975) and its properties have been examined. It is relatively stable in hexane and exhibits minimum decomposition in aqueous solution at pH 3·40. In ice-cold water, it has a threshold of 1·0 ppm. At 10 ppm, it tastes pungent, long-lasting, fresh-onion-like. There is a positive correlation between lachrymator concentration and bulb weight (Freeman and Whenham 1975B). The alliin precursor of the lachrymator, *S*-(*trans*-1-propenyl)cysteine sulfoxide, has been synthesised (Nishimura *et al.* 1975).

A sensory and gas chromatographic investigation of onion oil (Galetto and Bednarczyk 1975) has drawn attention to methyl propyl di- and trisulfides, propyl trisulfide, and to 3 components not yet identified (see also Belo 1973). The formation of dimethyl sulfide, a volatile of secondary importance in onion, has been investigated (Hattula and Granroth 1974).

In spite of the unambiguous identification of thiosulfonates in onion by Boelens *et al.* (1971), Freeman and Whenham (1976B) consider, on the basis of analytical data and sensory properties of the pure compounds, that they are less significant than the thiosulfinates.

Freeman and Whenham (1975A) investigated the large losses of onion flavor components which result from postharvest processes, such as cooking by boiling, dehydration, and freezing. On the basis of the above scheme, these losses can occur by (a) complete or partial enzyme destruction, (b) partial nonenzymic destruction of precursors, and/or (c) partial enzymic hydrolysis of precursors with loss of volatile reaction products. As compared with fresh onion rated as 100, enzymic activities of processed products were: freeze-dried 45, laboratory-frozen 18, hot-air dried 9, commercially frozen

DISULFIDES

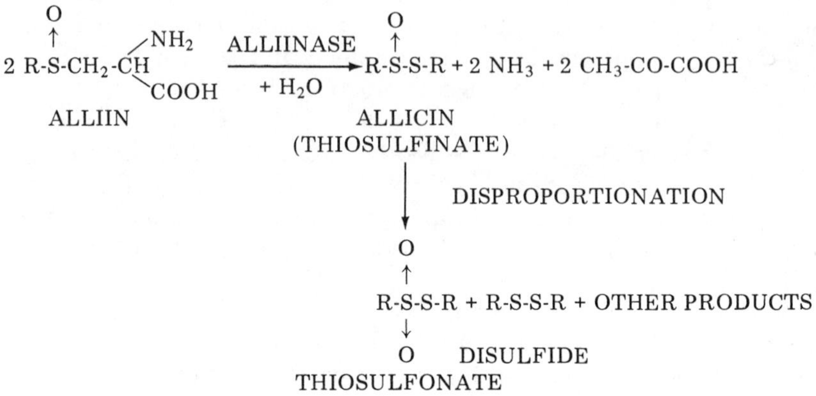

$$2 \text{ R-S-CH}_2\text{-CH} \overset{NH_2}{\underset{COOH}{\Big\langle}} \quad \xrightarrow[+\ H_2O]{\text{ALLIINASE}} \quad \overset{O}{\overset{\uparrow}{\text{R-S-S-R}}} + 2 \text{ NH}_3 + 2 \text{ CH}_3\text{-CO-COOH}$$

ALLIIN   ALLICIN
                              (THIOSULFINATE)

$$\downarrow \quad \text{DISPROPORTIONATION}$$

$$\overset{O}{\overset{\uparrow}{\text{R-S-S-R}}} + \text{R-S-S-R} + \text{OTHER PRODUCTS}$$
$$\downarrow$$
O   DISULFIDE
THIOSULFONATE

ADDITIONAL REACTIONS IN ONIONS

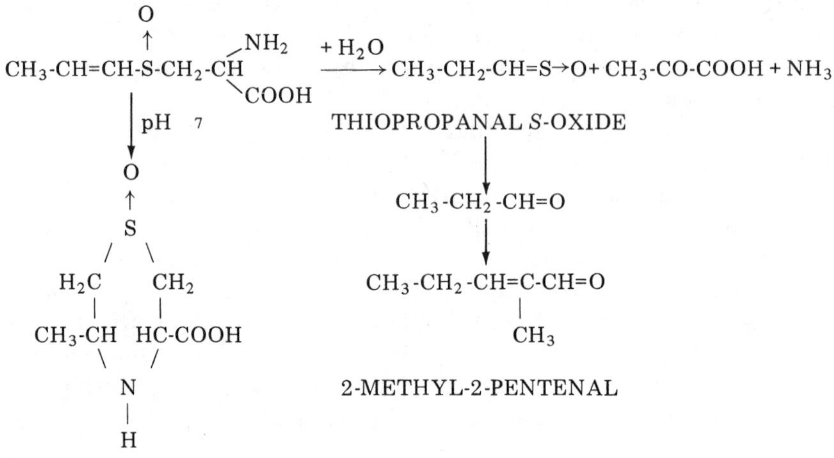

$$\text{CH}_3\text{-CH=CH-S-CH}_2\text{-CH}\overset{NH_2}{\underset{COOH}{\Big\langle}} \quad \xrightarrow{+\ H_2O} \quad \text{CH}_3\text{-CH}_2\text{-CH=S} \rightarrow \text{O} + \text{CH}_3\text{-CO-COOH} + \text{NH}_3$$

↓ pH 7   THIOPROPANAL *S*-OXIDE

↓

$$\text{CH}_3\text{-CH}_2\text{-CH=O}$$

↓

$$\text{CH}_3\text{-CH}_2\text{-CH=C-CH=O}$$
|
CH₃

2-METHYL-2-PENTENAL

CYCLOALLIIN

FIG. 29.2. ENZYMIC FORMATION OF S-CONTAINING FLAVOR COMPOUNDS

6, boiled 5, and pickled 0·01. Nonenzymic destruction of precursors was greatest in hot-air dried (about 80%) and least in freeze-dried (2%). Irradiation does not affect lachrymator production in onion immediately, but does so after storage, probably due to destruction of the enzyme rather than of the alliin (Nishimura and Mizutani 1975). The volatile components of several allium species have been surveyed in terms of their *S*-alk(en)yl-ʟ -cysteine sulfoxide precursors (Freeman and Whenham 1975C).

Akashi *et al.* (1975) have studied the mechanism of formation of unsymmetrical disulfides, which are important in the flavor of caucas, *A. victorialis* L. Whereas garlic alliinase does give some unsymmetrical disulfide from the caucas flavor precursors, *i.e.*, *S*-2-propenyl- and *S*-methyl-cysteine sulfoxides, it does not from mixtures of 2-propenyl and methyl disulfides or thiosulfinates. Unsymmetrical disulfides are therefore formed enzymically from the sulfoxides and there is some evidence of radical involvement. Using mixtures of synthetic *S*-alk(en)yl-ʟ -cysteine sulfoxides and alliinase, Freeman and Whenham (1976C) have been able to simulate very successfully their chromatograms (1975C) of the sliced vegetables, provided reducing power in the form of sodium borohydride was supplied also.

Overwinter storage is known to increase significantly the flavor intensity of onions. This trend has been monitored with varieties at 3 temperatures by Freeman and Whenham (1976A) by determination of (a) pyruvate, (b) lachrymator, and (c) total headspace GC areas. The increase ceased after about 190 days and was usually followed by a steep downturn, connected with sprouting and changes in respiration rate. The increase in flavor intensity is thought to be due to formation of additional precursors by enzyme-mediated release from its γ-ʟ -glutamyl peptide (see Schwimmer 1971; Schwimmer and Austin 1971). The use of such γ-glutamyl transpeptidases has been patented (*e.g.*, U.S. Secretary of Agriculture 1975) and they have been isolated from edible mushrooms (Iwami *et al.* 1974).

An area of current interest to flavor chemists is tissue culture. Freeman and his colleagues have looked at this aspect of onions also (Davey *et al.* 1974; Freeman *et al.* 1974). Undifferentiated callus produced only very small amounts of flavor components, but this is due to a lack of precursors rather than of alliinase. Callus which had differentiated roots exhibited flavor intensities approaching that of fresh onion, though lachrymatory potency was low.

Sensory Aspects

Here the concern is with flavors, not merely the chemistry of flavor. However sophisticated chemistry becomes, here its usefulness is constantly limited by the relative lack of understanding of the psychophysical aspects.

Multidimensional scaling has already been mentioned by Reymond in Chap. 28; and in connection with carrot flavor (see Alabran *et al.* 1975). There are two related directions in which distinct advances have been made recently: the definition of vocabularies and the contribution of the flavorist.

As regards vocabularies, most progress has been achieved with fermented beverages, with beer by Clapperton *et al.* (1976) and with cider and perry by Williams (1975). Such work is setting the pattern which those studying other commodities are well advised to follow (see also Harper 1975; von Sydow *et al.* 1974).

In the face of the progressively more sophisticated techniques of the flavor chemist, the flavorists have tended to retreat even further into their closely guarded expertise, but instances are beginning to occur of flavorists willing to contribute from their store of wisdom (see, for example, Broderick 1974A-C, 1975A-E, 1976A,B; Tong 1975). The resultant collaboration is greatly to be welcomed and valued and will hasten the advance of the subject, both directly and indirectly.

SUMMARY

The starting point at the beginning of this chapter was with the outstanding contributions which flavor chemists made in the last decade or so in the discovery of a whole series of character-impact compounds of fruits and vegetables, some of quite unanticipated structure.

In the immediate past, however, progress has only come in part from increasingly refined methods of analysis. Much of it has come from interdisciplinary areas, in particular those bordering on biochemistry and sensory analysis, including the art of the flavorist.

Perhaps it is appropriate to end with the happy thought that the centrifugal force of individual expertise is fortunately being more than matched by the centripetal ones of the interdisciplinary and international character of flavor research.

BIBLIOGRAPHY

ACREE, T. E., LEE, C. Y., BUTTS, R. M., and BARNARD, J. 1976. Geosmin, the earthy component of table beet odor. J. Agric. Food Chem. *24*, 430–431.
AKASHI, K., NISHIMURA, H. and MIZUTANI, J. 1975. Precursors and enzymic development of caucas flavour components. Agric. Biol. Chem. *39*, 1507–1508.
ALABRAN, D. M., MOSKOWITZ, H. R., and MABROUK, A. F. 1975. Carrot-

root oil components and their dimensional characterization of aroma. J. Agric. Food Chem. 23, 229-232.

BAYONOVE, C., CORDONNIER, R., and DUBOIS, P. 1975. Study of an aroma fraction characteristic of Cabernet Sauvignon grapes; identification of 2-methoxy-3-isobutylpyrazine. C. R. Acad. Sci. Ser. D, 281, 75-78. (French)

BELO, P. S. 1973. Enzymatic development of volatile components in onions. Diss. Abstr. B: 33, 4327.

BOELENS, M., DE VALOIS, P. J., WOBBEN, H. J., and VAN DER GEN, A. 1971. Volatile flavor compounds from onion. J. Agric. Food Chem. 19, 984-991.

BRODERICK, J. J. 1974A. Apple—has research helped? Flavours 5, 184-185.

BRODERICK, J. J. 1974B. Apricot—a diminshing art. Flavours 5, 231, 233.

BRODERICK, J. J. 1974C. Banana—a feeling for nature. Flavours 5, 284-285.

BRODERICK, J. J. 1975A. Blackberry—for a reasonable facsimile thereof. Flavours 6, 41, 44.

BRODERICK, J. J. 1975B. Cherry—common denominators. Flavours 6, 103, 109.

BRODERICK, J. J. 1975C. Grape and preference. Flavours 6, 171-172.

BRODERICK, J. J. 1975D. Peach—dynamic adaptations. Flavours 6, 243, 247.

BRODERICK, J. J. 1975E. Pineapple—some distance to go. Flavours 6, 351.

BRODERICK, J. J. 1976A. Raspberry—a case history. Flavours 7, 27, 30.

BRODERICK, J. J. 1976B. Do-it-yourself strawberry. Flavours 7, 51-52.

BUTTERY, R. G. 1975. Nona-2,4,6-trienal, an unusual component of blended dry beans. J. Agric. Food Chem. 23, 1003-1004.

BUTTERY, R. G., GUADAGNI, D. G., and LUNDIN, R. E. 1976A. Some 4,5-dialkylthiazoles with potent bell pepper-like aroma. J. Agric. Food Chem. 24, 1-3.

BUTTERY, R. G., GUADAGNI, D. G., and LING, L. C. 1976B. Geosmin, a musty off-flavor of dry beans. J. Agric. Food Chem. 24, 419-420.

BUTTERY, R. G., and LING, L. C. 1973. Earthy aroma of potatoes. J. Agric. Food Chem. 21, 745-746.

BUTTERY, R. G., and LING, L. C. 1974. Alkylthiazoles in potato products. J. Agric. Food Chem. 22, 912-914.

BUTTERY, R. G., SEIFERT, R. M., GUADAGNI, D. G., and LING, L. C. 1969. Characterization of some volatile constituents of bell peppers. J. Agric. Food Chem. 17, 1322-1327.

BUTTERY, R. G., SEIFERT, R. M., and LING, L. C. 1975. Characterization of some volatile constituents of dry red beans. J. Agric. Food Chem. 23, 516-519.

CHASE, T., JR. 1974. Flavor enzymes. In Food Related Enzymes. J. R. Whitaker (Editor). Am. Chem. Soc. Ad. Chem. Ser. 136, 241-266.

CLAPPERTON, J. F., DALGLIESH, C. E., and MEILGAARD, M. C. 1976. Progress towards an international system of beer flavour terminology. J. Inst. Brew. 82, 7-13.

CREVELING, R. K., and JENNINGS, W. G. 1970. Volatile components of Bartlett pear. Higher boiling esters. J. Agric. Food Chem. 18, 19-24.

CRONIN, D. A., and STANTON, P. 1976. 2-Methoxy-3-s-butylpyrazine—an important contributor to carrot aroma. J. Sci. Food Agric. 27, 145-151.

DAVEY, M. R., MACKENZIE, I. A., FREEMAN, G. G., and SHORT, K. C. 1974. Studies on some aspects of the growth, fine structure and flavour production of onion tissue grown in vitro. Plant Sci. Lett. 3, 113-120.

DRAWERT, F. 1975. Biochemical formation of aroma components. In Aroma Research: Proc. Int. Symp. Aroma Res. Zeist, H. Maarse and P. J. Groenen (Editors). Cent. Agric. Publishing and Documentation, Wageningen, Holland, The Netherlands.

FLATH, R. A. et al. 1967. Identification and organoleptic evaluation of compounds in Delicious apple essence. J. Agric. Food Chem. 15, 29-35.

FORREY, R. R., and FLATH, R. A. 1974. Volatile components of *Prunus salicina*, var. Santa Rosa. J. Agric. Food Chem. *22*, 496–498.

FORSS, D. A., DUNSTONE, E. A., RAMSHAW, E. H., and STARK, W. 1962. The flavor of cucumbers, J. Food Sci. *27*, 90–93.

FREEMAN, G. G., and WHENHAM, R. J. 1975A. The use of synthetic (±)-*S*-1-propyl-L-cysteine sulphoxide and alliinase preparations in studies of flavour changes resulting from processing of onion (*Allium cepa* L.). J. Sci. Food Agric. *26*, 1333–1346.

FREEMAN, G. G. and WHENHAM, R. J. 1975B. A rapid spectrophotometric method of determination of thiopropanal *S*-oxide (lachrymator) in onion (*Allium cepa* L.) and its significance in flavour studies. J. Sci. Food Agric. *26*, 1529–1543.

FREEMAN, G. G., and WHENHAM, R. J. 1975C. A survey of volatile components of some Allium species in terms of *S*-alk(en)yl-L-cysteine sulphoxides present as flavour precursors. J. Sci. Food Agric. *26*, 1869–1886.

FREEMAN, G. G., and WHENHAM, R. J. 1976A. Effect of overwinter storage at 3 temperatures on the flavour intensity of dry onion bulbs. J. Sci. Food Agric. *27*, 37–42.

FREEMAN, G. G., and WHENHAM, R. J. 1976B. Thiopropanal *S*-oxide, alk(en)yl thiosulphinates and thiosulphonates: simulation of flavour components of Allium species. Phytochemistry *15*, 187–190.

FREEMAN, G. G., and WHENHAM, R. J. 1976C. Synthetic *S*-alk(en)yl-L-cysteine sulphoxides-alliinase fission products: simulation of flavour components of Allium species. Phytochemistry *15*, 521–523.

FREEMAN, G. G., WHENHAM, R. J., MACKENZIE, I. A., and DAVEY, M. R. 1974. Flavour components in tissue cultures of onion (*Allium cepa* L.). Plant Sci. Lett. *3*, 121–125.

GALETTO, W. G., and BEDNARCZYK, A. A. 1975. Related flavour contribution of individual volatile components of the oil of onion (*Allium cepa*). J. Food Sci., *40*, 1165–1167.

GHOLAP, A. S., and BANDYOPADHYAY, C. 1975. Contributions of lipid to aroma of ripening mango. J. Am. Oil Chem. Soc. *52*, 514–516.

GILLIVER, P. J., and NURSTEN, H. E. 1976. The source of the acyl moiety in the biosynthesis of volatile banana esters. J. Sci. Food Agric. *27*, 152–158.

GROSCH, W., and LASKAWY, G. 1975. Differences in the amount and range of volatile carbonyl compounds formed by lipoxygenase isoenzymes from soybeans. J. Agric. Food Chem. *23*, 791–794.

GULAN, M. P., VEEK, M. H. SCANLON, R. A., and LIBBEY, L. M. 1973. Compounds identified in commercial blackberry essence. J. Agric. Food Chem. *21*, 741.

HARPER, R. 1975. Terminology in the sensory analysis of food. Flavours *6*, 215–216.

HATANAKA, A., KAJIWARA, T., and HARADA, T. 1975. Biosynthetic pathway of cucumber alcohol. Phytochemistry *14*, 2589–2592.

HATTULA, T., and GRANROTH, B. 1974. Formation of dimethyl sulfide from *S*-methylmethionine in onion seedlings (*Allium cepa*). J. Sci. Food Agric. *25*, 1517–1521.

HIIRSALMI, H. *et al.* 1974. Ionone content of raspberries, nectarberries, and nectar raspberry and its influence on their flavour. Ann. Agric. Fenn. *13*, 23–29, Food Sci. Technol. Abstr., 1975, *7*, 1J, 139.

IWAMI, K., YASUMOTO, K., and MITSUDA, H. 1974. γ-Glutamyl *trans*-peptidase in edible mushrooms and its application to flavor enhancement. Eiyo To Shokuryo *27*, 341–345; Chem. Abstr. 1975, *82*, 153999a.

JENNINGS, W. G., and SEVENANTS, M. R. 1964A. Volatile esters of Bartlett pear. Part 3. J. Food Sci. *29*, 158–163.

JENNINGS, W. G., and SEVENANTS, M. R. 1964B. Volatile components of peach. J. Food Sci. *29*, 796–801.

JENNINGS, W. G., and TRESSL, R. 1974. Production of volatile compounds in ripening Bartlett pear. Chem. Mikrobiol. Technol. Lebensm. *3*, 52–55.

KADOTA, R., and NAKAMURA, T. 1972. Food chemical studies on passion fruit, *Passiflora edulis* Sims. I. General chemical composition and volatile compounds of the juice. J. Food Sci. Technol. *19*, 567–572 (Jap.); Food Sci. Technol. Abstr. 1975, *7*, 1H 148.

KALLIO, H., and HONKANEN, E. 1974. An important major aroma compound in arctic bramble, *Rubus arcticus* L. Abstr. IV, I.U.F.S.T. Mtg, Madrid, 6–7.

KALLIO, H., and LINKO, R. R. 1973. Volatile monocarbonyl compounds of arctic bramble (*Rubus arcticus* L.) at various stages of ripeness. Z. Lebensm. Unters.-Forsch. *153*, 23–30.

KATAYAMA, O., and KANEKO, K. 1969. Aroma components of fruits and vegetables. V. Volatiles of watermelon. Nippon Shokuhin Kogyo Gakkai-Shi *16*, 474–479 (Jap.); Chem. Abstr. 1971, *74*, 2847.

KEMP, T. R. 1975A. Characterization of some new C_{16} and C_{17} unsaturated fatty aldehydes. J. Am. Oil Chemists' Soc. *52*, 300–302.

KEMP, T. R. 1975B. Identification of some volatile compounds from *Citrullus vulgaris*. Phytochemistry *14*, 2637–2638.

KEMP, T. R., KNAVEL, D. E., and STOLTZ, L. P. 1972A. *cis*-6-Nonenal: A flavor component of muskmelon fruit. Phytochemistry *11*, 3321–3322.

KEMP, T. R., KNAVEL, D. E., and STOLTZ, L. P. 1974. Identification of some volatile compounds from cucumber. J. Agric. Food Chem. *22*, 717–718.

KEMP, T. R., STOLTZ, L. P., and KNAVEL, D. E. 1972B. Volatile components of muskmelon fruit. J. Agric. Food Chem. *20*, 196–198.

LAMPARSKY, D., and SCHUDEL, P. 1971. *p*-Menthane-8-thiol-3-one. A new component of buchu leaf oil. Tetrahedron Lett. *36*, 3323–3326.

LEDL, F. 1975. Investigation of the aroma of fried onions. Z. Lebensm. Unters-Forsch. *157*, 229–234. (German)

LEE, P. L., SWORDS, G., and HUNTER, G. L. K. 1975A. Volatile compounds of *Eugenia jambos* L. (Rose-apple). J. Food Sci. *40*, 421–422.

LEE, P. L., SWORDS, G., and HUNTER, G. L. K. 1975B. Volatile constituents of tamarind (*Tamarindus indica* L.). J. Agric. Food Chem. *23*, 1195–1199.

MAGA, J. A. 1975A. The role of sulfur compounds in food flavor. Part I. Thiazoles. Crit. Rev. Food Sci. Nutr. *6*, 153–176.

MAGA, J. A. 1975B. The role of sulfur compounds in food flavor. Part II. Thiophenes. Crit. Rev. Food Sci. Nutr. *6*, 241–270.

MAGA, J. A., and SIZER, C. E. 1973. Pyrazines in foods. CRC Crit. Rev. Food Technol. *4*, 39–115.

MACLEOD, A. J. 1976. Volatile flavour compounds of the *Cruciferae*. *In* The Biology and Chemistry of the *Cruciferae*, J. G. Vaughan, A. J. MacLeod, and B. M. G. Jones (Editors). Academic Press, London, England.

MACLEOD, W. D., JR., and BUIGUES, N. M. 1964. Sesquiterpenes. I. Nootkatone, a new grapefruit flavor constituent. J. Food Sci. *29*, 565–568.

MEIGH, D. F., FILMER, A. A. E., and SELF, R. 1973. Growth-inhibitory volatile aromatic compounds produced by *Solanum tuberosum* tubers. Phytochemistry *12*, 987–993.

MILLER, P. H., LIBBEY, L. M., and YANG, H. Y. 1973. Loganberry flavour components of commercial essence. J. Agric. Food Chem. *21*, 508.

MORITA, M., and FUJIMAKI, M. 1973. Minor peroxide compounds as catalysts and precursors to monocarbonyls in the autoxidation of methyl linoleate. J. Agric. Food Chem. *21*, 860–863.

MURRAY, K. E., BANNISTER, P. A., and BUTTERY, R. G. 1975. Geosmin: an important volatile constituent of beetroot (*Beta vulgaris*). Chem. Ind. (London) 973.

MURRAY, K. E., SHIPTON, J., and WHITFIELD, F. B. 1970. 2-Methoxy-pyrazines and the flavour of green peas (*Pisum sativum*). Chem. Ind. (London) 897–898.

MURRAY, K. E. SHIPTON, J., and WHITFIELD, F. B. 1972. The chemistry of food flavour. I. Volatile constituents of passionfruit, *Passiflora edulis*. Aust. J. Chem. *25*, 1921-1933.
MURRAY, K. E., and WHITFIELD, F. B. 1975. The occurrence of 3-alkyl-2-methoxypyrazines in raw vegetables. J. Sci. Food Agric. *26*, 973-986.
NEY, K. H., and FREYTAG, W. 1972. Dimethyl sulphide as essential component of asparagus flavour. Further investigations of the volatile components of boiled asparagus. Z. Lebensm. Unters.-Forsche. *149*, 154-155. (German)
NISHIMURA, H., MIZUGUCHI, A., and MIZUTANI, J. 1975. Stereoselective synthesis of *S*-(*trans*-prop-1-enyl)-cysteine sulphoxide. Tetrahedron Lett. *37*, 3201-3202.
NISHIMURA, H., and MIZUTANI, J. 1975. Effect of γ-irradiation on development of lachrymator of onion. Agric. Biol. Chem. *39*, 2245-2246.
NURSTEN, H. E. 1970. Volatile compounds: the aroma of fruits. *In* The Biochemistry of Fruits and Their Products. A. C. Hulme (Editor). Academic Press, London, England.
NURSTEN, H. E. 1972. The Flavour of Foods. Rep. Prog. Appl. Chem. during 1971. *56*, 622-635.
NURSTEN, H. E., and SHEEN, M. R. 1974. Volatile flavour components of cooked potato. J. Sci. Food Agric. *25*, 643-663.
NURSTEN, H. E., and WOOLFE, M. L. 1972. An examination of the volatile compounds present in cooked Bramley's Seedling apples and the changes they undergo on processing. J. Sci. Food Agric. *23*, 803-822.
PARELES, S. R., and CHANG, S. S. 1974. Identification of compounds responsible for baked potato flavor. J. Agric. Food Chem. *22*, 339-340.
PARLIMENT, T. H. 1972. Some volatile constituents of passionfruit. J. Agric. Food Chem. *20*, 1043-1045.
PARLIMENT, T. H., and KOLOR, M. G. 1975. Identification of the major volatile components of blueberry. J. Food Sci. *40*, 762-763.
SCHREYEN, L., DIRINCK, P., VAN WASSENHOVE, F., and SCHAMP, N. 1976. Volatile flavor components of leek. J. Agric. Food Chem, *24*, 336-341.
SCHINZ, H., and SEIDEL, C. F. 1961. Postscript to paper No. 194 by H. Schinz and C. F. Seidel, Helv. Chim. Acta *40*, 1839 (1957). Helv. Chim. Acta *44*, 278. (German)
SCHUTTE, L. 1974. Precursors of sulfur-containing flavor compounds. CRC Crit. Rev. Food Technol. *5*, 457-505.
SCHWIMMER, S. 1971. Enzymatic Conversion of γ-L-glutamyl cysteine peptides to pyruvic acid, a coupled reaction for enhancement of onion flavor. J. Agric. Food Chem. *19*, 980-983.
SCHWIMMER, S., and AUSTIN, S. J. 1971. γ-Glutamyl transpeptidase of sprouted onion. J. Food Sci. *36*, 807-811.
SECK, S., and CROUZET, J. 1973. Volatile constitutents of *Lycopersicon esculentum*. Phytochemistry *12*, 2925-2930. (French)
SEKIYA, J., NUMA, S., KAZIWARA, T., and HATANAKA, A. 1976. Biosynthesis of leaf alcohol. Formation of 3Z-hexenal from linolenic acid in chloroplasts of *Thea sinensis* leaves. Agric. Biol. Chem. *40*, 185-190.
SEVENANTS, M. R. 1972. Personal communication, reported by Buttery and Ling (1973).
SINGLETON, J. A., PATTEE, H. E., and SANDERS, T. H. 1976. Production of flavor volatiles in enzyme and substrate enriched peanut homogenates. J. Food Sci. *41*, 148-151.
STEVENS, K. L., LUNDIN, R. E., and TERANISHI, R. 1965. Volatiles from oranges. III. The structure of sinensal. J. Org. Chem. *30*, 1690-1692.
STONE, E. J., HALL, R. M., and KAZENIAC, S. J. 1975. Formation of aldehydes and alcohols in tomato fruit from U-^{14}C-labeled linolenic and linoleic acids. J. Food Sci. *40*, 1138-1141.

SUNDT, E., WILLHALM, B., CHAPPAZ, R., and OHLOFF, G. 1971. The organoleptic principle of cassis flavor in oil of buchu. Helv. Chim. Acta 54, 1801-1805. (German)
TAKEI, S., and ONO, M. 1939. Leaf alcohol. III. Fragrance of cucumber. J. Agric. Chem. Soc. Japan 15, 193-195. Chem. Abstr. 1939, 33, 6524.
TAKKEN, H. J., VAN DER LINDE, L. M., BOELENS, M., and VAN DORT, J. M. 1975. Olfactory properties of a number of polysubstituted pyrazines. J. Agric. Food Chem. 23, 638-642.
TEWARI, G. M., and BANDYOPADHYAY, C. 1975. Quantitative evaluation of lachrymatory factor in onions by TLC. J. Agric. Food Chem. 23, 645-647.
TONG, S. T. 1975. Trace components in flavours. Flavours 6, 350, 355.
TRESSL, R., and DRAWERT, F. 1973. Biogenesis of banana volatiles. J. Agric. Food Chem. 21, 560-565.
TRESSL, R., HOLZER, M., and APETZ, M. 1975. Biogenesis of volatiles in fruit and vegetables. In Aroma Research: Proc. Int. Symp. Aroma Res. Zeist, H. Maarse and P. J. Groenen (Editors). Cent. Agric. Publishing and Documentation, Wageningen, Holland, The Netherlands.
U. S. SECRETARY OF AGRICULTURE. 1975. Flavor enhancement of Allium products. U.S. Pat. 3,725,085. April 3.
VAN STRATEN, S., and DE VRIJER, F. 1973. Lists of Volatile Compounds in Food, 3rd Edition. Rep. No. R.4030. Cent. Inst. Nutr. Food Res. TNO, Zeist, The Netherlands. Suppl. 1-6. 1975.
VON SYDOW, E. et al. 1970. The aroma of bilberries (Vaccinium myrtillus L.). II. Evaluation of the press juice by sensory methods and by gas chromatography and mass spectrometry. Lebensm.-Wiss. Technol. 3, 11-17.
VON SYDOW, E., MOSKOWITZ, H., JAKOBS, H., and MEISELMAN, H. 1974. Odor-taste interaction in fruit juices. Lebensm.-Wiss. Technol. 7, 18-24.
WILLIAMS, A. A. 1975. The development of a vocabulary and profile assessment method for evaluating the flavour contribution of cider and perry aroma constituents. J. Sci. Food Agric. 26, 567-582.
WINTER, M., and KLOTI, R. 1972. On the aroma of the yellow passionfruit (Passiflora edulis var. flavicarpa). Helv. Chim. Acta 55, 1916-1921. (German)
WINTER, M., and ENGGIST, P. 1971. Researches on aromas. XVII. On the aroma of raspberries. IV. Helv. Chim. Acta 54, 1891-1898. (French)
YAMASHITA, I., NEMOTO, Y., and YOSHIKAWA, S. 1975. Formation of volatile esters in strawberries. Agric. Biol. Chem. 39, 2303-2307.

Section 5

Nutrition

30

Introduction

Lawrence Rosner

At this Centennial meeting we intend to review various aspects of nutrition, including historical development and current knowledge. Obviously, in a half-day session we can only touch on a sampling of subjects. However, this sampling will be broad in scope as are the backgrounds of our speakers.

Dr. Harry Prebluda, a prominent consultant, will speak on animal nutrition, tracing the development of the field and projecting its future.

An exciting recent development has been recognition of fiber as an important dietary factor. A discussion of these findings will be presented by Dr. James Scala, Director of Nutrition and Health Sciences, General Foods Corporation.

The impact of food additives on our food supply can hardly be overemphasized. Dr. Benjamin Borenstein, Corporate Director, Consumer Research and Development, CPC International, Inc., will bring to us the story of the status of this field and how it was arrived at.

Finally, Dr. Maurice Shils, Director of Nutrition, Memorial Hospital, will discuss the latest developments in the field of clinical nutrition.

I am sure that you will agree that the subjects and speakers are worthy of being part of our Centennial meeting.

31

The Newer Knowledge of Animal Nutrition: The Road Ahead

Harry J. Prebluda

It is fitting during this double celebration of the Bicentennial Year and the Centennial of the American Chemical Society to look back briefly at significant exciting moments in the development of animal nutrition and focus on some historical events which helped place our country in its present leading position in food and animal production. The knowledge of nutrition as a science has progressed with the advances in chemistry so that as we understand the chemical nature of animal feeding materials or foodstuffs and their changes in the body after digestion, then we could follow the pathways in the animal body or human tissue.

Today nutrition is playing an increasingly important role as an instrument of peace as well as relief of hunger and improved health around the world. Nutrition is not an isolated science but is closely related and interfaced with other disciplines. Animal nutrition has great international significance in applied agriculture and is just beginning to be utilized to understand the pathways of carbohydrates, fats, proteins, minerals and micronutrients in body processes.

The early pioneer chemists and physiologists were amazed that vegetable foods, different in appearance and makeup from body cells, could be transformed in digestion into body tissues for conversion to eggs, milk, wool and meat. The scientists in those days could not explain just what was going on, and since they could not conduct experiments to fully prove out their beliefs, they simply speculated as to what took place in the digestion process. Dr. George Fordyce, a prominent London physician toward the close of the 18th Century, was the first investigator using controls to note that canary hens given mortar along with the feed remained in good condition during the laying season whereas those eating only seeds were in poor condition.

It has been estimated that in the last few hundred years at least a million scientific publications have appeared having to do with some kind of research on foods, their various properties, utilization or effects on animals. The very early work was carried out by apothecary-type scientists who were anxious to learn about the makeup of plant materials. Physiologists then followed and made digestion trials to find out just how foods were absorbed. Animal studies became necessary when it was learned that chemical analysis failed to provide the true nutritive value of a feed. It has only been during the last half century that most of the well-controlled, animal-type feeding studies have been made.

The purpose of this overview is to compare past and present developments in animal nutrition so that one can obtain a profile of the future road ahead.

Chemists of the 17th and 18th centuries contributed much to the understanding of food utilization and energy. Robert Boyle (1627-1692), James Black (1728-1799) and Joseph Priestley (1733-1804) opened many doors to new knowledge in this field. Priestley, the Unitarian preacher who loved to experiment, was responsible for the discovery of oxygen. In 1757, Benjamin Franklin, a New Jersey farmer, found that living bodies of animals and humans kept warm by the combustion of consumed food. About 20 years later, the French scientist Antoine Lavoisier presented a paper before the French Academy of Sciences on his research pertaining to the release of feed energy to animals. He proved that animals can unlock the energy from feed to maintain body temperature above that of the environment. Lavoisier also found that animals kept warm by using heat from the combination of oxygen in the air with the feed ingredients consumed. The combustion was greatly influenced by temperature. Exposure of animals to cold increased combustion and high temperatures decreased this oxidation. When animals worked, oxidation was stimulated. It is interesting that the U.S. Bicentennial coincides with the Nutrition Bicentennial when Lavoisier discovered that feed utilization was a chemical reaction. He learned that respiration involved oxidation and that the carbon dioxide exhaled by the animal was an index of the heat production.

Lavoisier's breakthrough studies inspired many chemists and scientists during the early 19th Century to work on methods of analysis so that there would be more accurate Tables on the composition of feeds and foods. In 1811, Gay-Lussac and Thenard suggested the first method for determining quantitatively amounts of carbon, hydrogen and nitrogen in organic materials. The method was not too accurate, but it helped point up the high nitrogen content in animal tissues other than fat. Another landmark of that period was Wöhler's synthesis of urea in 1828 from ammonium cyanate. He found that an organic substance could be made "without the aid . . . of any animal."

It was Mulder who first coined the term "protein" in 1839. His studies involved the examination of protein materials from both plant and animal sources. Mulder believed that protein materials are the most important of all the known sources in the organic kingdom.

It was Justus von Liebig who suggested in 1841 that the nutritive values of foods could be crudely estimated on the basis of the nitrogen content. Liebig was quite influential among chemists and his statements were held in high regard by livestock feeders. Liebig's main interest was the chemistry of agriculture. He first classified feeds and food into "albuminous" products, fats, carbohydrates and mineral salts in his book *Chemistry Applied to Agriculture and Physiology*. Liebig's composition Tables later became the reference tools of the animal nutritionist to design feeding experiments. Liebig called nitrogenous foods "plastic foods" (protein), and he thought that the main function of good nutrition was to replace destroyed tissue protein. He also proposed the idea that "the smaller the particles, the less resistance they meet in their diffusion in living tissue." This statement gave rise to liquid supplement feeding for livestock in the 20th Century and contributed to understanding foliar spraying of crops with nutrients. Liebig's ideas also laid the foundation for the fertilizer industry and soil enrichment as we know it today. The development of the steel-faced plow at the time of Liebig led to the mass production of grain and allowed the soil to be turned over at a depth for killing the grass.

Liebig found lactic acid present in meat products and believed that muscle extractives had unusual physiological qualities. Although he did not have supporting evidence, he recommended beef extract for patients who were not only weak and exhausted but depressed and despondent. Beef extract was quite popular for a number of years after Liebig's time but was later shown to have very little value except some flavor and a small amount of protein.

In those days American chemists beat a path to Liebig's laboratory and other universities on the European continent. The first U.S. agricultural experiment station was established in Connecticut about 100 years ago and it became the style for the leading experiment stations which followed to staff their laboratories with chemists who had worked in Liebig's laboratory.

There soon was a growing awareness and concern among American chemists and physiologists about the reliability of agricultural advice printed in foreign as well as U.S. Experiment Station Bulletins having to do with the feeding of farm animals. Many of the problems were directly related to the inadequacies of the existing analytical methods and their interpretation. Because of the general disappointment in the results of using chemical analysis for calculating the best nutritional rations for farm animals, a committee was formed in 1889 to consider better ways for more thorough chemical analysis of foods and animal feed. The group consisted of prominent chemists such as W. O. Atwater, G. C. Caldwell, E. H. Jenkins, W. H. Jordan with Dr. H. W. Wiley as Chairman.

The 50-page report which followed recognized that the estimates of the nutritive value of feed failed to agree with practical animal feeding experience. The committee did not make any recommendations; it did not indicate that the chemical approach to nutritive value was not fully the answer. Wiley did not sense the real importance of studying the effect of simplified animal rations having known compositions and yet nutritionally inadequate so that the test diet could be supplemented step-wise with known natural extractives or specific nutrients to have optimum nutrition. In the period after the Wiley report, physiologists and chemists built up the foundations of the newer knowledge of nutrition by isolating and purifying various proteins, carbohydrates, fats and inorganic salts. Later they looked into the physiological performance of these nutrients and their relationships with specific amino acids isolated from proteins. In 1898 the first edition of *Feeds and Feeding* by Dean William A. Henry of the University of Wisconsin was published. It became a landmark in animal husbandry and nutrition circles. Later it became a handbook for students and stockmen when rewritten in its many editions by Dr. F. B. Morrison at Cornell.

Liebig's work on proteins had attracted physiologists and bacteriologists who wanted to find out what happens in the alimentary tract when too much protein is consumed. It had been noticed by German workers as far back as 1868 (Senator) that intestinal putrefactive breakdown of protein produced toxic amines thought to be harmful. Early in the 20th Century, Metchnikoff, the head of the Pasteur Institute, published a book called *The Prolongation of Life.* He thought that body protection against premature aging could be brought about by encouraging growth of lactic acid organisms in the intestine to overcome the bacteria responsible for putrefaction. While all this was going on agronomists were busy doing analyses and studying the cost of producing protein from different crops so that farmers could grow nutrients and obtain energy at the lowest cost.

In 1871, the French chemist J. B. A. Dumas became the first individual to recognize that a diet which furnished only protein, carbohydrate and fat was not quite adequate to support life. He thought that shortages of milk and eggs accounted for the high death rate of children during the siege of Paris by the Germans. Efforts to feed milk substitutes using an emulsion of fat in a solution of sugar and "albuminous substance" brought on infant mortality. No one could explain why this happened. Dumas thought that there was something in the milk still unknown that had great nutritional significance. However, he did not carry out any experimental work to verify his hypothesis.

Lunin, in his dissertation study in 1880, was the first to restrict animals to purified diets. He fed a ration of purified protein (casein), milk fat, milk sugar (lactose) and a salt mixture to imitate the ash of milk. The adult test mice died after being on this ration for a short period. When Lunin provided milk to his test animals, they remained in good health. He concluded that there were substances other than casein, fat, lactose, and salts present in milk which were needed for good nutrition.

In 1897, a Dutch East Indies prison medical officer, Dr. Christian Eijkman, noticed that chickens in the dusty hospital yard developed a stiff condition of their feet and legs, never seen before in East Indian poultry. Eijkman thought that this condition was somewhat related to some of his beriberi prisoner patients. It was quite coincidental that the prisoners and the chickens ate the same food— polished rice. His hunch was fortified when he fed rice hulls to the sick chickens and they made a dramatic recovery. Eijkman believed that the starch in the polished rice produced a toxic condition which, in turn, was responsible for the nerve condition of his animals. When Eijkman's work was published in Europe people could not quite believe it since it came from "East of the Suez." It took many years before his work was accepted. It is interesting that almost 40 years went by before Eijkman was recognized to receive part of a Nobel Prize for this work. In the meantime, British and American nutrition researchers were not aware of Eijkman's work where one could produce under experimental conditions in fowl a disease resembling beriberi in humans.

It was Dr. Elmer V. McCollum of the University of Wisconsin who came to grips with the unidentified factors bothering chemists in the nutrition field. He was asked by Professor E. B. Hart to do some feeding work with young cows restricted to rations from single sources such as wheat, corn and oat plants. The objective was to see if the accepted methods of food and feed analysis provided a reliable basis for evaluating the nutritive value. After working a while with these animals, he decided that it would be much better to use smaller animals, especially if they had a short life span. McCollum reasoned that with smaller animals he could cut down making large quantities of purified ingredients for feeding studies. By using a rat as the experimental animal, he could accumulate data on growth and reproduction within a short time. It would take many years to carry out the same work using animals having a longer life span and a long gestation period. McCollum's rat colony for nutritional investigation was the first of its kind in the United States and was started in January 1908. The original rat breeding stock, which he left to the University of Wisconsin after he moved on to Johns Hopkins University, brought fame to Professors Hart, Steenbock and others working in the animal nutrition field. McCollum was still engaged in experiments with cows while he was getting started with his work using rats. In 1911, he had the help of Marguerite Davis, a University graduate student, who volunteered to take care of the colony. After some disappointing results with rats, using rations of improved palatability, McCollum had a breakthrough in his test work. He used an ether extract of butterfat or egg yolk to supplement the basal diet of purified protein, carbohydrates and inorganic salts. To his surprise, the animals grew and stayed well. When lard or olive oil was added as a source of fat to the basal diet, the animals lost their ability for well-being and seemed to fail rapidly. It was apparent that certain fats carried an essential factor for growth. It was soon found that extracting leaves of plants with ether produced a nutrient that really has the same effect. This unusual discovery upset previous nutritional thinking, since it had been assumed that fat serves only for the purpose of being an alternative source of food energy. Now it was possible to open up Pandora's box and determine just which essentials were lacking in the feed or food by finding out the nature of the supplements to make the diet fully complete for maintenance and well-being. It should be mentioned that the university administrators did not publicize McCollum's activities with his rat colony to farmers for fear of criticism. The Wisconsin legislature could hardly be expected to look with favor on tax monies being used to provide room and board for the farmer's pests. However, when Dean Russell learned of McCollum's breakthrough in December 1912, he insisted that the discovery be publicized immediately since the findings would be of great interest to Wisconsin farmers.

It was not until April 1913 that the first manuscript was sent in for publication to his former teacher, Professor Mendel of Yale, since he was the editor of the Journal of Biological Chemistry. It was coincidental that five months after McCollum's publication, Mendel published confirmations of the findings in the same Journal.

Although for a time there was a counterclaim concerning credit for the discovery of Vitamin A, within a few years it became quite clear that the credit belonged to McCollum alone.

The decade of McCollum's activities following his appointment to the University of Wisconsin was one of the most productive of his entire career insofar as it helped orient the thinking of physiologists and nutritionists around the world. This period brought forth many new ideas for using animals to discover hitherto unsuspected nutrients, especially when chemical procedures could not identify the deficiencies limiting the physiological well-being of the animal. McCollum's initial discovery influenced and inspired many people in the field of nutrition. The McCollum and Davis team was concerned that the fat-soluble A findings did not harmonize with that of Casimir Funk's observations. He found butterfat has no beneficial influence on pigeons having a nerve disease called "beriberi" when they were on a polished rice diet. Funk's research indicated that the normal diet needed at least one other substance, the absence of which brought on polyneuritis. While Funk was doing his experiments with pigeons, it took the Wisconsin group several years to find out through systematic investigation why young animals cannot grow well if restricted to single grains such as wheat, maize (corn), oats, peas, beans, rice, etc. In 1912 Funk proposed the "vital amine theory." He thought that special substances, having the nature of organic bases (amines) called "vitamines," are needed for maintenance of animal life. McCollum's butterfat factor did not quite fit into the Funk scheme of things. However, in 1915, McCollum and Davis discovered that water-alcohol extracts of wheat germ or rice polishings quickly improved the poor performance of animals fed polished rice. These extracts also cured the polyneuritis nerve condition in animals. The McCollum experiments became the basis for suggesting alphabetical terms to designate the two unidentified nutrients, fat-soluble A and water-soluble B. With these exciting developments a door was opened for an era of newer knowledge of nutrition, starting with vitamins A and B. It is interesting that McCollum addressed the American Feed Manufacturers Association in 1915. He predicted a great future for the use of vitamins in animal feeds. The following year McCollum found that the combination of a cereal grain with the leaf of a plant (alfalfa) provided a nutrient mixture which nourished experimental animals far better than any combination of seeds. This discovery explained the sound practice of pig farmers who feed their animals corn while allowing them at the same time to have daily access to forage plants as pasture. McCollum presented his new ideas in a lecture before the Harvey Society in 1917 and received national recognition.

The growing fame of McCollum for his originality and research achievements in animal nutrition brought him an invitation and an appointment in 1917 as a professor heading up the Department of Chemical Hygiene (this was later changed to Biochemistry) in the newly-endowed Johns Hopkins School of Hygiene and Public Health. In the spring of the following year, McCollum was invited by Dr. Milton Rosenau to present the Cutter Lectures at Harvard University. These lectures made such a profound impression on those attending that it was suggested that they be published as a book called *The Newer Knowledge of Nutrition: The Use of Food for the Preservation of Vitality and Health.* Many tens of thousands of copies were sold throughout the world. The book had many later editions and appeared with some of his associates, Nina Simmonds, Elsa Orent-Keiles, and Harry G. Day as co-authors. The "Newer Knowledge of Nutrition" era fired the creative minds of many American scientists and chemists in agricultural colleges, universities, medical schools, government research laboratories and also industrial laboratories. As the chemists unraveled the vitamin mysteries, it became clear that the newly-discovered vitamins were all needed for growth; however, it was amazing to learn that each vitamin was also essential for the prevention of a specific disease or pathology. From a chemical standpoint, the vitamins had very little similarity to each other. One vitamin after another was discovered, isolated, identified and later synthesized. Many are now commercially produced in large tonnage quantities.

Along with the vitamins came discoveries in the use of amino acids, minerals, trace elements, enzymes, antibiotics, coccidiostats, hormones, etc. Animal nutrition emerged as a science when these major findings began to interface with newer technological developments from other life science disciplines.

In this very brief review on animal nutrition, one does not have time to pay tribute to the hundreds of research scientists who have contributed to the explosive growth in this field, especially in the last 50 years.

The manufactured feed industry has been greatly benefited by the many scientific advances in animal nutrition. It should also be kept in mind that the feed industry has furnished many new research and development ideas to the colleges and the pharmaceutical industry.

About a century ago, the first formula feed mill was built in Waukegan, Illinois. At that time, the U.S. annual corn crop was only a billion bushels per year. Today it is at least six times as great. Animal nutrition has helped the feed industry use corn rations and come into being primarily to utilize wastes and by-products for which there were limited markets. In the "Gay Nineties," before the turn of the century, flour mills were dumping wheat bran in the rivers. By-products from breweries, distilleries, cotton ginning and meat packing found their way into streams or were burned. The best market for cottonseed meal in those days was the fertilizer industry. It was a time when agriculture was undergoing a change from a system of muscles and men, horses and mules, to one based primarily on mechanized operations. Vitamins came into the animal feed picture soon after Dr. McCollum's talk to the American Feed Manufacturer's Association in 1915. Chicken was considered a luxury item and politicians promised a "chicken in every pot." American farmers have been quick to recognize the expense and wasted energy involved in mixing their own feeds. Commercially manufactured feeds have grown in acceptance and have been helped by modern transportation methods.

The USDA started keeping production records in 1934. Only 34 million chickens were grown that year, and consumption was close to 16 lb per person. Today, the annual use of poultry and turkey is greater than 50 lb per person. Present costs compared to beef are less than during the 1929 Depression, without taking into consideration the smaller value of the dollar. This was made possible not only by progress in manufacturing but by research discoveries in animal nutrition which lowered the live weight price of broilers. Prior to McCollum's discovery of Vitamin D, there was no broiler industry as such because during the winter, chickens had a tendency to develop rickets (leg weakness) and die. Before Vitamin D products became available, the ultraviolet light of the sun was the only defense chickens had against rickets. Now, more is known about chicken nutrition than any other animal, including man. It takes over 4 billion broilers a year to satisfy present U.S. appetite and export demand. Figure 31.1 points up the feeding cost reduction in the last 50 years as brought about by nutrition research. Our country should be proud of this accomplishment since it is the result of dedicated teamwork involving industry, our colleges, experiment stations and our farmers. A hundred years ago, half our labor force produced the food and fiber needed by our nation. In 1910, about 35% of the U.S. population was engaged in farming. Today, we feed our country with

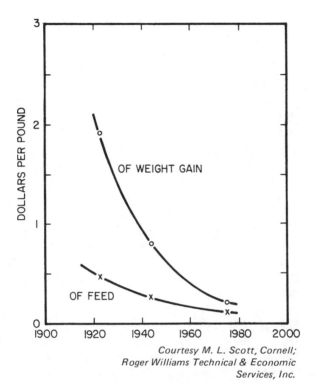

Courtesy M. L. Scott, Cornell;
Roger Williams Technical & Economic
Services, Inc.

FIG. 31.1. BROILER FEED COSTS (1975 DOLLARS)

less than 5% of the working force on our farms. Farmers gained tremendously when crop and animal production scientists developed superior strains of plants such as hybrid corn and improved genetics of farm animals. Much of the progress in breeding poultry, hogs, and beef animals would have been lost if nutrition advances had not kept up with the new developments. The impact of mechanization and inventive application brought great savings to the farmer and the consumer. An era of relatively low-cost food production was ushered in as a result of the improved feed efficiency. Better dairy animal breeding and feeding practices provided us more milk with fewer animals. The scientific breakthroughs in the last 20 years enabled us to almost double milk production with half as many animals.

The American feed industry is an intensely competitive industry and has constantly been subjected to technological and market changes because of new developments in animal nutrition. The profits are normally quite small and yet considerable gains or losses can come about from price fluctuations of basic ingredients.

In looking ahead to new developments in animal nutrition, here are some important changes which can be expected:

(1) We are faced with objectives of increased food production and interrelationships which run into serious energy shortages. Alternative energy technologies are gathering momentum to discourage the prophets of gloom that the world is winding down. Our food, feed and fiber position relies a great deal on the renewable energy from the sun. Greater understanding of photosynthesis and proper use of new growth regulators as well as better use of trace elements will help us do more with less to keep crop production costs down. Genetic improvement of grain crops and grasses will be coupled with greater efficiency for root systems and their associated bacteria to use nitrogen from the air and lower fertilizer costs.

(2) Regulatory agencies are expected to exert their usual constraints to make it difficult for agriculture to operate. There is a growing trend that the public is very much dissatisfied with having government agencies chase after environmental purity to prove 100% safety of animal feed additives which have had successful use for over 25 years. The FDA has in many instances over-reacted to uninformed groups which feel that the use of chemicals in feed and food production is bad. We now see that the nonsense of nonproductive defensive research without commensurate benefits to the taxpayer can be blamed for triggering higher inflation. World hunger problems may lead to changes in the American diet. The present world population of 4 billion is expected to almost double by the year 2000. Food shortages will help change attitudes of regulatory agencies. We must take a more critical look at environmental regulations controlling agriculture.

(3) Advances in synthetic and fermentation chemistry will lower cost of the amino acids used by both monogastric animals and ruminants. In 1950 the world use of amino acids by the animal feed industry was only a few million pounds per year. By the year 2000, the world daily use should be several million pounds. Methionine, lysine, and tryptophan will be important amino acids for use in feeds. The days of cheap fish meal for the feed industry will not return because of higher energy costs. Both fish meal and fishery wastes will also be upgraded for human use; this means greater use of amino acids in animal feeds to supplement vegetable proteins.

(4) It will be some time before single-cell protein (SCP) plays an important role in the U.S. animal feed nutrition picture. Single-cell proteins will have greater acceptability in other countries which are less efficient in growing plant proteins. There will be continued research in using municipal wastes to make protein from recycled

cellulose by fermentation. Eventually, these raw materials will have to compete with recovered heat values for industrial purposes. Federal and local governments will encourage joint municipal-industrial use of sewage sludge as starter material for fermenting other wastes intended for feed use. Economic pressures will force the animal feed industry to make better use of animal wastes. Nutritionists will also have some unusual opportunities to salvage worthwhile by-products from the clean-up of our waterways and make them available to the feed industry.

(5) Animal nutritionists will team up with chemists to look for improvements in the efficiency of low-cost synthesis of nonprotein nitrogen compounds for ruminant rations. Presently, the American feed industry uses close to 1 million tons of urea per year primarily for ruminant feeding. Ways will be found to have chickens and hogs make better use of nonprotein nitrogen derivatives. Advances in chemical engineering will affect the animal feed industry by the year 2000. Reactions in systems unheard of today will utilize new catalysts and immobilized enzymes to streamline costs. Ammonia synthesis at atmospheric pressure is likely before the year 2000. It means that simple peptides, custom tailored proteins, and even synthetic calories will be made with relatively low investment cost.

(6) The study of the microorganism population in ruminants and "by-pass" techniques in polygastric animals will lead to a better understanding of the role of microorganisms in monogastric nutrition. Encapsulations and slow release mechanisms will be used for greater nutrient effectiveness in monogastric animals. These same mechanisms will find their way to improve sanitation products needed to control unfriendly bacteria when raising farm animals. Microbiologists and animal nutritionists will pay greater attention to the role of microorganisms in the intestinal tract of monogastric animals, especially under disease conditions and other stress factors. The feed industry and nutritionists will turn to growth promotants and additives not used for human medication to avoid entanglement with regulatory agencies. The objective will be to emphasize preventive treatment. In the targets for such prophylaxis the emphasis will be against a broad range of harmful bacteria rather than a single type of organism. Previous observations that strains of *Lactobacillus plantarum* help enhance nutrient utilization and also can inhibit pathogenic intestinal bacteria will receive greater study. It is felt that improved methods of keeping toxic microorganisms under control in the digestive tract of single-stomach animals or ruminants could improve efficiency of feed utilization and well-being of farm animals.

(7) To further help nutritionists achieve greater feed efficiency

when disease factors upset the raising of farm animals, chemists will improve production of vaccines. Immunization in the future will involve the use of synthetic compounds. Combinations of chemically derived vaccines will be used along with antiviral agents and immunopotentiators to immobilize the invaders causing disease without being harmful to the animal.

(8) Nutritionists will find new interrelationships in the field of trace elements. Boron, chromium, cobalt, molybdenum, nickel, silicon, tungsten and vanadium will be in the limelight. Many of these trace elements will be found effective in the activation of enzymes. The role of zinc in nutrition will point up a functionality tied in with a taste acuity of feeding farm animals. Improved analytical methods for monitoring cobalt and molybdenum in feeding will encourage control agencies to approve limited use of these elements for animal feeding.

(9) Great things are expected from a new group of chemicals called "prostaglandins." These compounds are quite potent in that a billionth of a gram can provide some measurable effects. They are available from natural sources such as the sea whip. Several dozen natural compounds have been identified. Some of these compounds are being used overseas for estrus and pet population control. Prostaglandins will find markets for ova transfers where the fertilized egg is transferred from a valuable donor cow to a less valuable recipient. This will enable better cows to produce several dozen calves in a lifetime rather than a half dozen or less.

(10) Look for some unusual changes in nutrition of energy calculations. The feed industry will be examining this area in terms of spectral energy density as well as calories. These new perspectives have been announced recently. Spectral energy density concepts will put a new sophisticated dimension into present computerized feed formulations so that there can be a better predictability of nitrogen retention and palatability.

(11) Close to 4½ million tons of various kinds of pet foods were sold in this country this last year, representing a 2.5 billion dollar market. As animal protein products become upgraded for human consumption, pet foods will include more plant proteins, fermentation products, and amino acids in formulations to provide good nutrition and maintain the functional characteristics of texture, odor, and taste. Urban problems affecting this industry may call for greater enforcement of leash laws to minimize the ecological and public health menace of free-roaming dogs. There undoubtedly could be efforts to control canine births through feed additives.

(12) The recently revised USDA beef grading standards eliminated

conformation since it does not have any effect on eating quality of meat. Consumers will continue to demand meat at the retail beef counter that will be consistently tasty and tender on cooking. A new concept has been developed at the University of Nebraska whereby quick, low-cost procedures can be used for predicting tenderness of cooked beef from representative samples of the working muscles of the carcass. The evaluation is based upon a handful of trace elements and nitrogen analyses as factors for predicting the tenderness score. This new breakthrough could very well revolutionize the selling of meat products in domestic and international trade channels. Venders of meat products who wish to control the quality of meat sold can write into their specifications a range of predicted tenderness scores as measured by this special method. This discovery will be helpful to the animal nutritionist testing new supplements in feeding rations. Measurements can determine the influence of the supplement as it affects the tenderness of the cooked carcass meat.

(13) We have learned a great deal about interrelationships of Vitamin D and calcium levels of rations since McCollum's discovery of this vitamin. DeLuca at the University of Wisconsin has provided us recently with some new interpretations of the function and metabolism of Vitamin D. His findings will require new evaluations for optimum calcium levels when new analogs and Vitamin D-3 derivatives are fed to farm animals. The problems of leg weakness, cage fatigue, fragile egg shells, etc., may well disappear when we have a better understanding of the new Vitamin D metabolism.

As we look beyond the newer knowledge of animal nutrition we wonder whether the standards for human nutrition even approach those designed by professionals for animals and practiced by the American feed industry. There are many millions of people within our country's borders whose functional capacities could be improved with proper nourishment. Most nutrition troubles stem from a lack of the right kind of food. Surveys are showing that many have over-weight problems and there are deficiencies in a segment of our population. The average American's understanding of nutrition is most appalling because of a lack of education and misinformation. While many complain about the quality of air and water they are indifferent in speaking out about our poor eating habits.

Education in nutritional rudiments should be emphasized to young and old in all income brackets. For people to live beyond retirement age and enjoy the benefits of social security, there should be greater emphasis on good nutrition during middle life. It is during this period or earlier in life that the eating habits sow seeds for the degenerative diseases of old age.

SUMMARY

The terrific pace of technical accomplishment based on nutrition research burdens our society with social responsibilities which will be difficult to ignore. The contributions of the newer knowledge of nutrition greatly benefit mankind. Although we do acknowledge that good health is of primary importance, good nutrition practices require not only knowledge but self-discipline to change established eating habits. Should these things be rightfully used and not abused, we could then assure a better living and a richer, peaceful security for all people by the year 2000.

BIBLIOGRAPHY

ANDERSON, P. C. 1975. The spectral energy value system. Appl. Spectros. 29, 28–40.

DAY, H. G. 1974. Elmer Vernon McCollum, 1879–1967. National Academy of Sciences Biographic Memoirs 45, 263–335.

DeLUCA, H. F. 1973. Vitamin D metabolites in medicine and nutrition. Proc. Georgia Nutr. Conf. Feed Manufacturers. Atlanta, Febr. 14–16, 1973.

DUMAS, J. B. A. 1871. The constitution of blood and milk. Philos. Mag. 42, 129.

EIJKMAN, C. 1897. Ein Beri-Beri änliche Erkrankung der Hühner. Virchow's Arch. Pathol. Anat. Physiol. 148, 523–532. (German)

FORDYCE, G. 1791. A Treatise on the Digestion of Foods, 2nd Edition. J. Johnson Publisher, London, England.

LAVOISIER, A. L. 1777. Expériences sur la respiration des animaux et sur les changements qui arrivent à l'air on passant par leur poumons. Mem. Acad. Sci. Inst. Fr. 185. (French)

LIEBIG, J. von. 1840. Über die stickstoffehaltigen Nahrungsmittel des pflanzenreiches. Ann. Chem. 39, 129. (German)

LUNIN, N. 1880. Ueber die bedeutung der anorganischen salze für ernahrung des tieres. Inaug. Diss. Dorpat. (German)

McCOLLUM, E. V. 1957. A History of Nutrition. Houghton Mifflin Co., Boston.

McCOLLUM, E. V. and DAVIS, M. 1913. The necessity of certain lipins in the diet during growth. J. Biol. Chem. 15, 167–175.

McCOLLUM, E. V. and DAVIS, M. 1915. The essential factors in the diet during growth. J. Biol. Chem. 23, 231–246.

McCOLLUM, E. V. and KENNEDY, C. 1916. The dietary factors operating in the production of polyneuritis. J. Biol. Chem. 24, 491–502.

MULDER, G. J. 1839. Ueber die proteinverbindungen des pflanzenreiches. J. Prakt. Chem. 16, 129. (German)

OSBORNE, T. B. and MENDEL L. B. 1913. The influence of butterfat on growth. J. Biol. Chem. 15, 423–437.

PREBLUDA, H. J. 1974. The U.S. animal feed industry—a century of progress. Prof. Nutritionist, Winter Edition 6–9.

VAVAK, L. D., SATTERLEE, L. D. and ANDERSON, P. C. 1976. The relationship of cardiac sheer and trace element content to beef muscle tenderness. J. Food Sci. 41, 729–731.

WILEY, H. W. 1890. Report of the Wiley Committee. Exp. Stn. Rec. 2, 1850–1890.

WÖHLER, F. 1828. Ueber kunstliche bildung des harnstoffs. Ann. Phys. 12, 253–256. (German)

32

Food Additives and Nutrition

B. Borenstein

For purposes of this discussion, food additives will exclude staple commodities such as starch, salt and sugar and concentrate primarily on compounds produced by synthesis. The importance of functional additives, as well as nutritional additives, will be reviewed.

It is self-evident that the science of chemistry has been the foundation stone for nutrition since without techniques of isolation, identification and structure proof, the nutritional requirements of living tissue could not be identified, nor could the discipline of biochemistry exist. Although the basic scientists who pioneered in this field were probably not concerned with commercial availability of micronutrients, their work in structure proof and laboratory synthesis led rapidly to commercial synthesis of twelve vitamins and, more slowly, to commercial synthesis of several amino acids. It is not necessary to detail the well-known examples of chemical synthesis lowering the price of vitamins B-1, B-6, C, etc., as much as one hundredfold in a few years. Perhaps less is known about the commercial history of vitamin A. Shark liver oil was the major source of vitamin A in 1950 at an approximate cost of 30¢/million units. Complete synthesis of vitamin A from ethylene lowered the price of vitamin A palmitate oil to approximately 3¢/million units in 1976. Natural vitamin A is still available in very small quantities at 50¢ per million units for the nature food fadists.

It should be noted that these fadists are not convinced that 100 years of ACS have produced true, natural identicals. They still believe that natural is better and are paying enormous premiums for their beliefs. It is possibly a valid criticism of ACS that we do not yet know how to communicate with the lay public.

Equally important, the development of functional food additives—emulsifiers, microbial preservatives, antioxidants, flavors, etc.—made possible the development of totally engineered foods which require nutritional supplementation with vitamins, minerals and amino acids to maintain nutritional integrity.

Historically, margarine was probably the first designed food and was invented a few years before the ACS was founded. Modern margarine production uses a variety of precise hydrogenation steps, plus additives such as mono- and diglycerides, flavor compounds, plus vitamin A, beta carotene (provitamin A and color source) and D-2 to achieve desired technological attributes, PUFA (polyunsaturated fatty acids) balance, as well as to supply the micronutrients associated with butter.

A newer example of designed foods is vegetable-protein-based meat replacers. Here, sophisticated processing plus micronutrient and functional additives make possible meat extenders and custom-made "breakfast meats" minus cholesterol and saturated fat. Another group of designed foods developed in the United States primarily for overseas use are low-cost, blended dry foods such as CSM (corn soy milk) which combine staple foods to optimize protein at low cost plus added vitamins and minerals resulting in nutritionally complete foods.

This chemical version of the Green Revolution is, in effect, competing with the geneticist. Should we seek newer varieties of potatoes and tomatoes to increase vitamin C and vitamin A values? Attempting to increase vitamin C in potatoes when half the crop is processed into dehydrated potatoes and other products resulting in significant vitamin C degradation is wasteful. The demands of crop performance are so great—drought resistance, high yields, insect and disease resistance, texture for canning, desirable color—that only economically important problems should be brought to the geneticist. Increasing protein quantity and quality in crops is a valid challenge, but also a difficult one. To date, high protein rice has not solved problems of inadequate protein in the underdeveloped rice-consuming nations. I am not suggesting that agricultural research in areas such as nitrogen fixation and photosynthesis efficiency are not of the highest importance, but merely that food additives already offer solutions to specific aspects of nutritional shortfalls.

In the long run, our planet's population will probably necessitate a significant shift in the international distribution of foodstuffs to produce more equitable distribution among nations. This has already started in a small way with food costs rising in this country in the last three years. This could result in less freedom of choice of foods and increase the need for inexpensive, appealing, designed foods with nutritional quality domestically and abroad. Such demands would make functional additives essential to simulate conventional foods, and to produce palatibility plus variety. This would simultaneously require nutritional additives—minerals, vitamins and amino acids—to meet nutritional goals. The increasing body of regulations on all aspects of foods will, hopefully, not deter research and development on such projects.

The production of single-cell protein from petroleum is another example of chemical competition with the Green Revolution which may prove to be of critical importance in the future. Another potentially important finding is the recent report that glycine and diammonium citrate fed to humans has a sparing effect on protein requirement. The approach of increasing the world's useful food supply by adding synthetic essential amino acids plus a nonspecific nitrogen source such as diammonium citrate to available but poor quality protein sources around the world has to be the most important food technology-nutrition project on the horizon.

In the dental caries aspect of nutrition, perhaps the most exciting current development is the on-going research on xylitol use as a food additive to prevent caries. The possibility of using this new commercially available sugar alcohol as an anticaries sweetener in foods is a new and important possibility.

To sum up, it is reasonable to forecast that food additives will become even more important in the future.

33

Fiber–The Forgotten Nutrient

James Scala

It has become obvious in recent years that the major diseases of affluent societies are diet-related; consequently, nutrition has emerged as an important part of preventive medicine. Diet is recognized as a preventive factor in the cardiovascular diseases, dental caries, diabetes, osteoporosis, cancer, diverticular disease and anemia, to identify a few. Therefore, when a dietary factor is identified as either therapeutic or it exhibits a causal relationship to a public health problem, its role in nutrition must be thoroughly explored. Conversely, if a dietary component has undergone a major change in its dietary contribution, its role must be carefully re-evaluated to identify the effects of the change on health.

Against this background, fiber is a unique nutrient. Every society has, through folklore and "plain common sense," recognized its importance. At present, epidemiologists suspect that its depletion in our diet might be partially responsible for several public health problems. Against the epidemiological observations, nutritionists cannot identify with certainty the amount required for optimum health. Food scientists, chemists and research physicians are learning that there are several types of fiber and each type performs specific necessary functions. Consequently, fiber is being rediscovered by food and health scientists and recognized as a major input to well-being. Its re-emergence as a necessary dietary factor makes fiber the "Forgotten Nutrient."

FIBER

Portions of plant foods which are resistant to gastrointestinal digestion are referred to as fiber, roughage, and bulk. This portion of the diet was considered relatively inert until recent years when it has been associated with a number of diseases of Western civilization. Consequently, this "inert" part of the diet has received considerable attention in recent years and its importance is now being realized.

NOMENCLATURE

Crude Fiber, Dietary Fiber and Plant Fiber

In 1972, Trowell proposed the term "dietary fiber" be used to identify the remnants of plant cell walls not hydrolyzed by human alimentary enzymes (Trowell 1973). This physiologic definition does not limit fiber to specific polymers or materials; but only identifies its origin. In contrast, the definition "crude fiber" was developed to identify that portion of plant material which is resistant to digestion by weak acid followed by weak alkali (AOAC 1970). This classification was developed in the 19th Century for the grading of animal feed and has marginal value for human food.

The term "plant fiber" has been proposed as a collective term to classify the various substances of the plant cell wall (Spiller and Amen 1975). Plant fiber can be subdivided into "nonpurified plant fiber" for epidemiological discussions of physiological function and "purified plant fibers" for purified components of the cell wall such as cellulose or pectin.

A single precise term for fiber has not yet emerged. Since fiber is a collective term for many materials with both chemical and physiologic properties, no single terminology is likely to be developed. Therefore, for some years, scientists will be faced with conflicting terminologies.

CHEMISTRY OF FIBER

Most fiber is ultimately traceable to the plant cell wall which is a complex structure consisting of many substances. As foodstuffs, they are digestible to differing degrees by the alimentary enzymes of man and fermentable to varying degrees by the bacteria which inhabit his colon. Most have been sufficiently characterized to be classified.

Dietary fiber probably consists of six groups of substances (Trowell 1974):

(1) Structured polysaccharides consisting of cellulose, heteropolysaccharides and homopolysaccharides (formerly called hemicellulose).

(2) Lignins consisting of polymers of phenyl propane.

(3) Unavailable lipids such as waxes and cutin.

(4) Nitrogenous compounds.

(5) Trace elements associated with fiber such as zinc, silicon, chromium and manganese.

(6) Unidentified mineral salts and plant acids such as phytic acid and calcium.

Since the polysaccharide materials are most important to human nutrition they should be considered in more detail.

Cellulose—The major portion of crude fiber consisting of straight chain polymers of 1–4 linked D-Glucose in parallel by hydrogen bonding. These bundles of polymers are strong and insoluble.

Pectin—Chemically galacturonans, galactans and arabinans existing in the cell walls as polyuronic acid polymers. When the carboxyl groups have been converted to methyl esters, the resulting polymer is called pectinic acid; generally known as pectin. Pectin forms stable gels with sugar and acid.

Hemicelluloses—Are mixtures of linear and branch polysaccharides containing primarily xylose, arabinose, glucose and mannose and about 4% uronic acid. They include xylans, polyuronans, galactans, arabinoxylans and other similar compounds.

Lignin—Is usually considered a structural material acting as a cementing and anchoring agent in the cell wall. It consists of a group of complex aromatic polymers based on phenyl propane units; therefore, is not a polysaccharide.

Mucilages—Consist of mainly uronic acid polymers with a high water-holding capacity. They tend to form viscous gels when foods containing them are homogenized with water.

In summary, the plant cell wall is a dynamic and complex structure consisting of linear fibrils woven into a matrix of branched polysaccharides and in the lignified cell infiltrated with aromatic residues. These components appear in varying ratios according to the type and age of the plant, and conditions under which it is grown. When grain is processed and its various components separated, these cell wall materials become the fraction identified as "fiber."

PHYSICAL PROPERTIES

Plant fiber has four properties which form the basis of its physiologic effects. Each of these is operative to varying degrees for all of the components of fiber; however, no single source of fiber, nor every component of fiber elicits each characteristic (Eastwood 1973).

Water Absorption—Is characteristic of most of the polysaccharides which swell in water to form a gel-like mass. These same poly-saccharides are dispersed by mechanical or enzymic cleavage and the extent to which they swell in the intestinal tract is influenced by bile salts and ionic material. Many of them, especially pectins and musci-lages, can form gels which contain as much as 99% water. The lignins absorb little or no water.

Cation Exchange—Is characteristic of the uronic acid poly-saccharides. This property is related to the free carboxyl groups in the pectic materials and the hemicelluloses (Ershoff 1974).

Organic Absorption—Of surface active agents, such as the bile acids, is characteristic of fiber.

Gel Filtration—Is characteristic of most gel systems which have the capacity of molecular exclusion capacities. The same matrix can trap bacteria and enzymes which may, in turn, attach other molecules so entrapped or the gel itself.

ANALYTICAL PROCEDURES

Crude Fiber

The AOAC method of fiber analysis involves an hydrolysis of the sample by 0.25 N sulfuric acid followed by hydrolysis in 0.31 N sodium hydroxide. These hydrolyses are followed by incineration with "crude fiber" being the weight difference before and after igni-tion. This method does not account for 80% of hemicellulose, 50 to 90% of the lignin and 20 to 50% of cellulose (Van Soest and McQueen 1973). These errors make crude fiber values imprecise and it does not provide a good estimate of dietary fiber. Crude fiber is the value reported in food tables.

Southgate Method (Southgate 1969)

Southgate developed a system which employs organic solvents, enzymes, acids and alkali. This method permits differentiation be-tween cellulose, hemicelluloses, lignin and the water-soluble poly-saccharides. It is of value to the researcher seeking to quantify the effects of fiber on metabolism.

Detergent System (Goering and Van Soest 1970)

These methodologies consist of a series of treatments with neutral detergent, acid detergent, acid, permanganate and weighing to identify all components of cell wall materials. The determinations are not complex and can be conducted in most moderately equipped food laboratories. Although each determination identifies a specific cell wall fraction, the *neutral detergent fraction* is sufficient for most purposes.

This fraction consists of treatment with detergents at pH 7.0 for 1 hr (Goering and Van Soest 1970). As research in fiber continues, it is likely that the neutral detergent fraction will become more widely used and will eventually replace crude fiber. The pectins are not identified by this method and must be evaluated by one of a variety of methods (Kertesz 1950).

Critique of Analytical Methods

The determination of crude fiber is obsolete and has little value in evaluating dietary fiber in human food.

Southgate's method is very time-consuming, but allows detailed knowledge of the materials. It is probably best in a research environment where detailed information is necessary.

The detergent system is rapid and, for the foreseeable future, will provide sufficient knowledge about food fiber. The detergent system produces data which is very similar to the method of Southgate; therefore, it is adequate for most nutritional evaluation (McConnell and Eastwood 1974).

In conclusion, the detergent system appears to be the most practical for purposes of information, ease of evaluation and general utility.

DIETARY LITERATURE DESIGNATION

In nutritional literature and information, fiber is identified as crude fiber. However in research, dietary fiber, neutral detergent fiber and more recently plant fiber are expressed for specific research purposes. Therefore, the reader is often faced with confusing terminology. In the future, the neutral detergent designation will probably become more widely used. A comparison of the fiber content of various foodstuffs as determined by different methods is reported in Table 33.1 to illustrate the extent to which the data vary.

Scientists should assume that reported fiber is *crude fiber* unless designated otherwise. Most, if not all, USDA information and infor-

TABLE 33.1

COMPARISON OF "NEUTRAL DETERGENT FIBER" AND CRUDE FIBER

	Dry Matter (%)	Neutral Detergent Fiber (%)	Crude Fiber (%)
Bread, white enriched	64.2	3.3	1.7
Bread, whole wheat	64.4	14.9	5.1
Kelloggs All Bran	97.3	34.0	9.2
Kelloggs Special K	96.2	7.4	1.1
Kelloggs Corn Flakes	96.3	7.9	1.4
Nabisco Shredded Wheat	93.5	22.4	3.6
Ralston Purina Wheat Chex	96.4	17.6	3.5
General Mills Cheerios	94.0	8.8	2.7
General Mills Wheaties	95.3	13.8	3.0
Post Grape Nuts	94.9	11.9	2.2
Quaker Puffed Wheat	94.3	8.9	3.5
Scotts Porridge Oats	91.1	10.4	2.0
Pet Heartland	97.4	10.5	6.0
Apples	11.7	7.6	3.7
Beans, green	10.2	22.0	10.6
Beet root	12.5	11.8	4.7
Broccoli	11.0	11.3	12.6
Cabbage	7.8	14.2	8.4
Carrots	10.4	9.2	5.7
Cauliflower	8.7	15.1	10.1
Celery	5.0	14.4	13.3
Cucumber (peeled)	3.2	12.7	13.4
Cucumber (skin)	8.5	35.5	34.1
Eggplant (peeled)	6.9	21.8	18.6
Lettuce, romaine	4.9	17.3	12.4
Onions	9.1	7.1	9.7
Oranges (peeled)	13.6	3.7	2.7
Pepper (seedless)	9.9	17.2	14.5
Potatoes (peeled)	22.2	4.7	0.5
Potatoes (skin)	18.2	23.6	9.8
Radishes	11.9	14.3	11.4
Rutabaga (peeled)	10.0	10.2	7.4
Squash	12.0	14.7	6.2

SOURCE: Amen and Spiller (1975).

mation supplied by food producers is crude fiber. Kraus (1976) has compiled a comprehensive catalog of the fiber content of most foods.

FIBER INTAKE

Accurate estimates of the level of fiber intake by prehistoric man are probably impossible. With the rise of agriculture, about 10,000 years ago, cereal, grain and vegetable consumption probably accounted for a large part of the caloric intake. Therefore, if rural African society serves as a model from which to infer prehistoric man's crude fiber intake, a value of about 16-24 g per day is obtained (Nelson 1975; Lubbe 1971).

In contrast to this, estimates of crude fiber in modern Western diets range from 0.8 to less than 10 g per day (Nelson 1975; Robertson 1972; Burkitt and Trowell 1975). Inasmuch as estimates of the crude fiber content of the diet vary, the exact value is probably unclear and less than 5 g. In addition, the data obtained by Robertson and Burkitt indicate that during the 20th Century, our source of fiber has shifted from cereals to fruits and vegetables while remaining between 3 to 5 g daily. Table 33.2 summarizes these data for the British diet which is similar, if not identical, to the U.S. diet.

Since crude fiber cannot be quantitatively translated to dietary fiber, the data in Table 33.2 should be used as indicative of trends. In this light, the data indicate that cereal fiber has declined by over 50% and possibly by up to 90% in this century while total fiber intake has remained about 4 g per day.

EFFECT OF GUT FLORA ON FIBER

Dietary fiber is not broken down by the enzymes of the human alimentary tract, but is fermented by the microflora of the gut. The gut microflora can digest hemicelluloses, pectin and cellulose to varying extents (Williams and Olmsted 1936).

Digestion studies by Williams and Olmsted have been reproduced by Southgate (Southgate and Durnin 1970) and are summarized in Table 33.3 showing the extent to which fiber from various sources is digested by the gut microflora. It shows that hemicellulose is

TABLE 33.2

ESTIMATED COMPARISON OF CRUDE FIBER INTAKES FOR THE YEAR 1900 AND 1970[1]

| | Crude Fiber Intake | |
	1900 (g/day)	1970 (g/day)
Total fiber	3.6	4.2
Vegetables	2.1	2.6
Fruits and nuts	0.6	1.1
Cereals[2]	0.9	0.5
Cereals[3]	1.0	0.3

[1] The greatest change (decrease) has undoubtedly occurred in cereal fiber.
[2] Robertson (1972).
[3] Burkitt and Trowell (1975).

TABLE 33.3.

DISAPPEARANCE OF LIGNIN, CELLULOSE AND HEMICELLULOSE

Residue Added to Basal Diet	Lignin (g)	(%)	Fractions of Indigestible Residue Cellulose (g)	(%)	Hemicellulose (g)	(%)	Total Residue (g)	(%)
Wheat bran	1.9	10	11.4	30	29.8	35	43.1	30
Corn germ meal	*	*	29.5	57	59.5	63	89.0	60
Carrots	*	*	34.5	67	39.5	85	74.4	74
Cottonseed hull	3.7	12	4.8	14	10.6	30	19.1	18
Cabbage	*	*	29.8	55	33.1	80	62.9	80
Sugar beet pulp	*	*	30.9	55	29.7	89	60.6	65
Alfalfa leaf meal	0.6	3	5.0	12	1.1	6	6.7	9
Canned peas	*	*	25.4	45	12.1	84	37.5	53
Agar agar	—	—	—	—	48.6	60	48.6	60
Cellu flour	—	—	4.2	7	3.2	29	7.4	10

SOURCE: Williams and Olmsted (1936).
*Lignin content too small for valid results.

broken down more extensively than cellulose and that only about 10% of lignin is hydrolyzed.

PHYSIOLOGICAL EFFECTS OF FIBER

Intestinal Transit Time

In general, diets rich in fiber, produce bulky, soft stools which traverse the gut rapidly while diets low in fiber (refined diets) produce small hard stools. An inverse relationship between stool weight and transit time has been reported by many researchers (Burkitt *et al.* 1972; Holmgren and Mynors 1972). These findings are summarized in Table 33.4 which summarizes data from studies by Burkitt (Burkitt *et al* 1974).

Although high fiber diets in general exhibit more rapid transit times, the addition of fiber, specifically bran, to low fiber diets does not always produce a shortening of transit time. People on a bland diet who exhibit a long transit time show a decrease and those with a short transit time exhibit an increase. Indeed, this reaction indicates that fiber can be useful in the treatment of both constipation and diarrhea (Harvey *et al* 1973).

DISEASES OF THE COLON

Epidemiologists have identified a variety of diseases of Western man which do not exist in underdeveloped countries. They ascribe these to the difference in the fiber content of the diet (Burkitt and Trowell 1975).

These diseases fall into two classes; those related to the effects of

TABLE 33.4

AVERAGE BOWEL TRANSIT TIMES AND STOOL WEIGHTS

Group	Diet	Avg Transit Time (hr)	Avg Stool Weight (g)
African villagers	Unrefined (high residue)	20	510
Vegetarians and boarding school	Mixed diet	30	270
English boarding school	Refined (low residue)	55	90

SOURCE: Burkitt *et al.* (1974).

fiber on the consistency and frequency of the stools; those which are the result of the ability of fiber to bind bile acids and other dietary components.

The former relate to colonic problems such as constipation, diverticulitis and hiatus hernia while the later relate to the chemistry of the blood and toxic effects on the colon (the most devastating of which is cancer).

CONSTIPATION

Cowgill, documented the laxative effect of wheat bran and identified the requirement for dietary fiber (Cowgill and Anderson 1932; Cowgill and Sullivan 1933). He calculated that the 70-kg man is constipated on a daily diet containing 2.5 g of crude fiber, borderline on 4 g and 6.3 g for good colonic function. In addition, his studies indicated that wheat bran is superior to vegetable fiber.

Although there have been more recent studies, they tend to support Cowgill's findings and cannot be quantified until a more universally accepted method of evaluating dietary fiber is established. In addition, opinions, anecdotal evidence and folklore support these observations.

DIVERTICULAR DISEASE

Origin

Diverticular disease refers to the development of out-pouching of the colonic wall. Diverticula occur by obstruction following which the muscle fibers are separated permitting the mucosa to protrude. Diverticulosis is thought to exist in about 1/2 of those over 40 and as many as 2/3 of those over 80. It is, therefore, one of the most common age-related conditions of the colon (Painter and Burkitt 1971). Two hypotheses have been developed to explain the origin of the disease.

Diets low in fiber are felt to be conducive to diverticula formation because the food residues are more compact requiring greater pressure to propel them along. In addition, the less bulky stools require more straining to achieve ejection; hence the formation of diverticula (Painter and Burkitt 1971). In contrast, the soft, bulky stools of a high fiber diet are propelled easily and require little pressure for ejection.

More recently, it was proposed that flatus is the main factor in diverticula formation. This hypothesis proposes that the flatus is prevented from leaving the anus and is forced back up the rectum. This movement results in muscle changes similar to those already described and overexertion of the colonic musculature is the cause (Wynne-Jones 1975). This hypothesis would also explain the prevalence of diverticulitis in Western countries where the release of flatus is not accepted.

Clearly, the cause-effect relationship of dietary fiber diverticulosis remains to be firmly established. However, its absence among people on a high fiber, unrefined diet supports the refined diet, low residue hypothesis.

Treatment

Bran supplement to the diet of sufferers of diverticular disease provides symptomatic relief and clinical improvement (Painter et al. 1972; Findley et al. 1974). In the clinical studies bran shortened transit time, increased stool weight and produced regularity.

Treatment consisted of from 6 to 20 g of bran per day spread over three meals. These results are summarized in Table 33.5 and support very strongly the observation of Cowgill that it eliminates both constipation and laxation (Cowgill and Anderson 1932).

In recent years, the practice of using high fiber diets for treatment of diverticular disease has gained wide acceptance. To date, no adverse reports have been published and the method appears to be gaining support.

HIATUS HERNIA

Often called a "sliding hernia," hiatus hernia is an acquired defect in which the stomach protrudes upward through the diaphragm. It is

TABLE 33.5

BOWEL HABITS OF 62 PATIENTS BEFORE AND AFTER BRAN[1]

Frequency of Movement	Before Bran	After Bran
Irregular	13	—
Every 3 days	7	—
Every 2 days	8	—
Every day	28	31
Twice daily	3	25
3 times daily		6
Frequently	3	

SOURCE: Painter et al. (1972).
[1] The influence of increased fiber (about 10 g per day) on bowel movements in diverticular patients.

almost completely absent in Africa; therefore, assumed by Burkitt to be the result of a refined, low fiber diet. Burkitt's hypothesis maintains that hiatus hernia is the result of straining at stools during which the abdominal pressure forces the abdominal components against the diaphragm, resulting in the protrusion of the stomach upward through the diaphragm (Burkitt and James 1973). No evidence is available to show that a high fiber diet helps the hiatus hernia which is usually asymptomatic.

APPENDICITIS

An inflammation of the appendix is believed to be the result of an obstruction of the lumen or a muscular contraction. The disease is widespread in the United States, and almost nonexistent in Africa (Burkitt 1971A). Its incidence has been inversely correlated to crude fiber in the diet (Walker *et al.* 1973).

The epidemiological evidence appears to support the hypothesis that a low fiber diet induces appendicitis. However, there is virtually no evidence suggesting that fiber will relieve appendicitis once it has developed.

IRRITABLE BOWEL SYNDROME

This is a common clinical condition associated with disordered and excessive muscular activity of the colon. Chronic constipation exists and brings on straining at stools which often produces small, firm stools; at other times diarrhea is present associated with intermittent discomfort and pain. The disease is observed in fiber-deficient diets; therefore, it is attributed to a low fiber diet (Painter 1972).

The irritable bowel syndrome can be successfully treated with dietary fiber such as unprocessed bran (Piepmeyer 1974) or preparations containing refined hemicelluloses (Lieberthal 1955).

DISEASES RELATED TO THE ABSORPTIVE
PROPERTIES OF FIBER

Detoxification

Absorptive properties of fiber were identified as the result of its chemical composition (Eastwood 1973). This property makes fiber an effective detoxicant and Ershoff has demonstrated that fiber acts as an absorbant which can reduce the toxic effects of various food additives (Ershoff 1974). Therefore, the indirect effects of the removal of ingested materials should be two-fold. One, to change blood chemistry by preventing the absorption (or reabsorption) of materials; two, to prevent materials toxic to the intestine from exerting their effects while adsorbed to fiber. Thus, we have a basis on which the observed cardiovascular effects rest and the colon cancer-dietary fiber hypothesis stands.

COLON CANCER

Colonic cancer, second only to lung cancer as a killer among cancers, occurs least frequently in populations with a high residue, unrefined diet (Burkitt 1971B). This is supported by epidemiological studies on ethnic and racial groups which have migrated and assumed the dietary habits of the surrounding population. These studies are summarized in Table 33.6 which identifies colonic cancer as environmental and related to acquired dietary habits as opposed to physical location or origin (Doll 1969). From these and other observations, a hypothesis has been developed by Burkitt (Burkitt 1971B).

Since fiber increases the velocity and volume of material through the gut, it reduces the exposure time of gut tissue to toxic components. In concert with volume, cation exchange and organic absorption, it increases the potential for detoxification. Consequently, the detoxification property of high-fiber diets on known toxicants supports the development of this hypothesis (Ershoff 1974).

A number of observations support Burkitt's hypothesis. Deoxycholate, a by-product of gut flora on the bile acids is a weak, carcinogen (Booher et al. 1951). Further, people on high-fiber unrefined diets produce a substantially different group of bile acid metabolites than do people on a low-fiber refined diet (Senti and Dimler 1959). Hill supports these observations and has shown that the flora of

TABLE 33.6

COLON CANCER INCIDENCE IN MALES BY RACE
AND COUNTRY

Country	Race	Incidence
U.S. (California	Black	69.8
(Hawaii)	Caucasian	68.0
(Hawaii)	Japanese	66.4
Rhodesia	Black	18.2
Japan (Rural)	Japanese	11.8
South Africa	Black	10.8
Nigeria	Black	5.8

populations with a high incidence of colon cancer are different from those with a low incidence (Hill 1974). Thus, there is a growing body of evidence to show that quite possibly the cancer-causing agents are the by-products of the action of gut flora on bile acids and other components.

Epidemiologists attempting to relate fiber to cancer have in general been unsuccessful. However, Irving has shown a small negative correlation of colonic cancer with cereal consumption (Ruttloff et al. 1968). This finding would account for a small fraction of colonic cancer, but is both intuitively correct and in agreement with Burkitt's hypothesis.

In contrast to the Burkitt hypothesis and findings of Irving are studies which show that cancer correlates best to the high-fat diet in Western Countries (Wynder and Reddy 1975). The incidence of cardiovascular disease and colonic cancer is well established which also leads to some support to the dietary fat hypothesis (Allen and Whelan 1963).

Obviously, the issue has not been settled and both schools—the high-fiber low-cancer and high-fat high-cancer—contain elements of truth. An attempt at reconciliation is worthwhile.

Fecal fat does not change with dietary fat, therefore fat probably has its greatest influence on the amount of bile acids, dietary sterols and their metabolites which reach the gut. These are converted by the bacteria to many compounds, some of which are probably carcinogenic. Consequently, a high-fat low-fiber diet would have the best probability of introducing sterol metabolites into the gut with the longest residence time. Similar comment can be made with respect to any carcinogen in a high-fat low-fiber diet. In conclusion, it would appear that the addition of fiber to the diet and a reduction of fat would be beneficial and certainly not harmful, especially since accurate experimental or epidemiological data will be very long in coming.

CARDIOVASCULAR EFFECTS

The low incidence of ischaemic heart disease and low serum cholesterol among populations on an unrefined high residue diet has been attributed, in part, to the hypocholesterolemic effect of fiber (Dahlquist et al. 1965). However, the diet of these people is often, but not always, low in fat; long-term studies with South African white and Bantu prisoners on controlled diets have confirmed these observations (Reddy and Wostmann 1966).

Keys showed that pectin performs well as a cholesterol-lowering agent (Keys *et al.* 1960, 1961). Recent studies have confirmed Keys observations and have shown that wheat bran is not hypocholesterolemic (Jenkins *et al.* 1975; Palmer and Dixon 1966). Animal studies have, in general, confirmed the human feeding studies and have evaluated rolled oats, alfalfa, wheat, straw, cellulose and cellophane, rice bran and rice husk (Osman *et al.* 1970). The evidence appears to be very strong that some types of dietary fiber have a cholesterol-lowering effect. Attention should shift to possible mechanism.

Blood cholesterol is part of the total cholesterol pool and can be lowered by reducing the total pool. This reduction can be achieved by the elimination of neutral sterols and bile acids via the feces. Therefore, any material which absorbs these components and increases their movement into the feces will exert a cholesterol-lowering effect. In short, the absorbing agent prevents absorption of dietary cholesterol and reabsorption of the bile acids. This explanation is consistent with and dependent upon the absorption and chelation properties of fiber.

Epidemiologists have, in the main, emphasized fat and refined carbohydrate in their correlative studies on CVD. Consequently, vegetable and fruit fiber which experimental studies show is effective have been neglected (Klevay 1974). Indeed, at least one alternate hypothesis proposes that the low-fiber diets produce a copper: zinc imbalance which results in increased risk (Klevay 1974).

At this stage in human understanding, the studies of Keys and others suggest that an increase in fruit would be beneficial; indeed, it could hardly be detrimental associated with a reduction in fat consumption, and the general effect would probably do much good and no harm.

SUMMARY

There is little doubt that more intensive research is required on dietary fiber. However, fiber can be successfully used for some therapeutic purposes; it is equally evident that an increase in dietary fiber would be desirable and not detrimental.

The food scientists should pursue a method for accurate fiber determination, or take action to have the neutral detergent method accepted as the standard. Epidemiologists and medical scientists should pursue the relationship of dietary fiber to the colonic diseases with cancer as a primary consideration. The extent to which dietary fiber can alter blood chemistry should be pursued as one means by which food can become a sophisticated vehicle of preventive medicine.

It is in the introduction of fiber into fabricated foods that it has its greatest potential. However, how much fiber, what type of fiber, and in which foods remains to be discovered.

BIBLIOGRAPHY

ALLEN, P. Z., and WHELAN, W. J. 1963. The mechanism of carbohydrase action 9 hydrolysis of solep mannan by preparation of a-amylase. Biochem. J. 88, 69.

AMEN, R. J., and SPILLER, G. A. 1975. Dietary fiber in human nutrition. Crit. Rev. Food Sci. Nutr. 7, 39.

AOAC. 1970. Official methods of analysis of the Association of Official Analytical Chemists. Washington, D.C.

BOOKER, L. E., BEHAN, I., McMEANS, E. and BOYD, H. M. 1951. Biologic utilization of unmodified and modified food starches. J. Nutr. 45, 75.

BURKITT, D. P. 1971A. The etiology of appendicitis. Br. Surg. 58, 695.

BURKITT, D. P. 1971B. Epidemiology of cancer of the colon and rectum. Cancer, 28, 3.

BURKITT, D. P. and JAMES, P. A. 1973. Low residue diets and hiatus hernia. Lancet 2, 128.

BURKITT, D. P., and TROWELL, H. C. 1975. Refined Carbohydrate Foods and Disease; The Implications of Dietary Fibre, Academic Press, London, England.

BURKITT, D. P., WALKER, A. R. P., and PAINTER, N. S. 1972. Effect of dietary fiber on stools and transit times and its role in the causation of disease. Lancet 2, 1408.

BURKITT, D. P., WALKER, A. R. P., and PAINTER, N. S. 1974. Dietary fiber and disease. J. Am. Med. Assoc. 229, 1068.

COWGILL, G. R., and ANDERSEN, W. E. 1932. Laxative effects of wheat bran and 'washed bran' in healthy men. J. Am. Med. Assoc. 98, 1866.

COWGILL, G. R., and SULLIVAN, A. J. 1933. Further studies on the use of wheat bran as a laxative. J. Am. Med. Assoc. 100, 795.

DAHLQUIST, A., BULL, B., and GUSTAFSSON, B. E. 1965. Rat intestinal 6 bromo-2-napthyl glycosidase and disaccharidase activities. Arch. Biochem. Biophys. 109, 150.

DOLL, R. 1969. The geographical distribution of cancer. Br. J. Cancer 1, 23.

EASTWOOD, M. A. 1973. Vegetable fibre: its physical properties. Proc. Nutr. Soc. 32, 137.

ERSHOFF, B. H. 1974. Antitoxic effects of plant fiber. Am. J. Clin. Nutr. 27, 1395.

FINDLEY, J. M. et al. 1974. Effects of unprocessed bran on colonic function in normal subjects and in diverticular disease. Lancet 1, 146.

GOERING, H. K., and VAN SOEST, P. J. 1970. Forage Fiber Analysis, Agriculture Handbook No. 379, USDA, Washington, D.C.

HARVEY, R. F., POMARE, E. W., and HEATON, K. W. 1973. Effect of increased dietary fiber on intestinal transit. Lancet 1, 1278.

HILL, J. R. 1974. Steroid nuclear dehydrogenation and colon cancer. Am. J. Clin. Nutr. 27, 1475.

HOLMGREN, G. O. R., and MYNORS, J. M. 1972. The effect of diet on bowel transit times. S. Afr. Med. J. 46, 918.

JENKINS, D., NEWTON, C., LEEDS, A., and CUMMINGS, J. 1975. Effect of pectin, guar gum and wheat fibre on serum-cholesterol. Lancet 1, 1116.

KERTESZ, Z. I. 1950. The Pectic Substances. Chap. 8. Interscience, New York.

KEYS, A., GRANDE, F., and ANDERSON, J. T. 1960. Diet type (fats constant) and blood lipids in man. J. Nutr. 70, 257.

KEYS, A., GRANDE, F., and ANDERSON, J. T. 1961. Fiber and pectin in the diet and serum cholesterol concentration in man. Proc. Soc. Exp. Biol. Med. *106*, 555.

KLEVAY, L. M. 1974. Coronary heart disease and dietary fiber. Am. J. Clin. Nutr. *27*, 1202.

KRAUS, B. 1976. Guide to fiber foods. Signet *451*, W6794.

KRITCHEVSKY, D., and STORY, J. A. 1974. Binding of bile salts *in vitro* by nonnutritive fiber. J. Nutr. *104*, 458.

LIEBERTHAL, M. M. 1955. Irritable colon syndrome. Conn. State Med. J. *19*, 1.

LUBBE, A. M. S. 1971. Dietary evaluation. S. Afr. Med. J. *45*, 1289.

McCONNELL, A. A., and EASTWOOD, M. A. 1974. A comparison of methods measuring fiber in vegetable material. J. Sci. Food Agric. *25*, 1451.

NELSON, R. H. 1975. Role of unavailable carbohydrates in digestion. *In* Nutrients in Processed Foods—Fats—Carbohydrates. P. L. White, D. C. Fletcher, and M. Ellis (Editors). Publishing Science Group, Acton, Mass.

OSMAN, H. F., THEURER, B., HALE, W. H., and MEHEN, S. M. 1970. The influence of grain processing on *in vitro* enzymatic starch digestion of barley and sorghum grain. J. Nutr. *100*, 1133.

PAINTER, N. S. 1972. Irritable or irritated bowel. Br. Med. J. *2*, 46.

PAINTER, N. S., ALMEIDA, A. Z., and COLEBOURNE, K. W. 1972. Unprocessed bran in treatment of diverticular disease. Br. Med. J. *2*, 137.

PAINTER, N. S., and BURKITT, D. P. 1971. Diverticular disease of the colon: a deficiency disease of Western civilization. Br. Med. J. *2*, 450.

PALMER, G. H. and DIXON, D. G. 1966. Effect of pectin dose on serum cholesterol levels. Am. J. Clin. Nutr. *18*, 437.

PIEPMEYER, J. L. 1974. Use of unprocessed bran in treatment of irritable bowel syndrome. Am. J. Clin. Nutr. *27*, 105.

REDDY, B. S., and WOSTMANN, B. S. 1966. Intestinal disaccharidase activities in the growing germfree and conventional rats. Arch. Biochem. Biophys. *113*, 609.

ROBERTSON, J. 1972. Changes in the fiber content of the British diet. Nature *238*, 290.

RUTTLOFF, R. *et al.* 1968. Polysaccharide splitting enzymes in the brush border of the rat mucosa with particular respect to γ-amylase. Chem. Abstr. *68*, 111, 599a; Acta Biol. Med. Ger. *19*, 331 (1967).

SENTI, F. R., and DIMLER, R. J. 1959. High-amylase corn—properties and prospects. Food Technol. *13*, 663.

SOUTHGATE, D. A. T. 1969. Determination of carbohydrates in foods. II. Unavailable carbohydrates. J. Sci. Food Agric. *20*, 331.

SOUTHGATE, D. A. T., and DURNIN, J. V. G. A. 1970. Calorie conversion factors: an experimental reassessment of the factors used in the calculation of the energy value of human diets. Br. J. Nutr. *24*, 517.

SPILLER, G. A., and AMEN, R. J. 1975. Plant fibers in nutrition: need of better nomenclature. Am. J. Clin. Nutr. *28*, 675.

TROWELL, H. C. 1973. Dietary fiber, ischaemic heart disease and diabetes mellitus. Proc. Nutr. Soc. *32*, 151.

TROWELL, H. C. 1974. Definitions of fiber. Lancet *1*, 503.

VAN SOEST, P. J., and McQUEEN, R. W. 1973. The chemistry estimation of fiber. Proc. Nutr. Soc. *32*, 132.

WALKER, A. R. P., WALKER, B. F., RICHARDSON, B. D., and WOOLFORD, A. 1973. Appendicitis, fibre intake and bowel behavior in ethnic groups in South Africa. Postgrad. Med. J. *49*, 243.

WILLIAMS, R. D., and OLMSTED, W. H. 1936. The manner in which food controls the bulk of the feces. Ann. Intern. Med. *10*, 717.

WYNDER, E. L., and REDDY, B. W. 1975. Dietary fat and colon cancer. J. Natl. Cancer Inst. *54*, 7.

WYNNE-JONES, G. 1975. Flatus retention is the major factor in diverticular disease. Lancet *2*, 211.

34

Expanding Role of Nutrition in Clinical Medicine

Maurice E. Shils

During the period of incubation and genesis of the American Chemical Society a few investigators of those days were making basic observations which have led to advances in areas which remain of current interest in nutrition science.

ENERGY METABOLISM

A major problem then as now relates to energy sources and requirements. Liebig (1842) had stated that energy in muscular work was derived from the breakdown of protein. In the late 1860's evidence from various sources was presented challenging this concept. The most important study was that of Frankland (1866) professor of the Royal Institution of London who was the first to employ a combustion calorimeter for the measurement of the energy values of foods. As a result of these studies he found that muscles work at the expense of energy derived through the oxidation of non-nitrogenous foods, namely, fats and carbohydrates. These studies were followed by those of von Pettenkofer, Voit, Rubner and Magnus-Levy which opened the era of studies on energy exchange and metabolism in animals and men. They led to Cuthbertson's (1932) discovery of marked negative nitrogen balance following trauma which gave rise to the concept of a catabolic state in stress, fever and trauma, and to appreciation of the roles of various hormones in regulating metabolism in relation to energy needs (Hume 1974). Modern metabolic studies have given us information on energy and protein requirements during stress, influenced our treatment of disease and emphasized the need for efforts to prevent malnutrition in these types of patients. Today's patient undergoing a sophisticated metabolic study can be visualized as breathing under a plexiglas canopy with his or her respiration and energy exchange measured continuously with computerized printouts (Spencer *et al.* 1972). Protein, fat or carbohydrate metabolism is further measured by production of labeled carbon dioxide, with radio-immuno assays of blood levels of various hormones and with a large number of blood constituents monitored by continuous instrumentation with rapid printout of results.

AMINO ACID REQUIREMENTS

In 1870, Mehu discovered that proteins could be quantitatively precipitated from aqueous solution upon saturation with ammonium sulfate without being coagulated by this treatment. This gave rise to procedures for the systematic isolation and analysis of proteins for amino acids. By 1881, seven amino acids were known (glycine, leucine, tyrosine, serine, glutamic acid, aspartic acid, phenylalanine). When the last of the essential amino acids (threonine) was discovered by Rose, it became possible to do reliable systematic studies on the essential amino acid requirements of man. At present, the problem of amino acid requirements for growth, development and reproduction in health and in disease continues to occupy a central role in clinical nutrition. Examples of current interest include inborn errors of metabolism where manifestations of disease are associated with disturbances in the metabolism of amino acids; hepatic coma which is the result of the failure of liver function and which has a grave prognosis; renal failure where the end products of protein metabolism accumulate and are believed to cause undesirable metabolic changes; and amino acid needs in stress and trauma.

VITAMINS

Liebig (1842) believed that "albuminous" substance (e.g., protein) together with "fuel foods" (sugar and fats) and certain mineral salts were all that an animal needed for nourishment. However, clinical observations such as those of J. B. A. Dumas during the siege of Paris by the Germans in 1870-1871 on efforts to prevent starvation of infants with an artificial milk led to the conclusion that something essential to life was lacking in the artificial milk. Studies with experimental animals such as those of Lunin and others demonstrated that mixtures of purified proteins, fats, carbohydrates and minerals would not sustain life. These observations led Funk in 1912 to propound a theory that beriberi, scurvy, pellagra and possibly rickets were caused by deficiency or lack in diet of special substances, which he termed vitamines. In 1913–1914, McCollum and Davis and, shortly thereafter, Osborne and Mendel, reported the discovery of a growth factor in butterfat which McCollum called fat-soluble A and which was renamed Vitamin A. McCollum's technique for biologic testing of the nutritional value of foods initiated the vitamin era which has led to the discovery of 13 vitamins which are accepted as being essential for man. A major scientific achievement has been the elucidation of the biochemical basis for the essentiality of the water-soluble vitamins in various steps of intermediary metabolism of glucose, fatty acids and amino acids and in the synthesis of numerous organic compounds important in metabolic processes. A similar understanding of the basic role in biochemistry of each of the fat-soluble vitamins has not been achieved except for that of vitamin A in the visual process. This is an area of active investigation. The recent pioneer work of DeLuca and associates followed by the studies of others has established mono- and di-hydroxy cholecalciferols as biologically active metabolic products of vitamin D (DeLuca 1973) and has led to important clinical applications in the treatment of vitamin D-resistant disease in children and in chronic renal disease. As with the water-soluble vitamins, the patho-physiologic effects of deficiencies of the various fat solubles are known and these vitamins are useful in prevention of treatment of deficiencies.

In this country and in most technologically advanced countries rickets, scurvy, pellagra and beriberi have disappeared as significant diseases in terms of incidence in the total population resulting from inadequate intake. However, vitamin deficiencies continue to occur particularly in alcoholics, certain food faddists and patients with various malabsorptive syndromes.

A relatively new area of increasing interest has to do with the effects of drugs on vitamin metabolism and needs. For example, anticonvulsant drugs, such as diphenylhydantoin and certain barbiturates as well as oral contraceptives have been found to lower serum folate (Stebbins et al. 1973). Long-term anticonvulsant drugs given to patients in Europe have been reported to lead to vitamin D inadequacy (Richens and Rowe 1970).

Several decades ago "anti-vitamins," i.e., specific antagonists of individual vitamins, were actively investigated in the expectation that they would have efficacy as therapeutic agents. At present only several antifolates have significant clinical use; these are methotrexate (used in the treatment of certain neoplasia) and a few others active against bacteria and plasmodia. Current use of methotrexate includes acute high dosages followed by folinic acid as a "rescue" to minimize damage to non-neoplastic tissue.

In clinical practice multiple vitamins are widely prescribed.

Relatively few hospital laboratories are capable of measuring more than a few vitamins; even when such procedures are performed the lag time between sample collection and the laboratory report is discouragingly long. For this and other reasons there is relatively little objective information on the actual needs of vitamins in a variety of disease states. Fragmentary and intriguing data exist indicating that in conditions of hypermetabolism and trauma, significant decreases in circulating levels in certain vitamins may occur, particularly with respect to vitamin A and ascorbic acid. These vitamins are not lost in the urine and the reasons for their low circulating levels are unknown (Levenson *et al.* 1946). Whether this represents an increased need or whether there is increased tissue uptake or destruction is also unknown.

The discovery that a number of vitamins circulate in blood in protein-bound form has added a new dimension which requires further elucidation. It may well be that a decreased circulating vitamin A level in certain clinical situations is not related to a deficiency of this vitamin but rather to decreases in the concentration of its transport binding protein or in its release from the liver. We probably are entering a period in which measurements of protein binding are as important to understanding vitamin metabolism, at least for certain vitamins, as it is for an understanding of iron metabolism. There is no question that there is a need for basic studies on the effects of various disease states on vitamin requirements. Such information would undoubtedly lead to major modifications of presently available vitamin formulations produced for clinical use.

MALNUTRITION

The etiologies of disease-related malnutrition are numerous.

Malabsorption as an important cause of malnutrition can occur for many reasons. One category occurs with failure to digest food properly. There are defects in the digestion of protein, fats and carbohydrates as the result of pancreatic insufficiency. Overgrowth of intestinal bacteria in the upper small bowel ("blind loop" syndrome) leads to decreased amounts of conjugated bile salts (with resultant impaired fat absorption) and to impaired absorption of vitamin B-12. Deficiencies in certain hydrolytic enzymes in the brush border of epithelial cells can lead to malabsorption. Malabsorption occurs in a variety of disease states associated with abnormalities in the absorbing cells of the intestinal epithelium. Inflammatory bowel disease, such as regional ileitis, or infection such as Whipple's disease or the protozoan infestation Lamblia giardia may adversely affect the intestinal epithelium. Since the mucosa of the small intestine is second only to bone marrow in its sensitivity to damage by radiation, it is no surprise to find malabsorption in a certain proportion of patients who have been given fairly high dosage radiation therapy in the treatment of cancer.

Massive bowel resection as the result of arterial or venous occlusion, trauma or disease leads to serious malabsorption. Increased knowledge of and application of nutritional principles have permitted survival with progressively smaller amounts of residual bowel. Maintenance of good nutrition from the time of resection is necessary in order to permit maximum hyperplasia of the remaining bowel (Wilmore et al. 1971). The use of total parenteral nutrition (TPN or "hyperalimentation") or of certain defined formula diets which may be given by slow infusion through tubes into the remaining small bowel have been very helpful in permitting patients to survive who in former days would certainly have died. Protracted TPN at home is now a practical procedure (Shils 1975).

Poor absorption can result from failure in normal transport mechanisms from the intestinal cells into the lymphatics. This occurs when there is failure to synthesize neutral fats or to form chylomicrons which are essential for entry of neutral fat into the lymphatics. There may be fat malabsorption secondary to obstruction to lymphatic flow as in lymphangiectasia and in other diseases.

It has been known for a number of years that the long-chain fatty acids are synthesized in epithelial cells to neutral fat which go into the lymphatics in the form of chylomicrons. Fatty acids of chain length C6 to C12 have another route passing directly into the portal system in the same way as carbohydrates and amino acids where they are metabolized in the liver. This has given rise to a preparation called medium chain length triglycerides (MCT) made up of triglycerides of C6 to C10 fatty acids which can be absorbed and transported to the liver through the damaged intestinal epithelium and therefore may be useful in conditions of fat malabsorption (Greenberger and Skillman 1969).

The need for pancreatic amylase can be by-passed by the use of so-called oligosaccharides (C3–C9 polymers of glucose) or disaccharides (which can by hydrolyzed by brush border enzymes) or by monosaccharides.

It is now well established that the majority of the world's adult population tends to have restricted amounts of the lactose-splitting enzyme lactase (Kretchmer 1972). When there is lactase insufficiency, there are undesirable intestinal symptoms and signs upon ingestion of significant amounts of milk taken at one time. Damage to the intestinal epithelium can also lead to a shortage of this enzyme.

Anorexia which is severe loss of appetite is a hallmark of a number of chronic or acute systemic diseases. It creates serious problems for the patient with weight loss and other manifestations of malnutrition. While its cure depends upon successful treatment of the underlying disease, there are many things a physician can do to combat this, ranging from more palatable, more acceptable diets and liquid formulas to total parenteral nutrition. Such techniques allow the patient's nutrition to be maintained or improved while the underlying disease is being treated. Diseases or stressful situations in which there is increased metabolism such as extensive burns, multiple fractures, hyperthyroidism or serious infection impose an increased requirement for calories, amino acids and other nutrients. The increased energy expenditure and tissue breakdown is especially apparent in the immediate post-infective on traumatic state in previously healthy individuals but may have more dire consequences for the previously malnourished individual who has less reserves.

The term metabolic dysfunction designates those disease states leading to altered metabolism which creates nutritional problems. Increased secretion of certain corticosteroids by the adrenal gland or administration of high potency steroids on a chronic basis leads to abnormal retention of sodium and increased secretion of potassium and breakdown of muscle nitrogen and loss of calcium from bone. Certain hormones such as calcitonin, serotonin, prostaglandins or vasoactive intestinal peptides may cause diarrhea and large losses of water and electrolytes (Schmitt et al. 1975). Recognition of these resultant metabolic problems at an early stage enables physicians not only to try to treat the underlying disease but also to maintain the patient by provision of adequate nutrients and fluids so that serious debility does not result.

Organ failure includes failure of any organ system as the result of disease, resection, radiation or toxic agents which may lead to serious metabolic and nutritional disturbances. I shall limit discussion of this important area to some aspects of renal disease. Nutrition problems result from renal failure as a result of either inability to excrete end products of metabolism or failure to reabsorb nutrients from damaged renal tubules. When glomerular filtration is impaired there is a rise in accumulation of urea, the end product of protein metabolism; uric acid, the end product of purine metabolism; and various other derivatives of protein metabolism. Inorganic sulfates and phosphates cannot be excreted and the kidney may be unable to excrete hydrogen ion or manufacture sufficient bicarbonate with resulting metabolic acidosis in accordance with the severity of the renal disease and the nature of the dietary intake. Before the importance of potassium was well recognized in medicine, patients with acute and advanced chronic renal failure frequently died from abnormally high potassium levels and its effects on cardiac function.

Patients with borderline failure may be well maintained by restricting potassium, protein and phosphate intakes and by proper regulations of the amounts of sodium and water. Where kidney function is severely compromised, however, life can be maintained only by the use of chronic dialysis or by successful kidney transplant. It has been known for many years that patients with renal insufficiency tend to develop low serum calcium levels and this could be improved with increased amounts of vitamin D. However, the use of large amounts of vitamin D was only partially successful. The observation of Fraser and Kodicek (1970) that 1:25 di-hydroxy cholecalciferol is formed in the kidney has given clinicians a new insight into the causation and treatment of the calcium metabolism of renal insufficiency. The use of suitable isomers of vitamin D has been helpful in preventing and treating bone resorption.

Hemodialysis, although life saving, creates new problems for the patient including nutritional ones. There is increased loss of amino acids, minerals and certain vitamins in dialysis requiring close supervision of diets to eliminate deficiencies or excesses in order to maintain the patient with minimum number of dialyses. It has been shown that histidine which, heretofor, has not been considered an essential amino acid for normal adults is not properly synthesized in patients with chronic renal disease (Fürst 1972). It should be included in any amino acid formulation given by tube, orally or by intravenous means. Patients with acute renal failure who are unable to eat may be maintained in better condition with fewer complications when they are given infusions of essential amino acids in conjunction with adequate calories (Abel et al. 1973).

Several groups of Italian investigators established that restricted nitrogen intake in the form of diets containing their major source of nitrogen as essential amino acids promoted positive nitrogen balance in patients with chronic renal failure. While this approach has been clinically useful, especially in acutely ill patients, where amino acids can be given by tube or intravenously, a less restricted diet has often been found necessary for prolonged use by patients on oral intake. It was suggested, independently, by several groups in 1966 and 1967 that the keto analogues of the essential amino acids be used for the treatment of chronic renal failure. This proposal was based initially on the idea that abnormally large amounts of ammonium ion were available to uremic individuals from degradation of urea by bacteria in the gut which was then absorbed and this could be used to form the essential amino acids from the alpha-keto analogues. In this way a nearly nitrogen-free diet could in theory supply all the requirements of essential amino acids and permit synthesis of nonessential amino acids. This concept is now being put to test. In their experiments to date, Walser and associates (*cf.* 1975) have utilized a combination of keto analogues of valine, leucine, isoleucine, methionine and phenylalanine with smaller amounts of the essential amino acids lysine, tyrosine, threonine and histidine. This combination exerted a nitrogen sparing effect. In some patients it decreased the need for dialysis and in others temporarily improved renal failure. Their data suggest that, contrary to the original idea, the keto analogues may be working by reducing the rate of the appearance of urea by minimizing the diversion of nitrogen from muscle-derived amino acids into urea synthesis.

As a by-product of these efforts Walser *et al.* have broadened their studies into the possible utilization of these analogues and amino acids into treatment of so-called hepatic coma (portal systemic encephalopathy). Another aspect has to do with the impression that keto analogues are sparing nitrogen. It is known that branch chain amino acids are catabolized chiefly in muscle rather than liver where they provide the principle source of nitrogen release from muscle as glutamine and alanine. Such studies hold the promise of improved therapy for certain disease states by the use of these modified nutrients.

NUTRITION AND CANCER

Diet and nutrition interrelate with cancer in a number of ways (Shils 1973). There is increasing interest in the etiological role of diet in the causation and development of cancer. Rather crude epidemiological evidence suggests that diets high in meats, animal fats and refined carbohydrates tend to be associated with higher incidence of certain important malignancies such as those of breast and colon when compared to diets high in vegetable and cereal grains (Wynder and Shigematsu 1967; Drasar and Irving 1973; Armstrong and Doll 1975). Emphasis has been given to the possible role of "fiber" as a protective factor (Burkitt 1971) and that of fat as a predisposing factor (Wynder and Shigematsu 1967; Hill 1975). This is an important but very complex area. Case controlled studies in the United States are not in agreement and there are contradictory data about the relevance of fat (Enstrom 1975; Drasar and Irving 1973; Armstrong and Doll 1975), meat (Enstrom 1975) and fiber (Hill 1975). Less emphasis is given currently to the fact that other types of diets are associated with higher incidence of types of cancer other than breast or colon; i.e., a high incidence of gastric cancer in Japan and liver cancer in parts of Africa. In my opinion, the data presently available do not provide physicians with sufficient objective information to permit firm recommendations for major changes in diet in an attempt to prevent cancer. It is well known that significant restriction of total calories or of certain individual nutrients which interfere with growth will inhibit the development of spontaneous tumors in experimental animals and decrease the rate of growth of transplanted tumors (Shils 1973). Such observations at the moment do not have clinical applicability except perhaps in the presence of significant overweight in women.

Another area of great potential importance has to do with the role of dietary factors in the metabolism of carcinogenic compounds. Based on studies with experimental animals certain deficiency states do appear to make certain carcinogens more active, in terms of tumor development, presumably by decreasing the rate of normal metabolism and permitting more prolonged contact of the carcinogen with normal cells (Newberne et al. 1968). Another interesting aspect has to do with the drug metabolizing enzymes which are present in the greatest concentration in the liver but which also occur in other organs. Wattenberg (1975) has pointed out that such enzymes in the intestine and lung have low activity normally; however, when the animal is exposed to chemical inducers or to carcinogens, activities of these enzymes in these locations increase significantly. In addition, activity is induced by certain plant materials, the most potent being in the Brassicaceae family of vegetables. Indoles derived from such vegetables have inducing activity. Certain phenolic antioxidants also have this property. Such intestinal enzymes may serve as primary barriers to the entry of potential carcinogens to the body.

Cancer affects nutritional status by its systemic and local effects. The anorexia of cancer leads to poor food intake with deficiency of certain nutrients and progressive debility. The local effects are represented, for example, by obstruction of the intestinal tract so that the patient cannot eat or by fistula formation where the intestinal contents are deviated from their normal course. An important area concerns the effects of the treatments of cancer on nutritional status of the patients (Shils 1973). There is no part of the alimentary tract which is spared from the surgeon's scalpel in an attempt to cure cancer.

Such surgical resection or bypass of many of the areas lead to a variety of undesirable nutritional consequences. Radiation and chemotherapy may affect adversely appetite, gastrointestinal function and metabolism. This is a most challenging area to the clinician interested in maintaining good nutritional status in the patient and calls for significant knowledge of nutrition and of the effects of treatments.

Maintenance of good nutrition can play an important role as an adjunct in the treatment of cancer. Prevention or improvement of nutrition allows a better opportunity for successful completion of whatever treatment is indicated with a minimum of delay. Total parenteral nutrition or certain oral formulas may decrease the toxicity of certain chemotherapeutic agents.

One of the most intriguing areas of cancer concerns the reasons for the failure of the body's immune defense mechanisms in recognizing cancer cells as "foreign." There is some evidence that immunoglobulins known as "blocking antibodies" may coat cancer cells and thus foil the immune mechanisms. Jose and Good (1973) have made the intriguing suggestion based upon some animal experiments that certain clinical levels of protein deficiency may depress the formation of these "blocking antibodies" without suppressing the cell mediated immunity; the result is increased destruction of the cancer cells. They have also found that deficiency of certain amino acids may also have the same result (Jose and Good 1973). The clinical importance of this approach still remains to be determined.

SYSTEMIC EFFECTS OF MALNUTRITION

Malnutrition for the great majority of patients stems from inadequate caloric intake and/or inadequate absorption so that the patient develops progressive negative calorie balance, loss of body tissues including muscle wasting and weakness. In association with negative caloric balance deficits of other nutrients may occur which may have additional deleterious effects on the well-being of the patient. The patient tends to get out of bed less and prolonged bedrest in itself contributes to negative nitrogen balance as skeletal muscle atrophies; it also causes demineralization of bone. Although there is ordinarily a high priority for wound healing in malnourished patients, at some point this becomes impaired and there is slowing or actual failure to heal and resultant increase in infection of open wounds. Significant alterations often occur in fluid and electrolyte balances as malnutrition progresses (Moore 1959). While some of the fluid retention may be secondary to low albumin levels which may occur, this is usually a later occurrence in the accumulation of edema than are the changes in the hormonal control of salt and water with increased reabsorption of sodium by the kidney and retention of water. The normal mechanisms maintaining the cellular-extracellular potassium and sodium gradients are altered. With progressive weight loss there tends to be breakdown of skin especially at pressure points with ulcer formation. The loss of subcutaneous fat also leads to increased loss of body heat and temperature regulation problems.

It is now well-established that malnutrition of calories and of protein lead to depression of cell-mediated immunity (Vitale and Good 1974) and this may help explain the old observation of increased infection in malnourished individuals. Immunoglobulins may or may not be affected. There is recent evidence that immunoglobulin A, which is secreted in complexed form into the intestinal tract, may be decreased in malnutrition. Since this immunoglobulin is believed to play a significant role in resistance to infection, a decrease in concentration may explain the increased incidence of enteric disease in malnourished individuals.

Alterations in endocrine function occur in malnutrition, notably a depression of the output of thyroid hormones, decreased insulin, increased glucagon, and, in children with protein malnutrition, an increase in growth hormone (Gardner and Amacher 1973).

DIET-ASSOCIATED DISEASES

Diet and nutrition are associated with the development of the symptoms and signs of certain diseases. I emphasize the *development* of symptoms since the etiologics of these diseases do not appear to be nutritionally or dietarily determined. However, the outcome, in terms of manifest signs and symptoms of such diseases, i.e., the "risk," appears to be related, in part, to the dietary history of the patient.

Atherosclerosis

Atherosclerosis, with particular reference to coronary artery disease, is in large part a genetically determined disease, with hyperlipidemia, hypertension, obesity, diabetes and cigarette smoking as major risk factors. As a result of the work of Fredrickson and associates we know that there are four major plasma lipoprotein families (Frederickson 1974). Variations in abnormal lipoproteins concentrations have been classified into five types (Levy and Ernst 1973). Types II, III and IV appear to be associated with an increased risk of coronary artery disease. Dietary programs can modify these abnormalities toward normal. The ability to prevent or decrease coronary artery disease by dietary modification is much less clear and more data are needed to determine the effectiveness of such programs.

Diabetes Mellitus

Diabetes mellitus is associated with Type IV lipoprotein pattern. The dietary treatment of Type IV is not to decrease fat but rather to decrease carbohydrate while giving polyunsaturated fats. In late onset adult diabetes there is often overweight and reduction of weight can often improve glucose tolerance.

Obesity

The prevention of this syndrome is, of course, avoidance of consumption of excess calories and its treatment is persistent negative caloric balance to achieve desirable weight. However, just as with alcoholism, the reasons why certain individuals overeat to a point where weight becomes both a psychological and survival problem are complex and poorly understood. Important basic studies have yielded information on the development of adipose cells in relation to age and diet (Hirsch and Knittle 1970).

A series of weight-reduction diets have flashed in (and out) of lay magazines and books. They often bear the imprimatur of the name of a physician or a medical institution (the latter unauthorized) with justification based on claimed metabolic or nutritional "principles." The latest relates to diets restricted to amino acids or protein. The usefulness and problems associated with this approach remain to be seen.

Total starvation of the obese which has been studied in a number of medical centers is rarely used now. The relative ineffectiveness of voluntary weight reduction programs in the treatment of the obese has led to increasingly widespread use of the jejuno-ileo by-pass whereby the major part of the small bowel is by-passed and patient weight loss occurs because of the development of marked malabsorption. This treatment is potentially dangerous with the induction in a significant number of patients of a variety of nutritional problems which need to be closely watched for and treated. While undoubtedly helpful to many morbidly obese, it is a procedure best done and followed in centers versed in the necessary art and science including that of clinical nutrition.

Hypertension

Hypertension is another major syndrome which may have serious consequences if uncontrolled. As with diabetes mellitus, risks are increased with overweight, hyperlipidemia, sedentary life and cigarette smoking.

"Inborn Errors" of Metabolism

A large number of such genetic diseases are now known. Their existence has greatly increased our knowledge of certain aspects of intermediary metabolism. Some of these conditions may be treated by nutritional means. The classic examples are phenylketonuria and galactosemia which are treated, respectively, by reduction in the phenylalanine content of the diet and by omission or reduction of lactose in the diet. These diseases have resulted in a specialized branch of pediatrics and in the development of diet banks of special foods and food products which can be called upon by pediatricians to assist in the treatment of patients with specific metabolic errors (Scriver 1974).

There are other major diseases whose outcome is affected by diet and nutrition. Cancer has been mentioned with the possible relationships between its development and diet. Osteoporosis with its progressive loss of bone mass is believed by some to be related to diet but adequate evidence on this point is not yet at hand.

SUMMARY

The century of the existence of the American Chemical Society has seen dramatic and wide-ranging advances in our knowledge of those nutritional factors essential for normal growth, reproduction and maintenance in many species including man and of their basic roles in metabolism. This knowledge has given us the capacity to effectively treat and prevent the classical nutritional deficiency diseases. The thrust of clinical nutrition is in the direction of interrelating nutritional therapy with optimum treatment for a variety of other diseases and educating physicians in the importance of maintaining good nutrition in the care of the patient.

BIBLIOGRAPHY

ABEL, R. M., BECK, C. H., JR., ABBOTT, W. M. *et al.* 1973. Improved survival from acute renal failure after treatment with intravenous essential L-amino acids and glucose. N. Eng. J. Med. *288*, 695–699.

ARMSTRONG, B., and DOLL, R. 1975. Environmental factors and cancer incidence and mortality in different countries with special reference to dietary practices. Int. J. Cancer *15*, 617–661.

BURKITT, D. P. 1971. Epidemiology of cancer of the colon and rectum. Cancer *28*, 3–13.

CUTHBERTSON, D. P. 1932. Observations in the disturbance of metabolism produced by injury to the limbs. Q. J. Med. *1*, 233–246.

DELUCA, H. F. 1973. The kidney as an endocrine organ for the production of 1:25-dihydroxy vitamin D_3, a calcium mobilizing hormone. N. Eng. J. Med. *289*, 359–365.

DRASAR, B. S., and IRVING, D. 1973. Environmental factors and cancer of the colon and breast. Br. J. Cancer *27*, 167–172.

DUMAS, J. B. A. 1957. *Cited by* E. V. McCollum. A History of Nutrition, pp. 202–203. Houghton Mifflin Co., Boston.

ENSTROM, J. E. 1975. Colorectal cancer and consumption of beef and fat. Br. J. Cancer *32*, 432–439.

FRANKLAND, E. 1866. *Cited by* E. V. McCollum. 1957. A History of Nutrition, pp. 127–129. Houghton Mifflin Co., Boston.

FRASER, D. R., and KODICEK, E. 1970. Unique synthesis by kidney of a biologically active vitamin D metabolite. Nature *228*, 764–766.

FREDERICKSON, D. S. 1974. Plasma lipoproteins and apolipoprotein. Harvey Lect. pp. 185–237. Academic Press, New York.

FÜRST, P. 1972. [15] N-studies in severe renal failure. II. Evidence for the essentiality of histidine. Scand. J. Clin. Lab. Invest. *30*, 307–312.

GARDNER, L. J., and AMACHER, P. (EDITORS) 1973. Endocrine Aspects of Malnutrition. Kroc Foundation, Santa Ynez, California.

GREENBERGER, N. J., and SKILLMAN, T. G. 1969. Medium-chain triglycerides. Physiologic considerations and clinical implications. N. Eng. J. Med. *280*, 1045–1058.

HILL, M. J. 1975. Metabolic epidemiology of dietary factors in large bowel cancer. Cancer Res. *35*, 3398–3402.

HIRSH, J., and KNITTLE, J. L. 1970. Cellularity of obese and non-obese human adipose tissue. Fed. Proc., Fed. Am. Soc. Exp. Biol. *29*, 1516–1521.

HUME, D. M. 1974. Endocrine and metabolic responses to injury. *In* Principles of Surgery, 2nd Edition. S. J. Schwartz (Editor). McGraw-Hill Book Co., New York.

JOSE, D. G., and GOOD, R. A. 1973. Quantitative effects of nutritional essential amino acid deficiency upon immune responses to tumors in mice. J. Exp. Med. *137*, 1–9.

KRETCHMER, N. 1972. Lactose and lactase. Sci. Am. *227*, 70–78.

LEVENSON, S. M. *et al.* 1946. Ascorbic acid, riboflavin, thiamine and nicotinic acid in relation to severe injury, hemorrhage and infection in the human. Ann. Surg. *124*, 840.

LEVY, R. I., and ERNST, N. 1973. Diet, hyperlipidemia and atherosclerosis. *In* Modern Nutrition in Health and Disease. R. S. Goodhart and M. E. Shils (Editors). Lea & Febiger, Philadelphia.

LIEBIG, J. 1842. Animal chemistry or organic chemistry in its applications to physiology and pathology. Wm. Gregory (Editor). Taylor & Walton, London, England.

MOORE, F. D. 1959. The nature of starvation. The study of body composition in starvation and cachexia. *In* Metabolic Care of the Surgical Patient. W. B. Saunders Co., Philadelphia.

NEWBERNE, C. M. ROGERS, A. E., and WOGAN, G. N. 1968. Hepato-renal lesions in rats fed a low lipotrope diet and exposed to aflatoxin. J. Nutr. *94*, 331–343.
RICHENS, A., and ROWE, D. J. F. 1970. Disturbances of calcium metabolism by anticonvulsant drugs. Brit. Med. J. *4*, 73–76.
SCHMITT, M. G., JR., SOERGEL, K. H., HENSLEY, G. T., and CHEY, W. Y. 1975. Watery diarrhea associated with pancreatic islet cell carcinoma. Gastroenterology *69*, 206–216.
SCRIVER, C. R. 1974. Inborn errors of metabolism: a new frontier of nutrition. Nutr. Today, Sept./Oct., 4–15.
SHILS, M. E. 1973. Nutrition and neoplasia. *In* Modern Nutrition in Health and Disease. R. S. Goodhart and M. E. Shils (Editors). Lea & Febiger, Philadelphia.
SHILS, M. E. 1975. A program for total parenteral nutrition at home. Am. J. Clin. Nutr. *28*, 1429–1435.
SPENCER, J. L. *et al.* 1972. A system for continuous measurement of gas exchange and respiratory functions. J. Appl. Physiol. *33*, 523–528.
STEBBINS, R., SCOTT, J., and HERBERT, V. 1973. Drug-induced megaloblastic anemias. Semin. Hematol. *10*, 235–251.
VITALE, J. D., and GOOD, R. A. (EDITORS) 1974. Nutrition and immunology. Symp. Nutr. Immunol. Am. J. Clin. Nutr. *27*, 623–669.
WALSER, M. 1975. Nutritional effects of nitrogen-free analogues of essential amino acids. Life Sci. *17*, 1011–1020.
WATTENBERG, L. W. 1975. Effects of dietary constituents on the metabolism of chemical carcinogens. Cancer Res. *35*, 3326–3331.
WILMORE, D. W., DUDRICK, S. J., DALY, J. M., and VARS, H. M. 1971. The role of nutrition in the adaptation of the small intestine after massive resection. Surg. Gynecol. Obstet. *132*, 673–680.
WYNDER, E. L., and SHIGEMATSU, T. 1967. Environmental factors of cancer of the colon and rectum. Cancer *20*, 1520–1561.

Section 6

Food Safety

35

Introduction

Harold L. Wilcke

Food safety has been and must continue to be of paramount concern to all of us, and it is particularly fitting that this be recognized by the Agricultural and Food Division of the American Chemical Society in this Centennial program. As Emil Mrak pointed out in his discerning and thought-provoking Atwater Memorial Lecture, it was the concern of Dr. Wiley, an agricultural chemist in the U.S. Department of Agriculture, for food safety which resulted in the establishment of, as Mrak described it "the organization we now know as the Food and Drug Administration." The Food and Drug Administration, now a unit of the Department of Health, Education, and Welfare, represented by chapters in this section, has responsibility for interpreting and enforcing the laws and regulations designed to ensure the safety of much of our food supply, with certain other similar responsibilities being assigned to USDA.

However, it is the primary responsibility of those who produce, process, and deliver our foods to maintain the integrity of what has commonly become recognized as the safest and most nutritious food supply available to any population in the world.

These responsibilities are recognized and accepted by both the regulated and the regulators, with common mutual respect for the integrity of the people and the institutions involved, in spite of charges of collusion leveled against both USDA and Food and Drug when they feel that industry quality assurance programs are effective and, therefore, need not necessarily be changed.

Certainly there is no way to guarantee complete safety in our entire food supply—the perfect system has not yet been devised. We must recognize that some foods are quite safe when consumed at the usual levels, but may cause some problems when eaten in excessive amounts. George Irving (Chap. 39) and his group are uncovering some very interesting information about some of our long-time, well-recognized foods.

Much can be said about the pros and cons of some of the multitudinous regulations governing our food operations—regulations which restrict our choices and more often than not increase the cost of the food to the ultimate consumer. Are they worth the cost?

Unfortunately, questions raised about the safety of specific foods (or more commonly regarding additives to our foods) and whether those questions are justified or not (and more commonly are not), receive much more publicity than their significance might warrant. This creates questions in the minds of people who do not have all of the information which would be necessary to make proper judgments. Certainly, questions must be raised; but even more certainly, there must be some information regarding the validity of the question before it is brought out for public hearing. There must be more emphasis on balanced reasoning, and less on sensationalism with regard to our food supply.

This symposium on Safety of Foods is organized to present some of the real problems that have been encountered in protecting the safety of our food supply; and to give us some insight on what is being done, or has been done, to correct the problems; and some suggestions regarding the role of the chemist in this very critical area. Our speakers today are well equipped to answer any questions that should be discussed and you will all recognize them as pioneers and eminent authorities in this field.

36

Food Industry Concerns with Food Safety

Bernard L. Oser

The coincidence of the bicentennial celebration of the birth of our nation with the Centennial Anniversary of the American Chemical Society presents a propitious and unique occasion for contrasting the safety of our modern food supply with that of yesteryear.

The American revolution occurred during the very decade in which Priestley discovered "dephlogisticated air" and Lavoisier established its relation to the process of combustion and renamed it "oxygen." The preceding year (1769) marked an epochal event in the history of American chemistry since it was then that Dr. Benjamin Rush was appointed as the first professor of chemistry in the new world, at the medical school of my alma mater, the University of Pennsylvania. Also in that year James Watt patented his steam engine which may be said to have been the forerunner of the industrial revolution.

This being a Bicentennial Symposium on Food Safety, let's look back at the state of affairs in those not so "good old days." The "food industry" in prerevolutionary days, such as it was, had not progressed much beyond the ancient trades of brewing, milling, baking, and butchering, although cheese-making and wine production were also practiced commercially. When not homegrown, meats and vegetables came from local farms and some of the most prominent colonial figures, including George Washington and Thomas Jefferson, were farmers. The methods of food preservation then in vogue were principally variations of dehydration, salting, pickling and smoking, refinements of which are the major processes in use to this day.

The food industry as it is now constituted may be said to have begun with the pioneer contributions of two Frenchmen, Nicolas Appert, who in 1820 discovered the technique of "canning" (hermetic sealing) and especially Louis Pasteur who laid the biological foundations of fermentation (1857) and preservation by the controlled application of heat.

The hygienic conditions under which foods were produced in the days before refrigeration can only be imagined. Moldy bread was quite common and one may wonder about the condition of the loaves that young Ben Franklin carried under each arm as he trudged up High Street in Philadelphia. Even as late as the mid-19th century, milk was described as being obtained from "cows confined in ill-constructed, ill-ventilated, and improperly drained places—their milk being at the same time formed by such unnatural and improper food as brewers' and distillers' grains and distillers' wash."

During the middle ages purveyors of foods, particularly foods that were milled, such as grains and spices, or liquids, such as wine and honey, became adept in the knavish art of adulteration. The practice had become so widespread in both England and America in the latter 1700's that one could hardly be sure that *any* food purchased in a local shop was not adulterated, either fraudulently or in a manner actually hazardous to health and life.

Milk was watered, "cream" was thickened with starch paste, lard was mixed with beef tallow, bread was whitened with chalk or alum, and tea was blended with spent leaves. (There could have been more than one reason for the Boston Tea Party.) These are but a few examples of economic adulteration which depleted the purses but not the lives of the colonists.

However the more vicious forms of adulteration were characterized by a Massachusetts legislator in 1785 as those committed by "evilly disposed persons . . . from motives of avarice and filthy lucre . . . (who are) induced to sell diseased, corrupted, contagious or unwholesome provisions to the great nuisance of public health and peace."

An unsung warrior in the battle against unsafe adulterated foods was a young chemist who came to London from Germany in 1793, adopted the name Frederick Accum, found employment as a pharmacist's assistant and several years later established his own commercial laboratory. He pioneered in the application of analytical chemistry to the detection of adulteration of foods and attracted the attention of Sir Humphrey Davy who appointed him "operative chemist" at the Royal Institution. Accum met his nemesis for having the effrontery to identify in lectures and writings, the shops from which he purchased samples of grossly adulterated foods. The full title of his classic work, published in London and Philadelphia in 1820, is revealing: "A Treatise on Adulterations of Food, and Culinary Poisons. Exhibiting the Fraudulent Sophistications of Bread, Beer, Wine, Spiritous Liquors, Tea, Coffee, Cream, Confectionery, Vinegar, Mustard, Pepper, Cheese, Olive Oil, Pickles and

Other Articles Employed in Domestic Economy and Methods of Detecting Them." He reported foods colored with mineral pigments such as lead oxide and chromate, manganese dioxide, copper sulfate and arsenite, Prussian blue, indigo, and verdigris. Copper and lead contamination was also described as coming from the vessels in which foods were prepared. Extract of cocculus indicus (picrotoxin) was used to impart "an intoxicating quality to porter or ales." These and many other types of adulteration were disclosed as "illicit pursuits (having) assumed all the order and method of a regular trade." Consumer activism had yet to be a reality since, as Accum described it, "The practice of sophisticating the necessaries of life, being reduced to systematic regularity, is ranked by public opinion among other mercantile pursuits, and is not only regarded with less disgust than formerly, but is almost generally esteemed as a justifiable way to wealth."

It was not until 1860, 40 years later, that the first comprehensive food law was promulgated in England thanks to the attention focused on the prevalence of unsafe foods by several writers, particularly Dr. A. H. Hassall, who added microscopy to chemistry as a tool for revealing adulteration of foods.

The nefarious and hazardous practices of adulteration of foods and drugs continued nevertheless, both in our country and in England and, in fact, were intensified with the emergence of the chemical industry in the mid-19th century when more sophisticated chemical means of adulteration became available, such as salicylic acid, formaldehyde, boric acid, and coal tar derivatives which were used originally as textile dyes.

It is of interest in this centennial year of the American Chemical Society to recall that Dr. Harvey W. Wiley, the father of our first federal food law, was one of its founder members and played an active role during its formative years when membership and control was wrested from the "New York clique" and ACS became a truly national society. Much as he inveighed against adulteration in his crusade, Dr. Wiley disclaimed being an alarmist. He believed chemicals ought not be added to foods unless it could be shown that they bestowed qualities to the benefit of the "health, prosperity, and honesty of the community." He said the problem was mainly ethical and "injury to public health . . . is the least important question in the subject of food adulteration, and it is the one which should be considered last of all. The real evil of food adulteration is deception of the consumer."

Against this historical background let us consider present-day concerns of the food industry with the specific problem of food safety (as distinguished from economic and esthetic aspects of adulteration). Despite the clamor of professional activists, it stands to reason that it is to the best interest of manufacturers and purveyors of food to keep their customers alive and well, and not to jeopardize their own profit-making potential and their reputations by selling foods that are unwholesome or unsafe. Today's laws and regulations provide a rigid framework within which food safety is controlled. Voluntary compliance is the order of the day and cooperative quality assurance programs are gaining ground.

Outbreaks of food poisoning are rarely encountered nowadays and even then are mainly of microbial rather than chemical origin. In fact, cases of poisoning result more often from foods poorly prepared or stored in homes, restaurants, or institutions than from those produced, processed or distributed through normal commercial channels. Standards of good manufacturing practice covering plant location, sanitation, and operation have been developed which are applicable to the food industry generally and many are being adopted or proposed for specific target areas such as the frozen or prepared food industries.

Annual statistics published by the Center for Disease Control demonstrate that, compared to other aspects of the environment, the incidence of morbidity due to chemically contaminated food is extremely low although occasional accidents due to carelessness or ignorance generate wide public concern. Episodes of toxicity, like methyl mercury poisoning from the ingestion of denatured seed wheat or fish from polluted waters, have been largely responsible for the institution of increasingly tighter control over the safety of foods.

Public attention was directed to potential chemical hazards after the adoption of the Food, Drug and Cosmetic Act of 1938. It provided that definitions and standards of identity for foods be developed through public hearings. This aroused greater awareness of the legitimate uses of chemicals in formulation and processing. The introduction of new agricultural pesticides during the post-World War II period raised the issue of potentially hazardous residues in food crops. Except for individual allergies, known instances of human illness attributable to the presence of functional, direct or indirect, additives have been virtually nonexistent.

The amendments to the Food, Drug and Cosmetic Act covering pesticide residues, food additives and color additives were designed to require prior approval of their safety under conditions of intended use and so placed the burden of proof on the food industry. The degree of safety of foods consumed by the American public is greater by many orders of magnitude than was the case a century ago. It is not the threat of acute poisoning that now concerns the regulatory agency and the regulated industry so much as suspicions of subtle, cumulative, or chronic toxicity, difficult, if not impossible to correlate epidemiologically with the etiology of disease. For instance, it is well known that highly toxic substances occur naturally in a great variety of foods, not to mention the mycotoxins that are formed by molds in grains, nuts and fruits, both before and after harvest. The poisonous nature of oxalates, hemagglutinins, goitrogens, cyanogenetic glycosides, solanin, fish and shellfish toxins, etc., was discovered largely through human experience, long before toxicological studies were conducted in laboratory animals. The rarity of widespread food intoxication from the incorporation of these substances in man's diet is due to the low levels generally present which are tolerated without apparent harm. Nevertheless were their natural toxins first discovered today it is doubtful whether these foods would stand the tests for safety. It will be recalled that several natural sources of food flavors (e.g., tonka bean extract and oil of sassafras) have been proscribed on toxicological grounds.

In sharp contrast with the law regulating naturally occurring poisonous substances in food, added substances are subject to prior approval of safety under conditions of intended use, unless qualified scientists regard them to be safe on the basis of either common experience or "scientific procedures." Many problems were created by the provisions for exemption contained in the definition of food additives in the 1958 Amendment, particularly when it became necessary to arrive at the GRAS ("generally recognized as safe") status of the many natural and "synthetic" substances then in use. It was not long before it was realized that natural occurrence does not, ipso facto, confer safety on food components, and furthermore, many so-called "synthetic" substances were either derived from natural sources or chemically identical to their natural counterparts.

The current reviews of GRAS substances authorized by the Food and Drug Administration, pursuant to the 1969 White House Conference on Foods and Nutrition, and the redefinition of the conditions for eligibility for GRAS status, have involved extensive surveys of industrial usage as well as of U.S. food consumption patterns. It may be predicted that relatively few significant changes will be required pursuant to the review of the GRAS list by the Select Committee of the Federation of Societies for Experimental Biology and Medicine and the independent review of flavoring substances by the Expert Panel of the Flavor and Extract Manufacturers' Association.

More serious difficulties for both the FDA and the food industry have arisen from the application of toxicological procedures to the

safety evaluation of food chemicals. Among the principal compli-
cating factors are (a) the intrinsic lack of precision in the determination of quantitative "no effect" dosage in experimental animals;
(b) the arbitrary safety factors used in extrapolating these levels to
arrive at permissible dietary intakes and tolerance levels in human
food; and (c) the interpretation of tumorigenic responses at the
grossly exaggerated conditions employed in animal feeding studies,
particularly with relatively nontoxic substances for which high doses
can be tolerated by animals without reduction in life span. Some of
the most controversial rulings of the Food and Drug Administration
have resulted from the application of the concept embodied in the
Delaney Clause which proscribes the use in food of any substance
found to induce "cancer" when ingested by man or animal. The lack
of general agreement on reasonable specifications limiting the experimental conditions of feeding tests with respect to the size, frequency
and duration of test doses, have led to the removal from the market
of foods containing cyclamates, amaranth (Red No. 2) and trichloro-
ethylene.

The unanticipated difficulties that may be encountered in the administration of Congressional enactments are abundantly illustrated
in the case of the laws regulating food safety. Perhaps more attention has had to be devoted by both FDA and the food industry to
substances which past experience indicated to be safe and to incidental additives present in hitherto undetected amounts, than to
direct or newly-introduced additives. This has been the combined
result of several factors, viz., (1) the application of new analytical
instrumentation and methodology of exquisite sensitivity, capable
of revealing the presence of substances in the parts per billion and
lower ranges; (2) advances in toxicological procedures involving an
increase in the number of critical parameters of adverse biological
effects, such as more animals, more observations, mutagenic, teratogenic, and multigeneration reproduction tests; and finally (3) the
lack of an official policy which formally recognizes that minute
amounts of even toxic chemical residues can be inconsequential from
the standpoint of human safety. Thus, much to the concern of
industry and the embarrassment of FDA, "no-residue" substances
(many pesticides, DES and migrants from packaging materials) which
were latterly discovered to come under the Delaney Clause, have
been reassessed, banned or threatened to be banned, even though
prolonged use has revealed no indication of potential hazard.

For years many scientists have contended that the literal construction of the Delaney Clause precludes the exercise of scientific or
regulatory judgement. Agreeing with this view, former FDA Commis-

sioner, Dr. Alexander Schmidt recently stated that the agency is "struggling to perfect a regulation stating how sensitive the approved analytical method used to detect chemical residues must be, to insure safe use of a carcinogen in livestock feed."

The precipitate action of FDA in banning such food chemicals after years of apparently innocuous use has aroused considerable concern not only within the food industry but among consumers who feel that FDA has been remiss in not recognizing sources of potential harm. Fearing adverse market reaction, several major companies, hypersensitive to consumer reaction, have prematurely abandoned the use of substances whose safety has come under question in the media, even before the issues have been resolved or regulatory action has been instituted. Repeated situations of this sort have added greatly to loss of public confidence not only in the safety of foods but in the judgement of the administrative agency.

Stimulated by biased and frequently sensational accounts in the press and over the air, the public has come to expect the ultimate in food safety, i.e., absolute proof regardless of what present knowledge may hold in store. The social philosopher Erich Fromm, in a different context has stated that "Paranoid-like thinking is based on the assumption of a logical possibility and wants to have absolute certainty that something could not happen even in the most remote circumstances."

The FDA has come in for a great deal of undeserved criticism from Congressional committees, the General Accounting Office, and pressure groups, not to mention certain segments of the food industry itself. In defense of FDA, it may be appropriate to quote one of its severest critics, Senator Edward Kennedy, who pointed out at a recent hearing that Congress has given it "an extraordinarily wide range of responsibilities" yet "never provided the funds necessary to develop the Agency's scientific capability and to assure that it can carry out its enforcement responsibilities." The time to consider such contingencies is before rather than after enacting what the Senator called the "paradox of an overextended mandate." In this connection, a recent editorial in Science has stated that "Congress is perhaps the branch of government least suited to receive, process, and use scientific information, not because of intellectual incapacity, but because of its organization and protocol and the nature of the legislative process."

It is interesting to note the relative expansion of the Food and Drug Administration during the past 25 years encompassing the period since passage of the Pesticide Chemicals, Food Additives and Color Additives Amendments to the Food, Drug and Cosmetic Act.

The number of authorized positions has risen to 6264 from 1000 in 1951, while the appropriations for these fiscal years has increased from $5,467,000 to $201,800,000. That present civil service salary limits are insufficient to attract distinguished personnel of outstanding scientific calibre has been recognized not only by the professional community but by Senator Kennedy and his colleagues who are seeking to reorganize the Food and Drug Administration. It might also be mentioned that the physical separation in the Washington area of the regulatory and scientific branches of the Food and Drug Administration does little to add to the efficiency of its operations.

In conclusion I would like to offer a few recommendations which I believe are in the public interest:

(1) It may be wishful thinking but I would hope for better recognition of the tremendous advances that have been made in the present century toward ensuring the safety of the food supply despite expansion in the functional uses of pesticides and other chemicals.

(2) I would urge that consumers and professional advocates alike consider the entire question of food safety in proper perspective giving highest priority to hygienic and microbiological considerations rather than overstressing the alleged hazards of food additives.

(3) I would caution regulatory agencies to await adequate investigations and deliberate scientific judgement rather than reacting precipitately to unconfirmed suspicions and contrived "public concern."

(4) I would urge that Congress provide FDA with sufficient resources to enhance its scientific competence and stature so that its discretionary decisions will justify more general scientific support and engender greater public confidence.

(5) I would advise that the food industry be more effectively organized to support toxicological investigations and safety evaluations of substances in common use throughout the industry, so that it might be better able to meet such crises as have arisen in the case of food colors, packaging materials, solvents, etc.

(6) I would recommend that food scientists in both industry and academia support the effort of oncologists and biostatisticians to devise new approaches to the goal of "virtual," as contrasted with "absolute," safety since this appears to offer the possibility of a rational scientific solution to the Delaney dilemma.

37

A Centennial Look at
Food Regulation

Howard R. Roberts

The Commissioner of Food and Drugs on behalf of the entire Food and Drug Administration congratulates the American Chemical Society on the occasion of the Society's Centennial observance, and offers best wishes for another 100 years of professional contributions to American science and society. Chemistry is basic to FDA operations. The present-day Food and Drug Administration evolved from a laboratory within the Bureau of Chemistry of the U.S. Department of Agriculture in the late 19th Century. More than 900 chemists help staff the testing, research, and analytical laboratories of FDA at headquarters and in the field; thus, chemists represent the largest category of scientific employees in the agency. The equally important position of chemistry in the Bureau of Foods is illustrated by the composition of our staff and by our close relationship with both the American Chemical Society and the Association of Official Analytical Chemists. In another type of relationship between chemistry and Bureau of Foods, the Stauffer Chemical Company recently achieved the milestone of becoming the first producer of chemicals to join our Cooperative Quality Assurance program. The cooperative agreement covering Stauffer's monosodium glutamate plant in San Jose, California ensures that production will adhere to quality controls developed by Stauffer and approved by FDA. We are looking forward to similar participation by other manufacturers in the chemical industry. FDA is also aware of the recent formation of the Chemical Industry Institute of Toxicology and joins in the hope that it will contribute to more efficient and effective assurance of the safety of chemicals used in foods, drugs, and the environment.

The role of regulatory activity in food safety has become a matter of serious concern to both groups and individuals in our society. Alvin Toffler, the author of *Future Shock*, has illustrated this dramatically in a passage from Chap. 6 of his book, *The Eco-Spasm Report:*[1]

A few days later, the Food and Drug Administrator, Harold Whitwell, requests a meeting with the president on a matter of the 'utmost pressing urgency.' Whitwell begins by handing the president a 382-page blue-bound report stamped 'Top Secret' which proves beyond dispute that a disaster is in the making. Tens of millions of jars and cans of baby food now circulating in the market have been found to contain Arceon Yellow, a dye which tests now show to be the cause of serious mental retardation in infants. The president, after consultation with his election strategists and television specialists, moves swiftly.

That evening, in a special television appearance, he explains that while there is no cause for alarm he is, as a precautionary measure, asking all baby-food manufacturers using Arceon Yellow to suspend production, and all supermarkets and local stores to remove from their shelves all bottled or canned baby-food products until tests can conclusively show them to be pure.

He is also asking all producers of commercial Zeronacephon, a safe but expensive substitute for Arceon Yellow, to halt immediately all shipments except those to baby-food manufacturers, to guarantee that consumer supplies can be resumed as safely and speedily as possible.

Plastics manufacturers, who require Zeronacephon for many of their products, immediately protest and begin besieging Zeronacephon producers for stepped-up shipments and emergency supplies. The price of bulk Zeronacephon zooms from $12.20 per barrel to $33.82, and the shares of Texsyn Corporation, the prime manufacturer of Zeronacephon, leap from 4¾ to 12. Meanwhile, pediatricians' offices are jammed by mothers who think their children may have already been affected.

In the election year of 1976, this futurist's scenario has a special impact, but in any year it is an excellent illustration of the kind of real-world milieu in which FDA conducts its business. The public significance of that business is confirmed by FDA's continuing presence in the headlines and features of the national news media.

Food safety decisions are indeed made in the open, as public decisions today must be. These decisions continue to be increasingly complex and difficult because of a variety of factors. First, the expectations of the American public have risen sharply as a result of an increased sensitivity to the "right to know" and a heightened concern about environmental contamination. The food supply, more than any other facet of daily life, is subject to mounting attention and skepticism. Consumers expect as a matter of course that the food supply will be free of *any* risk. Demands for safety continue to rise as the chemist increases the sensitivity of analytical methods and as the toxicologist evolves stricter experimental safety protocols. To some extent, current expectations for food safety are the outgrowth of a world food supply which is still relatively abundant. This

[1] Toffler, Alvin. 1975. The Eco-Spasm Report. *In* Future Shock, published by Bantam Books, New York.

situation could change, however, and responsible regulatory policy must take this possibility into consideration.

Another factor is that FDA must deal with the safety of foods and food additives in the context of current uses and consumption patterns. That context is continually changing as eating patterns shift toward consumption of more than half of our meals away from home and toward increased dependence on fabricated foods. Moreover, regulatory problems must be solved within the limits of existing legislative authority and within the capabilities of present technology. Unfortunately, legislative change tends to lag considerably behind advances in technology.

The current environment in which regulatory decisions are made is further characterized by the public demand for cutbacks, or at the very least, stringent control, in Federal budgets and by legislative demands that sometimes compete with each other. Examples are the Privacy Act, which limits the information Federal agencies can disclose, and the Freedom of Information Act, which requires disclosure of certain information in the possession of Federal agencies. A few years ago, 10% of FDA's information was in the "public domain" and approximately 90% was held in confidential status; today, those figures are just reversed. At the present time we see a growing need for far broader policy input from scientists and experts in a number of fields. Yet the legal and moral demands for openness imposed by the Federal Advisory Committee Act, for example, may sometimes dilute the contribution of certain "experts" and delay accommodation that might well be achieved in confidential discussion. The future is not likely to become less complicated in view of current enthusiasm for Congress to fashion mechanisms, such as concurrent resolutions for one-House vetoes of regulations promulgated by administrative agencies, for exercising more direct oversight of regulatory agencies. Furthermore, consider the additional complications should bills such as so-called "sunset bills" be enacted, by which an agency would have a specified lifespan and would either go out of business or have to be reauthorized at the end of that period. While the intent is commendable, such an approach would have devastating consequences for long-range planning and the recruitment of capable people.

The reason for mentioning these matters is not to be negative and unduly pessimistic, nor to philosophize on the obvious fact that the world is very complicated today. Rather it is to emphasize, on the occasion of this historic celebration, that the next 100 years will require much more from both the ACS and the FDA than the last 100 years as we jointly contribute to American and world society.

Nonetheless we should be optimistic for the future as we contemplate the achievements of the past. Moreover, there is solid evidence that the American people, in spite of the mass of often contradictory information flowing to them, appreciate what both organizations are attempting to do on their behalf. Notwithstanding recent attacks on the Food and Drug Administration, FDA continues to rank high—perhaps highest—in public confidence among Government agencies concerned with consumer protection.

Rather than recount the history of the Food, Drug, and Cosmetic Act, let us consider three topics of current interest in assessing the Federal role in regulating food safety: first, specific programs that are calculated to achieve a higher degree of food safety in the United States; second, an appraisal of the "sociology of science" from the perspective of a science-based law enforcement agency; and last, some considerations in the efficient use of the resources given to the FDA by the American taxpayer.

The following are examples of specific programs in food safety in the areas of compliance, education and research.

A new major effort in the compliance area is the Food Inspection Strategy developed during FY'75 to provide for inspectional coverage of domestic food and associated industries. The strategy is based on an assessment of the risk associated with each food category and the selection of the type and frequency of inspection best calculated to identify risk and avoid it.

Firms producing high-risk products will be inspected in greater depth and at more frequent intervals than other firms. The type and timing of inspections will also depend on the compliance status of the firm and the length of time since it was last inspected.

Establishment inspections have been a mainstay of Federal enforcement ever since passage of the 1938 Amendments to the 1906 Act. Inspectional activities have varied in intensity and frequency according to such factors as resource constraints and compliance histories. Until recently, inspections of the food industry have primarily emphasized sanitation but it is now clear that technology requires more attention in inspecting products that may deviate significantly from manufacturing specifications and/or present a hazard to health. Because of the nature of modern food processing and distribution systems, it is possible that these products could be in distribution or even in the home before the plant's management is aware of any problem. Recognizing this possibility of danger, enlightened members of industry have instituted "early warning" quality assurance systems designed to monitor and control critical stages in production processes. It is the intent of our new inspection program to see that this effort is intensified. Our new strategy will utilize appropriate in-depth inspectional techniques (e.g., Hazard Analysis of Critical Control Points or HACCP) to gain insight into the details of a firm's continuous operations, rather than the performance only on the day that FDA inspectors are present.

Evaluation of the firm's quality assurance programs by specially trained inspectors is a fundamental part of the strategy. FDA believes that consumers are best protected when a firm has the internal capability to monitor and control its own production processes. Of course, food safety is not achieved by compliance activity alone; rather, it is achieved most fully through compliance activities supplemented by public information and industry education. Such education is typified by our Good Manufacturing Practices (GMP) regulations and the Cooperative Quality Assurance (CQA) program mentioned above. Currently, 75 firms are enrolled in our CQA program and agreements with 31 other firms are under development. GMP's, which specify the "shalls" and "shoulds" for food processing, have been issued for smoked fish, low-acid canned foods, bottled drinking water, and confectionery products. The GMP for bakery products has recently been published as a proposal, and GMP's for tree nut products, fermented foods, and frozen foods will appear next.

The last six years of data compilation and scientific literature evaluations of the substances that are Generally Recognized as Safe (GRAS) have produced an increasing awareness that the safety of FDA-regulated food additives must be substantiated to a greater degree than was possible in the past. In many cases safety studies do not exist. Moreover, as scientific and technological improvements continue in the fields of toxicology and pharmacology, it becomes increasingly more obvious that some earlier research studies were not adequate by today's standards to assure the safety of food ingredients. Thus for certain substances a considerable safety research effort will be required to justify continued use. In addition to the GRAS list items there are about 10,000 food packaging materials, 400 functional food additives and 1600 flavors for which a reevaluation of the safety data and scientific literature is required to ensure their continued safe use. How quickly this review can be accomplished depends on the available resources. The same is true of the full implementation of the new food plant inspection strategy and the timely issuance of GMP's.

Let us briefly consider the effects of the "sociology of science" on regulatory activities. Although neither Congress nor the American public necessarily appreciates the fact, FDA decisions are professional judgments based on the best scientific information available at a given time. Yet even its most devout supporters must admit that science has its limitations. Then too, science is not static but is, in fact, sometimes dramatically dynamic. All those now working in scientific fields can readily agree that no organizations nor individuals need apologize for changing their minds or re-thinking an earlier decision if new scientific evidence is found. As a matter of fact, this is the essence of good science. However, the need to review decisions on the basis of new scientific evidence is only one facet of the overall question of United States science policy.

As regulatory requirements grow increasingly more stringent, a number of ramifications must be thought through. First, what will a particular decision mean in terms of the supply of food and other essential materials? Second, how do we assure a source of trained leaders to initiate required programs in the private sector and to fill the evaluation and policy-making roles within the public sector? What will be the costs of the requirements and who will bear them, and how do these costs relate to the benefits achieved? An illustration of these general problems is the high cost of developing and obtaining approval for new products. If these costs are too high to be borne by small or medium sized businesses, what will be the effect on certain industries and on the ultimate costs and choices for consumers? Under current constraints are there sufficient realistic economic incentives to encourage the development of a product with limited market potential? What can we predict about the availability, for example, of an animal drug that might be fully effective against a disease which may appear infrequently and only in a particular region but nevertheless may destroy an entire animal industry in that region when it does occur? In short, we must look seriously at the impact of the profit motive on the development of science and technology in this country.

Let us consider an aspect of food safety which is at the very core of science-industry-government relations: namely, FDA's concern about safety testing in industrial and independent laboratories. This aspect was the subject of hearings before the United States Senate Subcommittee on Health in January and again in April 1976. The ramifications of this issue go far beyond FDA; they involve not only other units of government such as the Environmental Protection Agency and the National Cancer Institute, but also many industrial biological enterprises.

Under the Federal Food, Drug, and Cosmetic Act manufacturers carry the burden of demonstrating that their products comply with the safety provisions of the law. Thus, the FDA does relatively little toxicology testing of its own and no clinical testing. Instead FDA sets requirements, such as the type and extent of testing considered necessary for a determination of safety, and then reviews the data submitted by manufacturers to determine whether the data meet these requirements. For example, food additives undergo extensive testing in animals to determine potential teratogenicity and carcinogenicity whenever there is a likelihood that humans will be chronically exposed to such chemicals. In some cases a host of other toxicology tests may be required under a variety of laboratory conditions.

Inasmuch as food additive testing cannot feasibly be conducted in humans at this time, animal data are paramount for the determination of safety. It is, therefore, absolutely essential that such animal studies be conducted according to scientifically sound protocols and with close attention to quality control. Major toxicological studies are technically complex and can be conducted only through the combined efforts of a variety of scientists and technical personnel. A fundamental principle of both law and public morality is that FDA cannot in any way compromise its insistence that submitted data be accurate and be based on the best science available.

Studies involving repeated observations of many animals over the long term are not easy to carry out and it would be unfair to understate the technical demands of such studies, their tedious nature and the opportunities for inadvertent errors. Constant monitoring to ensure their quality is therefore essential. It is the responsibility of management in the food, cosmetic, pesticide and pharmaceutical industries to provide controls over the quality of testing conducted both in their own laboratories and by their contractors. They must remember that the data they produce are essential to the public health and are not mere arbitrary requirements of a capricious bureaucracy.

I do not believe that the structure of our safety testing is founded on incompetence or deceit but neither is it necessarily built on a sound foundation of scientific and managerial excellence. Therefore, several things need to be done in fairness to the public, the industry, and the agency. First, FDA must establish minimal guidelines for protocols to be used in safety testing and quality control procedures for conducting and reporting on animal experiments. Next, FDA must assure itself and the public not only that adequate protocols and quality assurance programs have been established but also, through periodic audit of testing laboratories, that these procedures are being followed.

FDA, in addition, must review submitted safety data with continuing vigilance. Any suspicion of irregularities will require an *ad hoc* investigation whose results could range from rejection of specific submissions to the institution of criminal proceedings. The certification of testing laboratories is also being considered. All these actions will be developed in coordination with other Federal agencies and with industry to minimize red tape and maximize good science.

These commitments, together with others that are required if the agency is to continue to function responsibly, require resources that we do not currently have. The same problems we have just been discussing have been reflected in budget amendments submitted to the Congress for increased personnel and operating funds. Basic FDA programs cannot be discontinued because the regulatory responsibilities are relatively fixed, even though new areas of effort are well justified. Yet the American public has the right to expect maximum effort and accountability to assure that the resources provided are being utilized as productively as possible.

The Cooperative Quality Assurance programs, the new inspectional strategy, and GMP's are examples of new kinds of approaches that protect the public health effectively and with more efficient use of resources than techniques used in earlier days of the FDA's history. Other developments should also be cited under the category of "efficient use of resources." In the recent past, FDA has increased its emphasis on internal capabilities for setting priorities. This function, after all, is the keystone of the appropriate deployment of our resources. A much more systematic approach to planning and priority setting has been developed by which all of the relevant decision-makers in the various organizational units become involved. The approach provides for a comparative weighting of the actual problems and their potential severity in terms of American public health, the public sensitivity to the problems, and, perhaps most important, the relative effectiveness of the technologies and resources that can be brought to bear on the identified problems. Decisions are then based on the "bottom line" of this evaluation.

Second, about four years ago, FDA embarked on a course that resulted in a substantial increase in regulations. Although the promulgation of regulations, with its numerous technical and legal steps, is a major user of Agency resources, this regulation-writing approach is believed to be a more efficient use of our resources. Previously, FDA had conducted its law enforcement activities through prosecutions on a case-by-case approach. Tremendous effort and time were required to prepare the cases properly and to prosecute them through the courts. Other organizations that might be violating the law in a similar manner were not a party to the criminal proceedings. Often long delays ensued before definitive opinions could be reached, and there was a great deal of uncertainty as to what the courts would say when they did speak. Therefore, it was reasoned, FDA could do a much more effective job by writing regulations which embodied the Agency's approach to the entire industry involved in a given area and with which the industry was expected to comply across the board. This approach has the added advantage of crystallizing and codifying Agency practice in a number of areas. Moreover, it gives consumers, industry, and all affected parties an opportunity to comment and to object so that their views and interests can be an essential part of the process. Certainly anyone who has consistently read the Federal Register will confirm the fact that FDA has been one of its chief users.

Related to the subject of efficient utilization of resources is the legal authority under which the FDA conducts its business. When this statutory authority is lacking or deficient in particular areas, it means that the agency is precluded from working in a way that would increase the efficiency with which the necessary task is performed. The Consumer Food Act of 1975 (S.641) recently passed the United States Senate and we are hopeful that the House of Representatives will also pass it during this session of the Congress. Although we vigorously oppose any possibility of restrictions on the traditional approach of strict liability for violations of the Food, Drug, and Cosmetic Act, we believe that this legislation in general provides crucial authority for the safety regulation of foods under today's conditions. The bill requires food processors to develop, implement, and maintain adequate procedures to assure safety, and empowers the Secretary of HEW to develop and enforce such safety assurance standards whenever processors fail to provide adequate protection against unreasonable risks of adulteration. Very significantly, the bill would extend FDA's inspection authority to include records maintained by food processors.

In addition, the proposed Act authorizes a national registration program for food processors, and requires complete—and in some cases quantitative—information on food ingredients, open dating, and the nutritional value of food products. Although additional resources will be needed to implement the legislation, the greater efficiency inherent in these approaches will mean that the provision of such resources will be one of the best investments the public can make in food safety.

Let us now look briefly at FDA's future. From a position of relative obscurity, the Food and Drug Administration has emerged over the last 15 years as one of the leading subjects for continuing national attention. FDA now stands at the threshold of still another new area. Its scope and structure, and even its name, are now being questioned in the form of two bills (S.2696 and S.2697) that would divide the present FDA into two agencies: the Food and Cosmetics Administration and the Drug and Devices Administration.

The main feature of these proposals is the creation in each new Administration of duplicate Bureaus and separate field resources. The Commissioner, Assistant Commissioners, and General Counsel would be appointed by the President with advice and consent of the Senate and for a term of 5 years. Several provisions of the bills address the need for additional competent scientists and the mechanisms thought best to provide the agencies with those resources. Lastly, "commerce" would be redefined so as not to limit action to interstate shipment per se, and the concept of "imminent hazard to health" throughout the present Food, Drug, and Cosmetic Act would be modified to the less confining concept of "serious and substantial risk of harm." We could all agree that a number of the proposed changes in these bills would benefit FDA. Moreover, the bills provide a timely vehicle to undertake the kind of public discussion that is essential to the controversial public issues that are so much a part of FDA's day-to-day operations. It is an anomaly of the day that many of those who believe in a strong and vigorous FDA are currently contributing to the substantial impairment of the agency. Likewise, it is ironic that, at the very time of general consensus that we need considerably greater resources to do our full job, proposals have been made for two separate but very closely related agencies with the undoubted consequence of even greater resource demands.

The Food and Drug Administration, like any other organization, is a living entity and it is highly questionable whether any living thing is helped by dismemberment. The present components of the Food and Drug Administration are far more interdependent than those outside the agency may realize and many examples arise every day of issues that require unified solutions. Interagency coordination can never be an adequate substitute for close working relationships between parts of the same agency.

The modern phenomenon of public criticism of important institutions of our society is certainly not unique to the FDA; therefore there is no cause to feel personal or organizational paranoia. This type of criticism is witnessed at every level of government and the causes of discontent are many and varied. A detailed analysis of this phenomenon goes beyond the scope of this discussion. We can say, however, that we must continue to seek solutions to the problems that result in discontent and hostility. As we find these solutions and implement them, the causes for discontent should lessen. In the meantime, it behooves each of us to see that an institution so vital to the public interest as the FDA does not suffer irreparable harm from the pessimistic mood of certain critics.

Public expectation of FDA's performance has far outpaced the provision of resources with which to meet those expectations. The public in general does not comprehend that the FDA must regulate products that account for roughly 30% of all consumer goods expenditures in the United States today on a yearly budget of less than one dollar per person. It is apparent that some agreement on the functions and priorities of the agency is essential to its future, and this agreement must involve the Congress, the Executive Branch, and the public at large. Finally, the most important principle on which we need public agreement is recognition of the enormous difficulty, and at times impossibility, of the tasks that face FDA today. Sometimes resolutions of issues by the agency simply seem to defy our best efforts. People must accept the fact that government officials, like the family doctor, cannot always provide a magic answer to every problem that arises. Officials must be candid and open about their potential for errors in judgment and the public must accept, and even nurture, an environment that allows agencies to reverse decisions when new information becomes available or when re-evaluation of old information has generated new conclusions. We must demand that our governmental officials display integrity, hard work, and the best judgment of which they are capable. We can also demand that they explain the reasons for their decisions and the basis for their actions. Scientific judgments are always open to question through well-established peer review systems and the legality of governmental action is a legitimate question for the courts. But we cannot allow ourselves to go beyond this. We must exercise sufficient self-restraint so that we do not demand infallibility or decisions which satisfy everyone, a goal that is clearly not possible.

In closing, I invite all chemists and other scientists to join in the public work that at this point is still unfinished, and to make their scientific expertise, both personal and organizational, available to groups who are pursuing the public interest. I applaud those instances where it is already being done. FDA welcomes the efforts of scientists to resolve the complexities of an ever-expanding body of scientific knowledge and its applications and legal controls. We welcome participation in the public debates about our public institutions in general and FDA in particular. Finally, let us all continue to recognize those higher goals, fundamental to the long-term stability of our government and indeed our entire society, which take precedence over our immediate individual concerns and needs.

38

Protection Against Biological and Chemical Hazards in Our Food Supply

Emil M. Mrak

This subject is indeed a broad one to cover in the time allotted, but I will try. First, I will discuss biological hazards and what the food industry is doing about them, then something about chemicals and the care taken to be certain they are safe.

BIOLOGICAL HAZARDS

Biological hazards and their effects on humans are more common than generally realized; this is well indicated by the reports of the U.S. Public Health Service Communicable Disease Center. Reports on chemical intoxications, on the other hand, are much less in number. In the case of chemicals, however, there is concern of possible long-term effects and eventual carcinogenesis. In the case of the biological hazards, the effects are usually immediate, of short duration, though in some cases they may cause death. Biological hazards are generally of two types, infections and intoxications. The latter include food-borne intoxications of microbial, plant and animal origin.

A very common infection is Salmonellosis, caused by Salmonella bacteria, which when ingested, multiply in the small intestines. They occur in water, animal products, and low acid foods, as the result of fecal contamination. They also occur in pelleted animal diets, and fertilizers made from waste products of slaughter houses, and are often carried by rodents.

Salmonella organisms are associated with the so-called paratyphoid group of bacteria that, at one time, we heard so much about. The prevention of such infections has involved, to a large extent, the protection of potable water by use of the chlorination process, which believe it or not, is now questioned—but more on this later. In the case of milk, pastuurization has been used. The application of hygienic principles at all stages of food animal production, from birth to its appearance on the dinner table, is important.

Industry is well aware of such preventive measures, and one may be certain that as a result of its own policing, inspection by the Food and Drug Administration and greatly improved techniques for processing and sanitation, the problem today is largely one restricted to the careless handling of foods after they have left the processing plant. This may be in the home, during institutional feeding, at picnics, and so on.

Two organisms involved in food infections and which cause dysentary and severe gastric distress are *Cl. perfringens* and *Cl. bifermentans*. Outbreaks caused by these organisms are fairly common in situations when foods are handled carelessly after leaving the processing plant. For example, one outbreak occurred in a school when students were fed day-old gravy on a chicken product which stood at room temperature the entire time.

The origin of these organisms is usually from human and animal fecal matter, though they may even occur in soil and on house dust.

The prevention of such infections involves education of those handling foods in the home, at school luncheons, and so on. I have said on many occasions, we require many courses in our schools such as driver training, physical education, foreign languages, history, etc., in order to make good citizens with strong bodies and safe drivers on the highways. Yet, seldom is there much, if any, teaching on foods and nutrition, though throughout life, an individual normally eats three or more times a day. Here, in my opinion, we certainly need a change in our education programs.

Viruses and Vibrios

Viruses and vibrios are little-known and seldom discussed, though extremely important and are becoming more important as time goes on and we learn more about them. You have certainly heard about hepatitis, an infection that results from the ingestions of contaminated food such as raw shellfish, water, and at times, even dairy products. Here again, industry avoids such contaminants by use of sanitary procedures which certainly involve avoiding untreated water.

Parasitic Infections

These infections are largely nematodes and include the commonly-known Trichonellá of pork. In order to avoid such infections, industry holds meats that may be contaminated at a low temperature storage sufficiently long to kill the organisms.

In other countries, but seldom, if ever, in the United States, there are a number of other nematode-like organisms that occur in raw, fermented or partially cooked fresh water fish, in raw mollusks, snails, land crabs, and pickled or raw smoked herring.

Tapeworms seldom, if ever, occur in this country any more, though they are common in other countries and, of course, in such cases, one should avoid eating raw and undercooked meats.

Protozoans and especially *Entamoeba histolytica*, organisms that cause diarrhea, can be picked up very easily in waters and foods in other countries, especially when water is not chlorinated.

Intoxications of Microbial Origin

The one most highly publicized intoxication of microbial origin is botulism. This is caused by an organism that grows in low acid foods under anaerobic conditions, and especially in canned foods. It has, however, been known to occur in certain other products.

In the case of canned foods, proper sterilization, under pressure, is a requirement and in order to be certain that this is accomplished, industry is policing itself and holding processor training schools throughout the country. In California, every canner is required to keep a record of the temperature of every batch of pressure cooked foods during processing.

Staphylococcal Infections

Staphylococci produce entero toxins that cause an illness involving diarrhea, nausea and fever. These organisms occur in low acid foods, water, milk and man himself. They may even occur in air and dust around the home. Careless handling of foods in the home, in institutions, at picnics, etc., is usually the cause of these distressful intoxications. There are sanitation training programs to avoid such occurrences, but they do not reach the home.

Mycotoxicoses

Intoxications, resulting from the production of toxicants by mold, came to light quite recently when flocks of turkeys and ducks died as a result of eating mold-contaminated peanut meal and grains.

Certain fungi growing on these feeds may produce extremely toxic substances termed aflatoxins. The fungus *Asperigillus flavus* produces at least two types of aflatoxins known to be toxic to turkeys, ducks, pigs, and calves. *Penicillium icelandicum*, growing on stored rice in Japan, has also been found to produce a toxin. These toxins are liver carcinogens.

Industry today is extremely careful to avoid the use of moldy materials in foods—this is done by machine sorting and even hand sorting if needed. Effective ultraviolet sorters, for example, have been developed for peanuts.

Here is another instance where courses in foods and nutrition in our schools would be most useful to the homemaker. She should be aware of the danger of moldy products, but at the same time be aware that blue cheeses, for example, are perfectly safe to eat and enjoy.

Poisonous Plants and Animals

Man has been exposed to poisonous plants since the beginning of time. The story of Socrates and hemlock poison, for example, is well known. Poisonous mushrooms are known to everyone, and so on. There are certain instances when the toxic substances can be removed by leaching as the Indians did with California buckeye, a technique they learned early in their history. There are other plant materials, such as the wild black cherry *(Prunus Serotina)* and lima beans that may, at times, contain large quantities of cyanide. In fact, some of the foods we eat may contain substantial amounts of cyanides in the seeds, not in the flesh—a situation quite true of apples and other commercially produced fruits such as apricots, prunes and plums.

In fact, in many cases, edible plants are closely related to others that are quite toxic. This is particularly true of the *Leguminoseae*, or bean family. We eat beans, peas and lentils with confidence and without worry, but there are highly toxic legumes such as the loco weed, and even soy beans which must be cooked before eaten.

Many edible plants contain toxins and, in fact, two books have been published on the subject—one by the National Academy of Sciences entitled *Toxicants Occurring Naturally in Foods*, and another by Irvin E. Leiner entitled *Toxic Constituents of Animal Foodstuffs*. Strange as it may seem, if an attempt were made to add these naturally-occurring toxins to foods today, they would not be permitted by the Food and Drug Administration. As a matter of fact, the Food and Drug Administration has proposed control of new hybrid plants created by geneticists, to be certain that naturally-occurring toxicants are not increased over 20%, and the nutritional value is not decreased substantially.

This has resulted, to a large extent, from the observation that a new variety of potato, introduced a few years ago, was relatively high in solanine, an alkaloid considered toxic. Likewise, erusic acid was observed to occur in substantial quantities in a new variety of rape seed, a substance said to cause the development of heart lesions. These observations, I am certain, prompted the Food and Drug Administration to propose standards for nutritional value and safety for new hybrids. A regulation of this type, however, would certainly require drastic changes in plant breeding programs. What will come of this remains to be seen.

There are a number of poisonous animals of which one of the most spectacular is the so-called Red Tide, a dinoflagellate which is responsible for mussel and clam poisoning. Apparently, the organism enters the livers of the shellfish, which when ingested by humans is deadly. In California, the State Department of Public Health issues warnings at certain times of the year when mussels should not be harvested because of the possible occurrence of Red Tide. In Japan, the producers of pearl oysters have Red Tide watchers so that when there is a possibility of it appearing, containers holding the oysters are moved. The oysters are not affected by the poison as such, but are deprived of oxygen in the water by the presence of the Red Tide organisms.

Certain fish are quite toxic, and this seems to vary with the location, the season, and what the fish eat. Ciguetara poisoning is one that has perplexed scientists for a long time. It occurs in certain types of edible fish in the South Pacific and the West Indies. Fortunately, it is not a problem in this country.

Puffer fish are also known to be toxic, although natives in certain areas remove and eat portions that are not toxic. We do not ordinarily eat these fish in this country. Tuna and mackerel have, at times, developed what is known as scombroid poisoning. It appears to be caused by the microbial conversion of histidine and histamine or saurine. Symptoms of scombroid poisoning develop rapidly and include headache, dizziness, heart palpitation, difficulty in swallowing, nausea, diarrhea and abdominal pain. It acts very much like an allergic response and recovery occurs in 8-24 hr. This is a perplexing problem of great concern to industry which is, at present, supporting an intensive research program to solve the problems.

I have spent some time on the biological aspects of hazards in our food supply, and have indicated that industry is exercising excellent control by the use of modern procedures of sanitation, careful quality control of raw materials, modern technology and processing procedures, self policing and training. The real problem, to a large extent, relates to the handling of foods after they have left the processing plant, and the solution to this is education and nutrition at all levels of education.

CHEMICAL HAZARDS

I should like now to consider chemical hazards including intentional additives, incidental additives, and unintentional additives.

I would like first to discuss intentional additives which are defined by the Food Protection Committee of the National Research Council as: "A substance or a mixture of substances, other than a basic foodstuff, which is present in a food as a result of any aspect of production, processing, storage, or packaging." This definition, however, is so broad that it includes incidental additives such as pesticides and unintentional materials that may be picked up during handling and processing.

Certain chemicals have been used in foods for thousands of years, although inadvertently so. For example, when man discovered the use of fire and the process of drying and smoking meats, he added chemicals such as formaldehyde, pyroligneous acid, and anthracenes which were absorbed from the smoke. Then again, salt has been used as a preservative since early times. The Chinese have used ethylene gas emitted from burning oil lamps in closed chambers for hundreds, if not thousands, of years, to ripen persimmons and citrus fruits. The gas which induces the fruit to ripen is now recognized as a natural hormone produced in plant cells as a ripening agent. At one time, the canners in California actually stacked boxes of ripe pears next to immature pears so the ethylene emissions from the mature pears would accelerate the ripening process in the immature pears.

Strange as it may seem, the Food and Drug Administration considered banning this material at one time, until it was learned to occur naturally in fruits.

Other substances used for many years are saltpeter, sugar, lactic acid, vinegar, spices, and so on.

The extensive use of food additives, however, is really quite recent in the history of man and can be related to great changes in our mode of living, involving urbanization, industrialization and more efficient agricultural production. One may ask why bother about using food additives at all; but they are important in improving the acceptability of many foods. If a person does not eat a food because it is unacceptable, then the food is of little value, even though it may be the most nutritious product in the world. Acceptability involves taste, odor, color, texture, feel, pain (characteristic of hot foods) and even noise (characteristic of crisp foods).

In recent years, with the change in our mode of living, there have been placed on the market a great many items that are termed convenience foods, which have a high utility value. These foods are easy to prepare because an enormous amount of labor has been taken out of the home and placed in the processing plant. Chemical additives play an important role in convenience foods.

Stability is another reason for the use of food additives. Preservation, of course, is most important, and it involves more than the prevention of microbial spoilage, for there are chemical and biochemical types of deterioration. For example, on the one hand, there may be the production of toxic substances such as aflatoxins if mold spoilage occurs and this, of course, is microbial in nature. On the other hand, the so-called browning reaction, a chemical type of spoilage, affects taste, flavor, color and may even involve the production of harmful substances. Stability, therefore, is an important factor and one for which food additives are indeed essential.

Safety, of course, is an all-important factor; for deteriorative processes may produce harmful substances, or there may be organisms that are infectious, or that produce enterotoxins and even carcinogenic substances. Safety, therefore, is a factor for which food additives are and must be used.

It is needless to say that nutritive value is important. I have placed this last because if the other factors prevent a person from eating a food, then the nutritional value of food is irrelevant and of no use to the individual.

Although there are a large number of food additives, relatively few are used in large quantities. The most commonly used additive is sucrose or ordinary cane or beet sugar, amounting to about 100 lb per person per year. The second most widely used additive is sodium chloride or table salt amounting to about 15 lb per person per year. After that comes corn syrup, about 8 lb, and dextrose about 4 lb per person per year.

Following these in descending order are 33 different food chemicals which together account for about 9 lb per person per year. It is interesting to note that of these 33, 18 are used either as leavening agents or to adjust the acidity of foods. They include such items as yeast, baking soda, citric acid, black pepper, mustard, and the much-discussed monosodium glutamate.

Next on the list are some 1800 additives of which we consume about 1 lb per person per year in total. These include many of the food flavors, certain food colors, and so on, and would amount approximately to the weight of one grain of table salt per person per year.

I am certain it will be of interest to define the functional and basic purposes for which additives are used in processing foods. These uses may be categorized as acids; alkalies; buffers; neutralizing, bleaching, thickening, sequestering, anticaking, firming, whipping, and curing agents; bread improvers; emulsifers; stabilizers; flavors; colors; nutrient supplements; preservatives; antioxidants; non-nutrient sweetners; humectants; glazes; and polishes.

This sounds like a tremendous variety of chemicals in use, but, as already indicated, the ones most extensively used, of course, are sugars, salt, materials to adjust acidity, yeast, baking soda, black pepper, mustard, and so on. The others in use are miniscule insofar as the average amount per person is concerned.

The incidental food additives are entirely different and are, to a large extent, pesticides for which the EPA and FDA have established tolerances. There are, of course, hundreds of pesticide formulations approved for use by the EPA. Of these, the ones that are currently of great concern, from the standpoint of safety, are the chlorinated hydrocarbons, organic phosphates, and certain herbicides, especially those containing dioxin as an impurity. Pesticides containing heavy metals, such as As, Pb, Cd, and Hg, have been banned or greatly restricted and, of course, likewise DDT, aldrin, dieldrin, chlordane and heptachlor.

Unintentional food additives are those that may be picked up from water, packaging materials, inks on packaging materials, equipment, waste disposal, and so on. A good example of such contaminants are the PCBs and Cl-containing compounds, such as chloroform occurring in water resulting from chlorination. The sale of PCBs in this country, however, has been restricted by the manufacturer for use in electrical equipment.

HCBs have been found in meat animals as a result of industrial waste disposal in areas where beef animals were grazing.

I could go on and discuss these in great detail, but of more importance is the matter of safety, so I should like now to review some of the facets of safety, where we are as I see it, what industry is doing, and perhaps what needs to be done to alleviate what I consider, at present, a chaotic situation.

We hear so much about the danger of food additives, whether they be intentional, incidental, accidental or unintentional, and yet there have been so few instances of harm to humans by ingestion of foods containing any of the common additives. Of course, there is always the unanswered question of the lifetime effect.

On the other hand, as already pointed out, recent reports of the Communicable Disease Center of the Public Health Service in Atalanta, Georgia, indicate that a very high percentage of the food intoxications reported in the United States are microbial and not chemical in origin. One of the recent reports showed that 93% of the food intoxications for a single year were microbial in origin. There is always the possibility, of course, that the reporting of chemical intoxications may be incomplete, but the same may be said of microbial intoxications. Nevertheless, they do point out the importance and seriousness of intoxications microbial in origin.

In spite of this, the publicity and emphasis certainly seem to be on chemical hazards and the possibility of their inducing carcinogenesis, teratogenesis, and/or mutagenesis.

With respect to carcinogenesis, it is of interest to give an example of the concern relating to an incidental additive—namely, DES, Diethylstilbestrol, and the actions that have been taken regarding the use of this chemical in the feedlot industry. When the Food and Drug Administration was asked to comment on the banning of DES and what it really meant, the reply was as follows, and I quote this from the FDA presentation to the Ways and Means Committee of the House of Representatives. Believe it or not, the testimony reads as follows: ". . . . based on findings that 5% of the average diet is beef liver, a person would have to consume 5 million pounds of liver per year for 50 years to equal the intake from one treatment of day-after oral contraceptives."

There is certainly an inconsistency here. It appears that we have different standards for the food we eat than we do for the society we live in—yet DES occurs naturally in beef liver.

The one thing that has become very clear to me, as a result of working on the Pesticide Commission for Secretary Finch, was, and still is, the need for generally accepted protocols to follow in testing for the safety of pesticides and food additives. Furthermore, closely related to this situation is the question of the reliability of the analytical procedures used in quantitatively determining the amounts of a chemical in a food. Equally important, if not more so, is the capability of the individual making the analyses. I am prompted to say this because I am aware of certain instances when serious errors have been made. These errors resulted from the use of inadequate procedures, and I cannot but wonder about the analyst. I am aware of at least two instances when these inadequacies were brought to light by industry chemists.

In another instance, there was much discussion about an analytical procedure for dioxin reliable to ppt. Apparently a collaborative effort by industry-government has developed a good method—good to ppt. I feel strongly that we need reliable methods of analyses and that we need people who have the talent and ability to use them. In line with this, a Committee of the Science Advisory Board of EPA, reporting on contaminants in drinking water, pointed out that it is likely that the majority of the drinking water purveyors do not have the sophisticated equipment or trained personnel to make such analyses.

Over the years, we have gone in our quantitative determinations from parts per million to parts per billion, and now parts per trillion (ppt). Can it be that parts per quadrillion are on the horizon and, if so, what does it all mean? Dr. William Stewart, former surgeon general of the United States, pointed out that the defensive posture has become commonplace for public health officials in the last few decades. We have the ability to detect foreign substances in the environment in increasingly smaller amounts. Yet, at the same time, our ability to understand the biological consequences of exposure to these small quantities, over long periods of time, has advanced slowly and with great uncertainty.

As I see it, therefore, we have confusion in two respects. One relates to the methods used for analyses and the reliability of the individual making the analyses. The second, what does it all mean with respect to safety? The chemists are developing methods to find smaller and smaller quantities, but have done little or nothing to clarify what such small quantities mean to the biological system. In his recent book, *Science of Survival,* Lord Ritchie-Calder has pointed out that we seem to be aiming at the detection of the last toxic molecule and, if this is the case, we will be busy for a long time without knowning what it means.

Let's consider the present philosophy with respect to safety testing. It is to administer to animals, massive dosages—very often under the most extreme conditions—to determine whether or not the chemical will cause an adverse effect. In fact, industry, I know, is pursuing safety testing in line with suggestions of FDA and EPA, but in view of the developments on analyses and the failure of the chemists to consider the meaning of ppb or ppt, industry finds itself in a complete state of frustration.

When serving as Chariman of the Commission on Pesticides for

Secretary Finch, I had the opportunity to listen to discussions concerning the effects of DDT on weanling mice. Doses of DDT were force-fed to weanling mice in amounts 30,000 times that which an average human might ingest in a day on a weight-per-weight basis. The panel was not in complete agreement as to whether or not the results were significant or meaningful.

There was a question as to whether the abnormalities observed in livers of some of the mice were nodules or tumors. After much discussion and argumentation, it was concluded that they were probably tumorigenic.

Then there was a question as to whether or not they would metastisize, but eventually it was pointed out that if they were considered tumorigenic, then they should be considered potentially carcinogenic. Finally, there was never agreement as to whether or not the number observed was statistically significant. This, I believe, is an example of the situation in which those doing safety testing in industry find themselves today.

But that is not the whole story. A variety of other methods have been used, such as injection into the intraperitoneal cavity, implantation in the bladder, injection into the base of the brain, rubbing on the skin and so on, as well as feeding enormous quantitites.

Furthermore, in view of new and evolving information, there seems to be a growing doubt as to whether or not the mouse should be used for more than screening purposes. Then, too, there are questions about the conditions under which the animals are grown and the possible influence of adverse environmental conditions on the animals. Apparently, mice grown in small cages in large numbers are under stress. Some have indicated that adverse conditions of growth can apparently affect the development and carcinogenesis. Other factors that tend to induce stress are light, darkness, nutrition, and whether or not the animal foods contain toxic substances such as aflatoxins. Apparently, dirty pens and the development of ammonia can have a very detrimental effect with respect to the tests. Noise, temperature, and palatability of the food may have an effect on the rate of growth. It has been shown, apparently, by experimental and clinical studies that chronic and acute starvation, as a result of caloric restriction or intake, and the quality and quantity of protein and also the vitamin and mineral deficiences alter endocrine gland functions. These, it has been said, might very well influence carcinogenesis. Then, too, there is the matter of water and the purity of the water.

There may also be an adverse response with respect to the size and location of cages and whether or not the male and female may be in adjacent cages.

Also, it comes down to the desirable animal to use. Some years ago, the National Research Council of the Food Protection Committee recommended the use of two animals. Since then, there has been an increasing body of evidence to indicate, in addition to the mouse, another animal should be used. In line with this, Dr. Adamson of the Cancer Institute fed 5 male and 5 female monkeys 500 mg per kg per day of sodium cyclamate for approximately 5 years. These animals apparently exhibited normal growth in their development as well as normal hemochemical and hematological analysis. This brings me to a comment that I made earlier—that we are in an uncertain and even chaotic situation with respect to testing for safety.

The chemists have gone along analyzing more and more for less and less. It seems to me that it is time for the chemists to team up with toxicologists and public health people to help solve the uncertainties they have created by the improved methods of analyses they have developed for less and less. They must work with toxicologists, confer with them, and stimulate research to establish the meaning of going from ppm to ppb to ppt from the standpoint of safety.

In a speech presented to the Chemical Marketing Association February 1976, Dr. Julius Johnson pointed out that industry is already doing an enormous amount of testing. Let us take, for example, the safety evaluation tests made by industry before submission for registration:

Toxicology: 90-day rat test ⎫
             90-day dog test ⎪ all for carcinogenesis
             2-year rat test ⎬
             1-year dog test ⎭
             Reproduction, 3 generations of rats
             Teratogenesis on rodents
             Effects on fish and shellfish
Metabolism: On rodent and/or dog
             On plants
Analytical: Food Crops to 0.01–0.05 ppm
             Meat 0.1 ppm
             Milk 0.005–0.05 ppm
Ecology: Environmental
             Stability
             Movement
             Spectrum
             Accumulation

In addition to these and as time goes on, the tests will most likely include mutagenesis and test animals such as birds, primates and, if possible, humans.

Nevertheless, there are no generally accepted protocols to follow in testing for carcinogenesis and, in fact, the principles for decision making are still in the stage of formation.

With respect to teratogenesis, if we take seriously the test procedures being used, we can start right now outlawing table salt, aspirin, and Vitamin A, for, under certain test conditions, these have been shown to be teratogenic and, in fact, strongly so, in some instances.

We know even less about the testing for mutagenesis, although a promising procedure involving the use of salmonella is on the horizon. It should be tested extensively before being used routinely, and especially so since some believe there is a close relationship between mutagenesis and carcinogenesis. If this proves to be true, it will greatly simplify certain aspects of safety testing. One can but wonder if those developing analytical procedures for ppb and ppt should not familiarize themselves with the test and try for it.

This is especially true since the government today is loaded with new and young regulators endowed with ideals, but deficient in scientific training. These regulators seem inclined to respond to pressures more than to sound scientific facts and, understandably, this has caused in some instances adverse feelings on the part of scientists in the same agencies. There is no question but what we must be careful when it comes to human harm and safety. We must consider hazards seriously, we must use good judgment, we must consider the cost-benefit ratio and, above all, we must consider the health risk-benefit ratio. No one has worked out a clear-cut simple formula for these, for it requires understanding and a breadth of thinking. It is indeed difficult to broaden our interests, activities, and understanding, and especially so by the young regulators and the chemists contributing to the problem; but it is important and I hope it will come to pass.

BIBLIOGRAPHY

LEINER, I. E. 1969. Toxic Constituents of Plant Foodstuffs. Academic Press, New York.

LEINER, I. E. 1974. Toxic Constituents of Animal Foodstuffs. Academic Press, New York.

NATIONAL ACADEMY OF SCIENCE. 1966. Toxicants Occurring Naturally in Foods. National Academy of Sciences, Washington, D.C.

RIEMANN, H. 1969. Food-Borne Infections and Intoxications. Academic Press, New York.

39

Evaluating Food Additives–
The GRAS List

George W. Irving, Jr.

The food laws enacted early in this century were designed to assure wholesomeness and safety of food in the United States. The food industry, operating in compliance with the regulations promulgated and administered by the Food and Drug Administration (FDA) to implement these laws, has afforded us not only safe, wholesome food but foods in abundance and variety. Since the early days however, the food industry and its surveillance have become increasingly complex. Population grew and concentrated in cities, skills and time for kitchen preparation of consumable items from raw foods dwindled, and expectancies were established for continuous supplies of non-seasonal foods. As a consequence, need arose for a greater volume and diversity of nutritious, attractive, and palatable processed foods that had the ruggedness and stability to permit long-haul transportation and longer shelf-life in stores and the home. To accomplish some of these objectives there was increasing use of additives. While many of these additives were familiar, natural substances, many were not, adding new dimensions to the terms wholesomeness and safety. These developments culminated in the passage in 1958 of the Food Additives Amendment of the Food, Drug and Cosmetic Act, a law designed to regulate the use of food additives.

DEFINITIONS

According to the Food Additives Amendment, the term food additive means "any safe substance the intended use of which results or may reasonably be expected to result, directly or indirectly, in its becoming a component or otherwise affecting the characteristics of any food, including any substance intended for use in producing, manufacturing, packing, processing, preparing, treating, packaging, transporting, or holding food."

In practice, any person wishing to add a substance to processed food must propose to the Food and Drug Administration that a regulation be issued prescribing the conditions under which it may be safely used. There were a number of exceptions noted in the law, among them the "generally recognized as safe" provision, which exempted from the requirement of premarketing approval those substances deemed (by experts qualified by scientific training and experience to evaluate safety) to be safe under the conditions of intended use. The exception included substances used in food prior to January 1, 1958 which had been adequately shown through scientific procedures or experience based on common use in food to be safe under conditions of intended use. These excepted substances came to be known as GRAS, the acronym for Generally Recognized as Safe. Thus, those proposing use of an additive in food were required to present evidence of its harmlessness when used as proposed and receive approval, unless the substance was considered GRAS, in which case no prior approval was required.

WHY RE-EXAMINE THE GRAS SUBSTANCES?

Since 1958, many food additives have received approval and conditions of safe use are recorded as separately numbered sections in Title 21 of the Code of Federal Regulations. The GRAS substances are also accorded a section in the Code (21 CFR Part 182) where some 600 are listed, subdivided according to the major functions in food they are intended to perform.

For a number of years, administration of these regulations proceeded relatively smoothly. However, within the current decade, attention, particularly public attention, has become increasingly focused on the safety aspects of all things to which the public is exposed. The adequacy of current regulations to protect the consumer is being questioned. As a result, many Federal agencies are critically re-examining their regulations, among them the FDA with respect to its regulations on food additives.

The impetus for such re-examination appears to stem from several converging factors. Improvements in toxicological, biochemical, pathological, and analytical methodology permit the detection of some biological effects previously overlooked, as well as the measurement of concentrations of substances at extremely low levels. While it is recognized that many of these biological effects are still at the fringes of our understanding of them, some scientists maintain that they are indicative of possible subtle harm to humans exposed chronically to minute amounts of substances formerly regarded as safe. Because there is frequently no unanimity concerning the degree of risk associated with a specific substance in this regard, public anxieties are understandably aroused, particularly when uncertainties seem to surround the safety of the foods we eat.

Food additives provide an inviting target for those who believe that reassessment of the hazard associated with all environmental factors is in order. Unlike salt added to food from the table shaker or sugar from the bowl, substances added to processed foods become unrecognizable as such. Label declarations inform the consumer, but once he recognizes that a food contains a substance which he believes may be harmful to him, he must either refuse to eat it or consume it despite his misgivings; he is unable to remove the portion that offends him. It is this involuntary consumption of those food additives he fears may harm him that leads him to insist that all additives be proved safe before they are permitted in processed foods.

HOW THE RE-EXAMINATION IS BEING CONDUCTED

It was in this atmosphere that FDA in 1969 began a re-examination of the safety of food additives, beginning with those classified as GRAS. This choice was presumably made because, unlike the regulated additives where evidence of harmlessness is presented and reviewed before approval is granted, the GRAS substances have never been subjected to such careful scrutiny. The re-examination process involved the following steps:

1. Compilation of a bibliography of all publications in the scientific literature from 1920 through 1970 concerning each of the GRAS substances. (FDA contract with the Franklin Institute Research Laboratories.)

2. Survey of the food industry to determine the extent of use of each GRAS substance in all processed foods and calculation of possible human intakes of each substance from these data. (FDA contract with National Academy of Sciences—National Research Council.)

3. Preparation of a Scientific Literature Review (monograph) on each GRAS substance or group of closely related substances, based on the bibliography provided, and including reprints of all relevant publications. (FDA contracts with several institutions.)

4. Evaluation of the safety of each substance, based on the human

intake estimates and the Scientific Literature Review, and submission of a report to FDA. (FDA contract with the Federation of American Societies for Experimental Biology.)

5. Revision of the regulatory status of each GRAS substance based on the evaluation report and other relevant factors. (Actions proposed by FDA in the Federal Register.)

Elaboration of experiences in steps 4 and 5 will illustrate the procedures being employed, some of the problems encountered, and the progress being made both in the evaluation of the health aspects of using the GRAS substances as food ingredients and in the appropriate revision of regulations concerning them.

The Evaluation Procedure

For the evaluation of the GRAS substances, the Life Sciences Research Office (LSRO) of the Federation of American Societies for Experimental Biology (FASEB) under contract with FDA, elected to utilize a selected committee of scientists and appropriate professional staff. The Select Committee on GRAS Substances (SCOGS) numbered 9 initially, lost 1 member, and has added 3 during the nearly 5 years of its existence. The present eleven members and their affiliations are listed at the end of this chapter. Members were chosen for their experience and judgment with consideration for balance and breadth in the professional disciplines needed to ensure necessary expert coverage of the factors involved. Breadth in knowledge was considered preferable to narrow depth in a restricted specialty. The problem of ensuring complete expert coverage in a small committee is formidable. Shortcomings in this respect are mitigated, as they arise, by the use of *ad hoc* consultants who serve only as information sources.

Since SCOGS' conclusions are serving as a basis for regulatory decisions, committee members and staff are required to declare initially and periodically thereafter, all of their affiliations. These declarations are reviewed carefully by FASEB and by FDA to assure that there are no conflicts of interest on the part of members of the Committee or staff. Members devote a minimum of 4 full days each month to the work of SCOGS, including 2 days for the regular, monthly sessions of the Committee.

A staff of 8 professional and supporting personnel is directly associated with the GRAS substances evaluation effort, augmented frequently by as many as 7 other professional and nonprofessional members of the LSRO staff. Staff members attend all sessions of the Committee, provide and discuss information as required, maintain continuous liaison with FDA, and are alert to disclosure of relevant, new information at symposia and other scientific meetings. However, debating opinions and conclusions, interpreting data and the arrival at final evaluations are the sole province and responsibility of the Committee members.

The evaluation process—from the raw data to a completed report—consists of several discrete steps.

1. A subcommittee of two members of the Select Committee, chosen with an eye to matching subject matter with professional backgrounds, prepares the first draft report based on information furnished in the monograph, supplemented by information derived by staff from a number of sources. These supplemental sources vary with the substance under study but always include recent references obtained through the computer retrieval systems of the National Library of Medicine. Full committee deliberations are then conducted with the first draft report as the point of departure. All committee members have had opportunity by then to study the same raw data. Discussion continues until there is understanding among members regarding the general adequacy of the draft opinion, conclusion, and supporting data as representing the thinking of all members.

2. Based on the approved first draft, staff prepares a second draft incorporating comments and changes made during committee discussion. The second draft is mailed to committee members for their emendation. Any substantive changes at this stage are reconsidered at a subsequent full committee session.

3. Based on the approved second draft, staff prepares a third draft to reflect any further changes, verifies every statement and figure against the original articles cited, and rechecks all calculations. Upon signed approval by all committee members, this draft becomes the tentative report of the committee.

4. The tentative report is submitted to FDA for announcement in the Federal Register that SCOGS is prepared to hold a public hearing for those who express interest in presenting data, information and views on the subject of the report. SCOGS' tentative report is made available for perusal in the Office of the Hearing Clerk (FDA). The Federal Register notice discloses SCOGS' conclusion and identifies all materials used by the committee in its deliberations. Copies of the monograph and many of the additional reports and materials cited are also available from the National Technical Information Service.[1] If a request is received a hearing is held, recorded, transcribed, and published. SCOGS considers the information imparted at the hearing and may revise its tentative report appropriately.

[1] National Technical Information Service, U.S. Department of Commerce, P.O. Box 1553, Springfield, Virginia 22161.

5. When the hearing is completed, or if no hearing is requested, the tentative report, revised as necessary, is submitted for review and endorsement to an Advisory Committee consisting of 1 elected member of each of the 6 constituent societies of FASEB. If substantive issues are raised, they are considered and resolved by the Advisory Committee and SCOGS. The final report, now the FASEB report, incorporating all agreed-upon revisions, is submitted to FDA by the Executive Director, FASEB.

FDA Action

FDA considers the FASEB reports together with other relevant factors, in affirming or changing the regulatory status of the substances concerned. It does this by announcing proposed action in the Federal Register and includes the essential elements and conclusions of the FASEB report. After a suitable period is allowed for comment and appropriate revisions in its proposal have been made, FDA announces its final action in the Federal Register.

FASEB reports come to one of four conclusions on each of the GRAS substances evaluated. Based on the several substances evaluated by SCOGS upon which FDA has proposed or taken regulatory action, SCOGS' conclusions translate to regulatory actions as follows:

Conclusion No. 1: There is no evidence in the available information on _____ that demonstrates, or suggests reasonable grounds to suspect, a hazard to the public when it is used at levels that are now current or that might reasonably be expected in the future.

FDA action: Affirmed as GRAS.

Conclusion No. 2: There is no evidence in the available information on _____ that demonstrates, or suggests reasonable grounds to suspect, a hazard to the public when it is used at levels that are now current and in the manner now practiced. However, it is not possible to determine, without additional data, whether a significant increase in consumption would constitute a dietary hazard.

FDA action: Affirmed as GRAS with limitations or restrictions indicated.

Conclusion No. 3: While no evidence in the available information on _____ demonstrates a hazard

	to the public when it is used at levels that are now current and in the manner now practiced, uncertainties exist requiring that additional studies be conducted.
FDA action:	Interim food additive regulation established; use continues while new data are accumulated.
Conclusion No. 4:	The evidence on _____ is insufficient to determine that the adverse effects reported are not deleterious to the public health when it is used at levels that are now current and in the manner now practiced.
FDA action:	Prohibited from use in food until safer usage conditions are established.

Philosophical Considerations in the Evaluation Process

In asking FASEB to evaluate the health aspects of using the GRAS substances as food ingredients, FDA sought advice that would enable them to revise the regulatory status of these substances. It was realized that the GRAS substances were being, and had for many years been used, in processed foods with no apparent reasons to doubt their safety. Yet, since critical examination of the world's knowledge concerning them had never been attempted, it was decided to conduct such a critical examination. There was considered to be no need or justification to deny continued use of these substances pending their evaluation.

Recognizing that rigorous proof of safety is unattainable, the evaluation process began with the concept that the GRAS substances were safe unless reason could be found to cast doubt on their safety. In making its evaluations, therefore, SCOGS relied primarily on the absence of substantive evidence of, or reasonable grounds to suspect, a significant risk to the public health in the continued use of GRAS substances. Moreover, it was realized that a reasoned judgment was expected even in instances where the available information was qualitatively or quantitatively limited. When uncertainties became apparent, the committee was expected to indicate the specific additional information needed. It was and is SCOGS' intention to provide FDA with scientific evaluations only, based wholly on existing published scientific data. Conclusions were not to be influenced by such considerations as the benefits of using a GRAS substance versus the risks of not doing so; the possible economic impacts of removing a substance from the GRAS list; the effects of forcing the exchange of one GRAS substance for possibly larger quantities of another. Resolution of such questions is the responsibility of FDA.

Proceeding in this manner, SCOGS expected within a relatively short period of time, to provide FDA with assurance that some of the GRAS substances were safe, with or without limitations on their use, and that some required additional information before a judgment of safety could be made, with or without removal from use while necessary data were being accumulated.

PROBLEMS IN EVALUATION

Noteworthy are a few of the problems, apart from the normal difficulties one would expect in a task of this nature and magnitude, that have arisen during the course of SCOGS' work.

Quality of Basic Inputs

The first basic input in the evaluation process—the Scientific Literature Reviews (monographs)—were found to be generally good but deficient in some respects. In assembling pertinent literature one aims at completeness and currency. As a practical matter, the degree to which this is achieved depends on such factors as the number and nature of abstract sources searched, the "key words" used in the search, synonyms searched for, and the degree to which "redundancy" searches are made. Thoroughness in these respects varied among monographers. In preparing the summaries, the monographer was expected to select papers so that the summary would be representative of the available data. Of 500 references in a bibliography, the monographer might require only 100 to document a representative summary of the salient facts. In this process choices were made, judgments were exercised, and errors of omission and commission sometimes occurred. Since most of the monographs were completed during the first year of the evaluation effort, they became progressively more out-of-date as the interval between their preparation and the evaluation became longer.

Because of these factors, SCOGS early adopted the practice of regularly searching beyond the monograph material to assure completeness, currency and representativeness of the assembled facts upon which it was to base its evaluation. These searches included bibliographies in specialized areas of toxicology, reports of international bodies, such as the Food and Agriculture Organization and World Health Organization of the United Nations, official statistical reports of imports and production of substances in the United States, the files of the FDA, and the combined knowledge and experience of the Select Committee and the LSRO staff. In addition, a notice was placed in the Federal Register inviting all who had relevant information on any of the GRAS substances to send it to the Committee; many have availed themselves of this opportunity. As a final safeguard, the Committee insisted upon relying only on original papers rather than on information gleaned from secondary sources and was often able to cross-check its information for completeness against references contained in authoritative reviews.

The second basic input in the evaluation process—the human intake estimates provided by a subcommittee of the NRC—also posed some problems for the committee. The NRC surveyed manufacturers by questionnaire concerning the addition of GRAS substances to foods and estimated the possible daily intake of these substances within various age groups. Based on information supplied by those manufacturers who reported adding the substance to at least one food product in a food category, a weighted mean was calculated for the usual percentage addition of the substance to these foods. The NRC subcommittee then estimated possible average daily intakes from Market Research Corporation of America data on the mean frequency of eating foods by food category, U.S. Department of Agriculture data on mean portion size, and the assumption that all food products within a category contained the substance at the mean percentage derived from the questionnaire data. The latter assumption is likely to lead to overestimates of intake. The NRC subcommittee has recognized that in most cases its calculations of possible intakes were overstated, often by considerable margins, and that their average estimated total dietary intakes were likely to be much higher than would be the intakes achieved through consumption of a diet consisting totally of processed foods to which the substance had been added at the maximum level. Thus, the NRC estimates were consistently and intentionally on the high side. The Committee found this reassuring in that minimizing possible human exposure was avoided. However, in relating these high estimates of human intakes to "maximum no-effect levels" in animal studies, the intake of some GRAS substances appeared to be closer to these levels than might actually be the case. For these substances, alternative means for estimating intakes were used by SCOGS. For example, official import and production statistics were used to calculate per capita intake. Such estimates were often considered as more typical of actual intakes than the NRC figures.

Currently, the NRC subcommittee is reviewing its original data and making additional surveys with the expectation that revised and more typical intake estimates can be made for some substances. In addition, the Market Research Corporation of America is supplying intake data to FDA on some GRAS substances based on the results of daily menu diary surveys of a large population. These improved figures are being considered by SCOGS and incorporated into their reports when they are available.

Practical means for obtaining data on how much of each GRAS substance is actually consumed daily remains a difficult problem. Realistic estimates in this respect are crucial to safety judgments in some cases.

Product Identity

Problems occur in relating published biological and consumption data on the GRAS substances to the product actually added to processed foods. Specifications for the food grade of many of the GRAS substances are not listed in the Food Chemicals Codex. More often than not, authors are casual about the nature or source of the chemicals or substances used in their experiments. The older literature is worse in this regard than some of the more recent publications, but not much worse. Clinical reports rather consistently fail to identify the source and quality of substances used. Many authors are prone to overlook the significance of identifying the isomer used when two or more are possible.

It is recognized that there are practical limits of "purity" in food chemicals and that rigorous purification to single, identifiable chemical entities is not feasible or necessary in all instances. However, there are occasions when better qualitative and quantitative specifications would be helpful. At present, one is forced to speculate whether any of the biological results reported are due to contaminants known to be possible in the product tested, or whether the absence of untoward results might have been due to the experimental use of a product of greater purity than that actually used in food. Such questions arise frequently in the consideration of complex, usually natural, substances where composition can vary depending on the plant or animal source, the producing country, and on whether it has been subjected to "purification."

Evaluation of the GRAS substances from available, published, scientific information is particularly vulnerable in these respects. In contrast to petitioned food additives, where submission of biological data is required on the actual substance to be used, the biological information on GRAS substances must often be pieced together from reports published over a period of half a century, by people whose purposes were not necessarily to prove safety, and where their samples of the substance were of the quality that was then available to them.

Chronic Toxicity

Acute and short-term feeding studies reveal that the GRAS substances at the levels used in food are not toxic by usual criteria. Frequently, however, very much less has been done to study long-term, let alone lifetime, chronic effects which would be more significant in evaluating the safety of substances consumed, as the GRAS substances are, in relatively small quantities over long periods of time. Studies of effects on reproduction are also often lacking. Much of this deficiency is undoubtedly due to lack of incentive to conduct expensive long-term feeding tests and special research on substances classified as GRAS where many, many years of use in food have revealed no obvious, researchable clues to potential harmfulness. Nevertheless, the Select Committee is aware that such long-term effects, if any, could be important in the quest for maximum margins of safety. The problem is that there no techniques, *in vivo* or *in vitro*, that are now commonly accepted as true indicators of such chronic effects as carcinogenicity and mutagenicity, and few that provide unequivocal evidence of teratogenicity. For example, scientists still debate the significance of experiments that show barely significant increase in the incidence of cancer in cancer-prone animals after administration of massive doses of a test substance; no resolution has been found of the differences that occur in the testing of a substance for mutagenicity by the dominant lethal, cytogenetic, host-mediated, and chick embryo techniques where one or more may show presumptive evidence of mutagenicity while the others do not. Until more indicative methods are devised and accepted, evaluators must rely, as the practical alternative, on conservative and reasoned judgment based on all of the credible scientific information available.

Total Body Burden

The present evaluation of the safety of GRAS substances involves only the safety to consumers of these substances at the levels now added to processed foods. Inherent toxicity of a substance at the human intake levels that would result, is a matter of first concern. However, many GRAS substances are also normal components of unprocessed foods, and the total exposure to a particular GRAS substance of an individual consuming a varied diet could be considerably greater than that indicated only by the amount added to processed food. For this reason evaluators tend to consider total body burden with respect to a substance as a relevant consideration in assessing safety. Several examples illustrate the problems involved.

A number of calcium compounds and phosphate salts, now considered GRAS, are without demonstrable toxicity at the human intake levels that result from the amounts now added to processed foods. Yet dietary calcium:phosphate balance is important to health and distortion of this balance is known to cause problems. How much do added GRAS calcium and phosphate salts contribute to distortion of this balance, considering that unprocessed foods contain both calcium and phosphate and the daily load will vary depending on the food choices of the consumer?

Sucrose, a GRAS substance of very low inherent toxicity, is consumed in greater amounts than any other additive. However, such added sucrose contributes less than 2/3 of the total body burden, the remaining 1/3 coming from voluntary consumption of natural sucrose-containing foods, additions to foods during preparation, and additions to foods from the sugar bowl.

Caffeine is a GRAS substance of demonstrated pharmacological activity. Its principal GRAS use is as a flavor additive in a few foods, including cola beverages. However, major daily caffeine intake for adults is due to voluntary consumption of such beverages as coffee and tea. How much does added caffeine in cola beverages contribute to the total body burden of caffeine?

In these and other similar instances, the problem of the evaluator is to decide whether there is an adequate margin of safety, considering the body burden of a GRAS substance from all sources, and if not, whether restriction of its additive use would significantly lessen the burden and decrease possible risk to the public health.

"Nonaverage" Populations

The safety concept, as generally administered with respect to the food supply, concerns normal, adult individuals. Occasionally, however, untoward effects are attributed to consumption of an additive or of certain foods by the very young, the growing child, the adolescent, the woman taking birth control pills, the pregnant woman, the person who is allergic or otherwise ill, the individual receiving continuing medication, the elderly and infirm, those with inborn errors of metabolism, the teenager who sometimes exhibits bizarre eating habits, and the individual who continues these habits into adulthood. Knowledge of the existence of these situations or combinations of them, poses problems for the evaluator in determining how extensively he should probe into possible interactions between the GRAS substance being evaluated and other additives, food constituents, or medications to which the individual may be concomitantly exposed. Do we err if we do not conclude that use of an additive should be withdrawn from or limited in the food available to the vast majority of consumers so that the most susceptible will be protected from possible harm?

SUMMARY

The foregoing sampling of the kinds of problems that have emerged during the GRAS substances evaluation effort, illustrates why committee criteria have had to be revised and amended regularly as the work has progressed. It also illustrates that safety evaluation is rarely a simple matter of summarizing numerical data, but rather a process involving interpretations and judgments at virtually every step.

The Select Committee believes that its "true-life" experiences in facing and handling the complex issues involved in the safety evaluation of the GRAS substances may have value beyond that of its conclusions with respect to specific substances. For this reason the Select Committee expects to publish soon a paper entitled, "Evaluation of Health Aspects of GRAS Food Ingredients: Lessons Learned and Questions Unanswered," which will embrace the many intangibles that have no place in individual evaluation reports. The purposes in presenting this paper will be fourfold: to illustrate the range of factors to be taken into consideration in the safety assessment of a given food ingredient; to offer estimates on the state-of-the-art and commentaries on the nature of the technical dilemmas that are encountered in rendering scientific judgments on food safety; to provide suggestions concerning the philosophical, procedural, and scientific ramifications of such an evaluation process; to point out needed research to improve the validity and meaningfulness of the associated data. The paper should be read together with a sampling of actual evaluation reports. With actual reports as specific examples, the lessons learned and questions unanswered take on the dimensions and significance the Select Committee believes it is important to convey.

Progress toward completing the evaluation of the GRAS food substances is presented in the following table. The evaluation experience is one that all perhaps would not enjoy, but those of us who have, nurse the hope that the effort will have immediate as well as lasting usefulness in moving closer to the resolution of the complex of problems involved in providing convincing assurance of the safety of the substances being added to foods.

GRAS Substances Evaluation Scoresheet, July 1, 1977

	Monographs		Substances	
1. Received from FDA	118		360	
2. Final reports submitted		72		230
3. Conclusions reached		40		122
4. Remaining to be considered		6		8
		118		360
5. Additional monographs expected	50		83	

SELECT COMMITTEE ON GRAS SUBSTANCES

Joseph F. Borzelleca, Ph.D.
Professor of Pharmacology
Medical College of Virginia
Health Sciences Division
Virginia Commonwealth University
Richmond, Virginia

Harry G. Day, Sc.D.
Professor Emeritus of Chemistry
Indiana University
Bloomington, Indiana

Samuel J. Fomon, M.D.
Professor of Pediatrics
College of Medicine
University of Iowa
Iowa City, Iowa

Bert N. La Du, Jr., M.D., Ph.D.
Professor and Chairman, Department
of Pharmacology
University of Michigan Medical School
Ann Arbor, Michigan

John R. McCoy, V.M.D.
Professor of Comparative Pathology
New Jersey College of Medicine and
Dentistry
Rutgers Medical School
New Brunswick, New Jersey

Sanford A. Miller, Ph.D.
Professor of Nutritional Biochemistry
Massachusetts Institute of Technology
Cambridge, Massachusetts

Gabriel L. Plaa, Ph.D.
Professor and Chairman, Department
of Pharmacology
University of Montreal Faculty of
Medicine
Montreal, Canada

Michael B. Shimkin, M.D.
Professor of Community Medicine
and Oncology
School of Medicine

University of California, San Diego
La Jolla, California

Ralph G. H. Siu, Ph.D.
Consultant
Washington, D.C.

John L. Wood, Ph.D.
Distinguished Service Professor

Department of Biochemistry
University of Tennessee Medical Units
Memphis, Tennessee

George W. Irving, Jr., Ph.D. (Chairman)
Life Sciences Research Office
Federation of American Societies for
  Experimental Biology
Bethesda, Maryland

Publisher's Note:
  Because of the FDA changes made in the subject matter of this chapter since
the ACS Centennial symposium was held, Dr. Irving has updated the data to
conform with the latest information up to press time (July 1977).

Index

Other AVI Books

AN INTRODUCTION TO AGRICULTURAL ENGINEERING
Roth, Crow and Mahoney

CARBOHYDRATES AND HEALTH
Hood, Wardrip and Bollenback

DAIRY TECHNOLOGY AND ENGINEERING
Harper and Hall

DIETARY NUTRIENT GUIDE
Pennington

DRUG-INDUCED NUTRITIONAL DEFICIENCIES
Roe

EVALUATION OF PROTEINS FOR HUMANS
Bodwell

FABRICATED FOODS
Inglett

FOOD COLORIMETRY THEORY AND APPLICATIONS
Francis and Clydesdale

FOOD COLLOIDS
Graham

FOOD QUALITY ASSURANCE
Gould

FOOD SCIENCE
2nd Edition   *Potter*

FRUIT AND VEGETABLE JUICE PROCESSING TECHNOLOGY
2nd Edition   *Tressler and Joslyn*

FUNDAMENTALS OF ENTOMOLOGY AND PLANT PATHOLOGY
Pyenson

IMMUNOLOGICAL ASPECTS OF FOODS
Catsimpoolas

INTRODUCTORY FOOD CHEMISTRY
Garard

LABORATORY MANUAL FOR ENTOMOLOGY AND
PLANT PATHOLOGY
Pyenson and Barke

POULTRY PRODUCTS TECHNOLOGY
2nd Edition   *Mountney*

PRINCIPLES OF ANIMAL ENVIRONMENT
Esmay

PRINCIPLES OF FOOD CHEMISTRY
deMan

TECHNOLOGY OF FOOD PRESERVATION
4th Edition   *Desrosier and Desrosier*

THE MEAT HANDBOOK
3rd Edition   *Levie*